Intelligent Medicine

Intelligent Medicine

Artificial Intelligence, Patient Safety, and Liability under an EU-Based Perspective

By

Vera Lúcia Raposo

BRILL | NIJHOFF

ISBN 978-90-04-54399-7
ISBN 978-90-04-54398-0 (PDF)
DOI https://doi.org/10.1163/9789004543980

Library of Congress Control Number: 2026930682

Brill Nijhoff and Brill Deutschland GmbH are part of De Gruyter Brill.
www.degruyterbrill.com

Questions about General Product Safety Regulation: productsafety@degruyterbrill.com

Contents

Preface

Looking forward, I anticipate AI playing a role in almost all medical procedures in the foreseeable future. This forecast is not solely based on observations but predominantly on the numerous advantages that AI is set to bring to healthcare services, as supported by various studies.

Given this trajectory, defining the standard of care in AI-based medicine is long overdue. Establishing these guidelines is crucial to maximise the benefits of AI, ensure patient safety and prevent potential instances of medical liability. At a later stage, it is crucial to clarify the assignment of legal liabilities, not only to ensure that patients can easily receive compensation but also to enable the various stakeholders (AI developers and manufacturers, hospitals, doctors) to fully understand their respective roles.

It is vital for professional bodies and scientific associations to establish operational rules for using AI, considering the specific technical aspects within each medical speciality. This responsibility primarily rests with healthcare professionals.

However, it has long been acknowledged that the existing legal framework significantly shapes the standard of care, particularly concerning issues like preserving medical confidentiality, obtaining informed consent, and organising collaborative medical practices. Navigating this area presents a challenge: legal professionals struggle due to the novelty and disruptive nature of AI technology, alongside insufficient current regulations. Legal advancements often lag behind technological progress. The case law is still developing, and prevailing opinions in doctrine frequently lack substantial foundations.

The questions to solve are numerous and complex: Should doctors disclose to patients that AI-supported decision systems were used in diagnoses? What additional information should patients be given? How should doctors proceed when they lack the necessary information due to AI's lack of transparency and/or explainability? Are doctors obligated to adhere to AI recommendations in diagnoses or treatments? When is it appropriate for doctors to disagree with AI recommendations? Does the use of AI undermine the trust between doctor and patient? Who is to 'blame' when something goes wrong: the doctor, the hospital, the AI developer?

The questions are numerous, yet comprehensive and legally sound answers are limited. This legal uncertainty poses an imminent risk to healthcare professionals, a pattern often observed during the initial stages of adopting new technology. However, it is crucial not to let this ambiguity hinder or discourage the advancement of AI in medicine. While doctors are irreplaceable, those who disregard AI may find themselves easily replaced.

Acknowledgements

To author a book is, above all, a solitary undertaking. In the darkest moments – when no words flow from the keyboard and the page remains obstinately blank – it is easy to believe we are alone in this daunting task. And yet, that is not quite true.

For this particular book, I owe a debt of gratitude to the colleagues who, through their own writing, our conversations, and their generous engagement with my drafts, helped shape and sharpen my ideas: Leonardo Vanneschi (NOVA IMS), João Leite (NOVA School of Science and Technology), José Fragata (NOVA Medical School), André Pereira and Carla Barbosa (Faculty of Law, University of Coimbra), Oliver Quick (University of Bristol Law School), Barry Solaiman (HBKU Law), and Sofia Palmieri (Harvard Law School).

I am also deeply grateful to my students – past and present – who, by persistently questioning me, pushed me to articulate more precise and thoughtful answers, often becoming unwitting co-authors of the conceptual framework this book advances.

To my institution, NOVA School of Law, thank you for embracing this project and for granting me the freedom – and the rare gift of time – to pursue it. Above all, thank you for believing in me when I said, 'trust me, I can do it'.

And to my family – Pedro, Madalena, and Eduardo – my deepest thanks for forgiving all the moments I should have been present and was not. This book is, in no small part, also yours.

This study is funded by the Foundation for Science and Technology within the scope of CEDIS – UID/00714/2025.

Abbreviations

AI	Artificial intelligence
AIA	AI Act
AILD	Artificial Intelligence Liability Directive
AMA	American Medical Association
CDSS	Clinical Decision Support Systems
CFR	Charter of Fundamental Rights of the European Union
CPOE	Computerised Physician Order Entry
CJEU	Court of Justice of the European Union
DL	Deep Learning
EHRS	Electronic Health Records
EU	European Union
GDPR	General Data Protection Regulation
IoT	Internet of Things
IT	Information Technology
LIME	Local Interpretable Model-Agnostic Explanation
ML	Machine Learning
MRI	Magneticresonance Imaging
NLP	Natural Language Processing
PLD	Product Liability Directive
SHAP	Shapley Additive exPlanations
FDA	U.S. Food and Drug Administration
TCM	Traditional Chinese Medicine
XAI	Explainable AI

Introduction

In the Introduction of this book, let me start from the end: Artificial Intelligence (AI) will not replace human doctors. This might kill suspense, but on the other hand will also operate as a kind of initial disclosure to prepare readers for one of the basic premises of this book: human doctors are irreplaceable. AI might substitute human decision-makers in some healthcare settings, but will be unable to replace human doctors *per se*.

Still, I predict that in the near future, AI will intervene in pretty much every medical act. This is not an ungrounded belief. On the contrary, it is substantiated by the added value and the many benefits AI can bring to healthcare delivery, as highlighted by several studies.[1]

A Johns Hopkins study based on the US reality in 2016 estimated that around 250,000 deaths occur each year as a direct consequence of medical errors, potentially making it the third leading cause of death in the country (though this figure remains debated).[2] One might assume that at least some of the fatalities could be reduced with the substantial integration of AI in healthcare delivery, as a sizeable portion of these deaths stems from inaccurate diagnoses and inadequate treatments. AI technology has the potential to effectively decrease the number of patient fatalities resulting from preventable medical errors.[3]

In sum, AI has the potential to revolutionise healthcare by improving diagnostic accuracy, enabling personalised treatment plans, enhancing operational efficiency, and driving medical research advancements. However, to fully take advantage of the benefits of AI in healthcare, it is essential to establish clear

1 Cf. Ali, Abdelbaki, Shrestha, et al., 2023: 100333; Harry, 2023: 326–335; Nikolinakos, 2023: 3–21.

2 James, 2013:122–128.

The current statistics represent a significant escalation compared to the figures stated in the Institute of Medicine's 1999 report, 'To Err is Human: Building A Safer Health System' (Institute of Medicine (US) Committee on Quality of Health Care in America, 2000), which projected an annual death toll of up to 98,000 individuals as a result of hospital errors. The stark disparity between these two sets of data underscores the growing magnitude of the issue at hand, emphasizing the pressing need for comprehensive measures to enhance patient safety and mitigate the occurrence of avoidable fatalities within healthcare settings.

3 For instance, using data from approximately 700,000 individuals, the AI system known as Deep Patient exhibited remarkable predictive capabilities for various diseases during testing on new records. Interestingly, without any specific expert guidance, Deep Patient autonomously managed to identify hidden patterns within the hospital data, indicating the likelihood of several medical conditions, including liver cancer. Cf. Miotto, Li, Kidd, et al., 2016: 26094.

 | HTTPS://DOI.ORG/10.1163/9789004543980-002

rules for its deployment and a specific and adequate medical standard of care. That is, to ensure patient safety and minimise the risk of litigation, several crucial factors need to be addressed.

The transformation that is currently taking place in healthcare due to AI is prompting a profound shift in the notion of the 'medical act', the content of the doctor-patient relationship and the very understanding of medicine. This shift aligns with the principles of P5medicine, which advocates for a proactive approach that incorporates the five P's: prediction, personalisation, prevention, participation, and precision.[4]

To begin with, a strong regulatory framework, able to consider the specificities of AI in healthcare, is imperative.[5] Such a framework should outline the lawful and ethical uses of AI, set standards for AI algorithms' validation and testing, and define the roles and responsibilities of the different stakeholders, especially healthcare providers and developers. It must also include clear rules on data protection and security, considering that AI systems in healthcare rely heavily on patient data (a type of especially sensitive data), which demands strict adherence to data privacy and security regulations.[6] Thus, clear guidelines must be in place to protect patient information from breaches and unauthorised access. The AI legal framework must also ensure algorithm transparency and explainability[7] (i.e., doctors should be able to understand how the

4 Anaya, Duarte-Rey, Sarmiento-Monroy, et al., 2016: 833–842, 834.

5 'Highlights that the use of AI in the health sector should be anchored in strong ethical requirements such as equitable access to healthcare, privacy, liability, transparency, explainability, reliability, inclusiveness and representability of data sets, and constant human oversight; stresses that the design of AI-based systems must address the risk of resources being wrongly allocated to individuals based on faulty or biased categorisation, prioritisation or malfunctioning technology, leading to misdiagnosis, maltreatment or no treatment at all; believes that the highest ethical standards should apply to all healthcare applications and that ethical rules should be established at a very early stage in their development and design, i.e. ethics by design; underlines that automated decision-making in healthcare applications may pose risks to patients' well-being and fundamental rights and stresses that AI must therefore have a supportive role in healthcare, where professional human oversight should always be maintained; calls for AI in medical diagnoses in public health systems to preserve the patient-doctor relationship and to be consistent with the Hippocratic oath at all times; notes, however, that AI improves the accuracy of screening and is already outperforming doctors' diagnoses in several instances; finds that the existing liability frameworks do not provide sufficient legal certainty and do not uphold the right of patients to legal redress in the event of misdiagnosis and incorrect treatment through AI; welcomes, in this regard, the upcoming legislative proposal on AI liability; notes that it is important to protect health professionals as users of AI systems, as well as patients as end recipients, providing them with sufficient and transparent information' (European Parliament, 2020, par. 22).

6 Paul, Ferrag and Almomani, 2023.

7 Bernal and Mazo, 2022: 10228.

AI arrived at its conclusions or recommendations, enabling them to trust and validate the system's outputs), as well as the continuous monitoring and evaluation of the AI system to identify and address potential biases, inaccuracies, or performance issues to ensure that AI remains up-to-date, safe, and effective.[8]

Proper mechanisms of risk management and liability are required.[9] This involves establishing guidelines for allocating liability in case of AI-related adverse events. Moreover, collaboration between stakeholders (healthcare providers, technology developers, policymakers, supervising health authorities, and patient advocates) is crucial in defining best practices, standards, and guidelines for AI implementation in healthcare.[10]

Training and education of healthcare professionals are crucial, so they can understand the limitations and capabilities of AI, know how to interpret its results and integrate it into their clinical decision-making process.[11] E-Skills and AI ethics are strange to most healthcare providers, but crucial in this regard.

This book will comprehensively address the most critical issues at the intersection of AI and healthcare delivery.

Firstly, it will explore the advantages that AI brings to the field, such as improved diagnostic accuracy and personalised treatment plans, while also addressing its limitations. The latter includes bias, the challenge of dealing with incomplete or insufficient data and the potential lack of empathy in AI systems.

Secondly, the book will deal with the definition of the standard of care for healthcare providers when using AI. It will dive into the complexities of establishing a clear and specific standard of care when incorporating AI into healthcare practices. Based on the different risks and disadvantages of AI medicine, the book aims to provide a comprehensive understanding of how to define a responsible and effective standard of care when using AI in healthcare.

Next, the book will delve into the legal consequences of when a patient suffers any form of harm (whether due to a violation of the medical standard of care or for some other reason). As the use of AI in healthcare becomes more frequent, the issue of assigning responsibility in the event of patient harm requires careful analysis. This book will explore the complexities of assigning liability to different stakeholders, including developers, healthcare institutions, and healthcare providers, in the event of adverse outcomes related to AI

8 Reddy, Rogers, Makinen, et al., 2021: e100444.
9 Gupta, Kamboj and Bag, 2021: 1–18.
10 Baig, Almuhaizea, Alshehri, et al., 2020: 253–256.
11 Canadian Agency for Drugs and Technologies in Health, 2025.

interventions, to shed light on best practices for assigning responsibility in a fair and effective manner.

Finally, the book will suggest some guidelines for avoiding patient harm and litigation. Prevention is crucial to ensuring patient safety and reducing the risk of lawsuits. The book will provide practical, evidence-based guidelines for healthcare providers (and indirectly developers and policymakers) to minimise the potential harms associated with the use of AI in healthcare delivery.

The content of this book reflects my perspective as both a lawyer and a legal scholar with extensive experience in medical law, particularly in matters concerning the medical standard of care and medical malpractice. Naturally, there are aspects – especially those involving technical guidelines for the clinical use of AI – for which I do not claim expertise, as they fall outside the domain of legal analysis. What I offer here is a legal-ethical account of the AI-supported medical standard of care, shaped by normative reasoning and legal methodology, and therefore likely to differ from the approach a medical practitioner might adopt.

Through an in-depth exploration of these core issues, this book seeks to contribute to the responsible and safe integration of AI in healthcare. It aims to provide valuable insights and actionable recommendations that can guide healthcare stakeholders in benefiting from AI while reinforcing patient safety and ethical standards, thereby fostering trust and confidence in the transformative power of AI in healthcare delivery.

CHAPTER 1

What Is AI

1 Understand AI

1.1 *From the Turing Test to Current Regulations*

Alan Turing, often regarded as the pioneer of AI, proposed an operational test of intelligence to assess whether a machine could exhibit behaviour indistinguishable from that of a human.[1] The crucial assessment of machine intelligence, in his view, lies in its ability to produce results indistinguishable from human-generated outcomes.[2] In essence, AI's purpose is to replicate human thought processes and decision-making, as its intelligence is modelled on ours. The Turing Test requires a computer to respond to written questions in a manner indistinguishable from a human, challenging the human interrogator to determine if the responses originate from a person or a computer.[3]

While Turing's approach offers a foundational, humancentric conception of machine intelligence, contemporary legal systems adopt far more operational and policydriven definitions. Within the EU, the definition that will take precedence is the one provided by the regulation on AI.[4]

Not surprisingly, the definition of AI has become a controversial one.[5] The challenge in defining AI lies not in the concept of artificiality, but in the complexity of defining 'intelligence'. The definitions of intelligence vary widely, highlighting many complex human traits that are difficult to define. Such attributes encompass a web of interconnected facets, including consciousness, self-awareness, linguistic proficiency, learning capacity, abstract thinking, adaptability, and reasoning ability.[6]

1.2 *AI, Machine Learning and Deep Learning*

AI is a broad term used to describe computer systems capable of executing tasks that conventionally demand human intellect, including visual perception,

1 Turing, 1936: 230–265.
2 Cf. Turing, 1950: 433–460.
3 Neufeld and Finnestad, 2020; 819–827.
4 The AI Regulation, also called AI Act, will be presented later in this book.
5 On the various definitions of AI, Samoili, Cobo, Gómez, et al., 2020: 7–28. On the definition of AI in the AI Act, Casaburo and Gugliotta, 2023.
6 Scherer, 2016: 359–360.

 | HTTPS://DOI.ORG/10.1163/9789004543980-003

decision-making, speech recognition, and language translation. It constitutes an umbrella concept encompassing various forms, each at distinct stages of development.[7]

A subtype of AI known as machine learning (ML) distinguishes itself by its capacity to learn from datasets, allowing the system to draw conclusions or predictions without being explicitly programmed for predefined responses. ML algorithms evolve through data exposure and mature with experience, not relying on pre-established rules. Much like human learning, ML evolves from the examples (the data) presented, enabling this form of AI to modify and enhance its algorithms without direct programmer intervention. The objective is to confirm the adaptability of the trained model beyond the initial dataset, diving into new applications that the algorithm has not encountered before. Resembling the human learning experience, ML begins with limited knowledge and skills, then accumulates expertise based on feedback, ultimately refining choices.[8] Recently, ML has become the prevailing approach for AI applications in healthcare. This approach enables computer systems to learn from data and enhance their performance without explicit programming.[9] It has been suggested that ML could revolutionise healthcare by aiding doctors in identifying patients prone to specific incidents, discerning those requiring greater attention, and identifying candidates for aggressive treatments.[10]

Deep learning (DL), a subset of ML, employs artificial neural networks with multiple layers to detect patterns within extensive datasets. While ML typically employs one or two intermediate layers, DL introduces additional intermediate layers, imitating the human brain, where each intermediate layer represents a nexus of connections between neurons. Central to the deep neural network lies the artificial neuron, which is nothing more than a mathematical model. Although initially inspired by the model of neurons in human and animal brains, the artificial neuron resembles a biological neuron within a computer framework. Its purpose is to replicate the biological neuron's capacity to learn from experience. Each neuron stands as a singular computational unit, with neuron groups assembled into layers, in such a way that each neuron in a layer is connected to the next layer. The greater the layer count, the more intricate the machine's representation becomes. Successive layers tend to grasp more

7 In this book, I will generally use the term AI to refer to the overarching field, without specifically distinguishing between ML or DL, as this is the approach commonly adopted in EU legislation.

8 Sarker, 2021.

9 Mehta and Devarakonda, 2019–2021.

10 Wong and Bressler, 2016: 2366.

abstract characteristics of network input. First-layer neurons observe raw input data, while the final-layer neurons generate output. A distinctive aspect is that no individual neuron encodes a discrete decision-making component. Instead, thousands or even hundreds of thousands of neurons collaborate to reach a decision.[11] Thus, it also requires significantly more computational power. Greater task complexity relies on the more abstract representations generated in the deeper layers.[12] Unfortunately, heightened layer counts render the machine's logic more challenging for humans to grasp and follow.[13] As these models become more complex, they become less comprehensible to users and even to the original developers of these neural networks.

Put simply, AI refers to computers programmed by humans to execute tasks demanding human aptitude. ML buils on this, with the added ability to enhance performance through the integration of new data inputs. DL, a form of ML, delves deeper into layers, necessitating more data, yet providing heightened precision in results.

1.3 *Algorithms*

1.3.1 Algorithms and AI

Even though all AI comprises algorithms, not all algorithms are a form of AI.[14]

An 'algorithm' is a mathematical construction that can be described as a 'control structure that accomplishes a specific purpose under certain provisions',[15] i.e., a set of steps designed to achieve a specific objective. Even a basic program, containing a concise set of 'if ... then' rules integrated into any kind of software, can be perceived as an algorithm. While some algorithms are simple and direct, others can be intricate and elaborate. Traditional algorithms, whether straightforward or complex, do not involve any learning process. In contrast, an AI model operates differently by updating its internal parameters over time, enabling its behaviour to adapt without altering the underlying algorithm.[16]

AI algorithms can acquire knowledge from the data they process. In the example provided by Maddox et al.,[17] in the case of breast cancer prediction, AI algorithms can use well-established genetic sequences like BRCA1 as the benchmark for accuracy. By leveraging relevant data, AI algorithms can discern intricate connections that conventional analytical methods cannot access. This

11 Yu, Beam and Kohane, 2018: 720.
12 Taye, 2023: 91.
13 Jorstad, 2020: 3.
14 Villasenor and Foggo, 2020: 301.
15 Hill, 2016: 35–59.
16 Villasenor and Foggo, 2020: 301.
17 Maddox, Rumsfeld and Payne, 2019: 31.

ability enables AI to detect subtle patterns, such as multiple minor changes in a chest computed tomographic image, which together may indicate the presence of pneumonia.

1.3.2 Functions of Algorithms

Algorithms are often used to either predict uncertain future outcomes or to identify and subjectively categorise unknown entities based on inference rather than direct measurement. As an example of predictions, consider a predictive model for patient readmission risk.[18] Hospitals and healthcare systems often employ algorithms to analyse a patient's medical history, current condition, demographic information, and other relevant data to estimate the likelihood of that patient being readmitted to the hospital within a specific timeframe after the initial discharge. Healthcare professionals can then use this predictive score to allocate resources effectively and provide tailored care plans for patients at higher risk of readmission. As an example of the second case, consider an algorithm used to diagnose rare diseases using symptom patterns.[19] Imagine a scenario where a patient presents with a set of symptoms that do not immediately match any known medical condition. In such cases, traditional direct measurement methods might not provide a clear diagnosis. However, algorithms can be developed to analyse databases of medical records, genetic data, and symptom patterns to infer potential rare diseases that could be causing the patient's symptoms. They are able to compare the patient's symptoms to patterns observed in previously diagnosed rare diseases. This type of algorithm assists medical professionals in making informed decisions, suggesting further tests, and narrowing down potential diagnoses for rare and complex cases.

Within medical applications, genetic algorithms stand out as the most widely utilised form of evolutionary computation.[20] Drawing inspiration from natural biological evolution, these search and optimisation algorithms function by generating numerous random solutions for a given problem. Over successive generations, this population evolves, leading to the emergence of satisfactory solutions. The process involves retaining the best solutions while eliminating the inferior ones. Repeating this cycle among the better elements results in population improvements, survival of the fittest, and the generation of

18 Artetxe, Beristain and Grana, 2018: 49–64.
19 Lee and Yoon, 2017: 6–9.
20 Sharma and Kumar, 2022.

novel solutions. Genetic algorithms find application in various tasks, including diagnosis and prognosis, medical imaging, and signal processing.[21]

1.3.3 Locked and Evolutive Algorithms

Algorithms can be locked in time, or they can evolve.

An algorithm that remains locked will consistently produce the same outcome when provided with identical data, maintaining its stability over time and with repeated use. However, this fixed nature does not inherently guarantee safety. In fact, locked algorithms can potentially be more harmful than 'unlocked' or 'adaptive' algorithms, particularly if they produce erroneous results based on outdated training data or perpetuate biases in patient care.[22] Considering this, locked AI health technologies might necessitate new regulatory approvals in real-world usage if significant and unexpected patterns of results arise, stemming from incorrect priors about the data fed into the AI. Additionally, updates to the algorithms might be deemed necessary to align with advances in medical knowledge.[23]

Unlocked or adaptive algorithms are computational processes that possess the ability to evolve, learn, and adjust their behaviour based on new data through updates to their internal parameters rather than changes to the underlying algorithm. These algorithms are designed to adapt to changing circumstances, patterns, or inputs over time, allowing them to improve their performance or accuracy. Unlike traditional algorithms that follow static rules or instructions, unlocked or adaptive algorithms have a degree of flexibility that enables them to modify their internal parameters or structures in response to new information. This adaptability allows them to refine their predictions, classifications, or outputs as they encounter more data, leading to potentially enhanced performance in various applications. For this reason, they represent the future of AI health technology, surpassing human performance.[24]

1.4 *The Development of AI across the Years*

Back in 1955, John McCarthy, a mathematics professor, introduced the term 'AI' in his grant application for the 1956 Dartmouth AI Conference. In his words, AI was the science and engineering devoted to creating intelligent machines.[25]

---

21 Gen and Lin, 2023: 635–674.
22 Prabhu, 2019: 621–622.
23 Gerke, Babic, Evgeniou et al., 2020.
24 Prabhu, 2019: 621–622.
25 McCarthy, Minsky, Rochester, 2006.

Little did he know that this seemingly modest proposal would lay the foundation for one of the most transformative fields of human innovation.

In that moment, a journey of exploration and discovery started, as researchers and engineers embarked on the mission of building intelligent systems able to perform tasks that traditionally require human intelligence. This has been a process of continuous development. Over the decades, AI has evolved from theoretical discussions to practical applications, increasingly invading various aspects of our lives.

The mid-1950s is considered the first AI boom,[26] a period when significant advancements and interest in the field of AI began. During this era, the concept of the 'perceptron' was introduced, which is a type of artificial neural network that receives multiple inputs and provides an output based on them. This perceptron idea laid the foundations for what would later develop into DL, a subfield of AI that has seen remarkable progress in recent years.[27] DL involves training artificial neural networks with large amounts of data to enable them to perform complex tasks and make sophisticated decisions. During the 1980s, early advances in neural network research laid important groundwork for later developments in DL, contributing to the idea that computers could improve performance through exposure to data.[28] DL is now a crucial component of modern AI systems and has contributed to numerous advances in various sectors.

Back in 2007, IBM introduced Watson, a supercomputer that some years later would win the Jeopardy contest. A core strength of Watson lay in its adeptness at understanding and interpreting natural language. It also possessed the remarkable capability to process vast amounts of documents within minutes.[29] These achievements can be attributed to Watson's capacity to read, synthesise, and retain research papers and clinical trials at an exceptional pace. Still, it was also heavily criticised and was not particularly successful.

The pursuit of AI has led to significant advancements, driving the development of smart assistants, recommendation systems, autonomous vehicles, medical diagnosis tools, and much more. These intelligent machines are revolutionising industries, improving efficiency, enhancing decision-making, and opening new possibilities for human progress.

---

26 Matsuzaki, 2018: 258.

27 Rosenblatt, 1958: 386, 387, 399–401.

28 See, 2021: 422.

29 Demonstrated by a study at the University of North Carolina School of Medicine, Watson's diagnosis matched that of oncologists in 990 out of 1,000 cancer cases tested (Galeon, 2016). Even more impressive, Watson identified treatment options that escaped oncologists' attention in 30% of cases (See, 2021: 423).

With the evolution of AI, the ability to tackle intricate tasks has been amplified due to recent breakthroughs in computing power and the abundance of data at our disposal. As information processing capabilities have soared, AI designers now possess the means to execute complex calculations, effectively transforming inputs into corresponding outputs. Consequently, the correlations identified and employed by AI models to generate classifications and predictions have become increasingly intricate and less intuitively understandable from a human perspective. This shift is attributed to the complex nature of these correlations, which have transcended the boundaries of human thinking.

However, the journey has not been without challenges and debates. As AI becomes more pervasive, questions regarding ethics, liability and respect for fundamental rights have surfaced. Striking a balance between technological advancement and societal well-being remains a critical consideration.

1.5 *The Basic Functioning of AI*

AI systems operate through three fundamental components: inputs, algorithms, and outputs.[30]

Inputs include the data with which the algorithms are fed, which in the medical domain encompasses, for instance, medical notes, clinical outcomes, records, and exam reports. The quality of training data significantly impacts AI performance. The more the better, that is, more data and better labelled information can lead to better results.[31]

Outputs are the results, decisions, or predictions produced by the algorithm based on its analysis of input data and its learned patterns. Examples of outputs in healthcare delivery involve incident predictions, patient diagnoses, the definition of appropriate drug doses, or tumour localisations. The accuracy and reliability of outcomes are crucial factors in evaluating the effectiveness of AI systems in various applications. However, AI outputs are not necessarily correct. Many algorithmic conclusions are based on statistical learning and computational theory, resulting in probabilities rather than certainties. Thus, some argue that such inconclusive evidence cannot form a solid foundation for decision-making, especially in critical areas like medical decisions.[32]

1.6 *Models of AI Learning*

In an advanced self-learning architecture, the system acquires knowledge and expertise in specific classification tasks through exposure to extensive

30 See Taulli, 2019.
31 Bright Data, 2025.
32 Mittelstadt, Allo, Taddeo, et al., 2016: 5.

annotated and classified datasets. For example, the system might learn to differentiate between healthy and diseased lung X-rays after studying a substantial dataset of pre-classified lung images. This process of learning from data empowers the architecture to make accurate and informed classifications, thus enhancing its performance over time. In the domain of learning, diverse approaches exist, including unsupervised, supervised, semi-supervised learning and reinforcement learning.[33]

Unsupervised learning involves the agent extracting patterns from input data without explicit feedback. The algorithms are supplied with untagged data, and it is the algorithms' responsibility to uncover latent patterns in the data without any guidance or reference to outputs.[34]

Supervised techniques function through the accumulation of a considerable dataset comprising training cases (inputs) and the subsequent examination of relationships between these inputs and their corresponding outputs. As the algorithm proceeds, it acquires the ability to produce accurate outputs for fresh inputs in novel scenarios, pinpointing specific parameters that reduce disparities between its predictions and the real outcomes observed within these cases. In doing so, algorithms develop predictive capabilities for new cases beyond their training datasets. This process mirrors human learning through trial and error. Picture teaching a young child a new task, such as using the potty: when the child performs the task correctly, it is rewarded; if the child errs, there is no reward. Over time, the child aims to perform successfully to receive more rewards.[35] In a similar vein, within AI reinforcement learning, the agent explores various actions, grasps which actions lead to the best outcomes, and fine-tunes its decision-making strategy through numerous interactions. This procedure encompasses a balance between experimenting with new actions and choosing actions known to yield greater rewards. Unlike unsupervised learning, where data are unlabelled, supervised learning employs labelled and annotated data.[36]

Semi-supervised learning involves a limited number of teacher-provided outputs to the agent, combining guidance and independent learning. This approach is designed to strike a delicate balance between leveraging the benefits of labelled data, where the model receives explicit guidance from a teacher or supervisor, and allowing the system to independently explore and learn from unlabelled data. By incorporating a limited number of labelled examples,

33 Banteka, 2020: 545; Matsuzaki, 2018: 261.

34 Zimmerman, 2017: 9.ff.

35 Hundt, Murali, Hubli, et al., 2022.

36 Zimmerman, 2017: 8 ff.

semi-supervised learning aims to optimise the efficiency of the learning process, capitalising on the advantages of both supervised and unsupervised learning paradigms.[37]

Reinforcement learning, on the other hand, is a distinct form of learning that is neither supervised nor unsupervised. It focuses on learning from interaction with an environment, using rewards and punishments to guide learning. It involves an ML agent taking actions within a given environment to achieve specific goals. These actions lead to particular outcomes, and the agent receives feedback in the form of rewards or penalties based on the outcomes of its actions. The agent aims to learn the optimal strategy that maximises cumulative rewards over time. Algorithms receive directives regarding inputs and corresponding outputs to understand how to act in the future.[38]

1.7 *Hardware and Software*

Within an AI system, there are two primary manifestations: software-based systems and hardware-embedded devices.[39] The software exists on the purely digital level, showcasing its skills through voice assistants, image analysis software, search engines, and speech and face recognition systems. Hardware finds its footing in physical objects, such as advanced robots, autonomous cars, drones, and Internet of Things (IoT) applications. This distinction leads to a subsequent distinction between non-embedded AI (purely software-based AI) and embedded AI, which is part of hardware devices.

In the specific healthcare domain, we can also find this dichotomy. On the one hand, we have electronic health record (EHR) systems, applications focused on text and data mining, and platforms designed to facilitate treatment choices, which are purely software. On the other hand, we are also faced with surgical assistance robots, ingenious prostheses and tangible support systems tailored for elderly care, that is, hardware devices.

1.8 *Different Types of AI Development and Autonomy*

When it comes to the scope of purposes, three distinct categories emerge. 'Weak AI' or 'narrow AI' – the only one achieved so far – is designed for specific tasks. 'General AI' or 'strong AI' systems, still in development, would possess skills and knowledge equivalent to humans, enabling them to perform a wide

37 Li and Liang, 2019: 669–676.
38 Shakya, Pillai and Chakrabarty, 2023: 120495.
39 Anastasia, Madonnab and Monica, 2021: 338–343.

array of tasks. Lastly, 'super AI' is nowadays a purely theoretical concept that would surpass human capabilities.[40]

Regarding AI's interaction with human users and its correlation with human intelligence, distinctions arise based on the extent to which machines emulate human tasks. Some machines repeat human tasks, aiming to enhance human efficiency and speed with fewer errors. This category is termed 'autonomous AI'. On the other hand, 'augmented AI' involves machines capable of performing tasks beyond human capacity.[41]

Another division exists between substitutive and complementary AI.[42] This distinction hinges on whether technology replaces or complements human expertise. Complementary AI, like prediction-oriented ML systems, furnishes outputs that inform professionals, enhancing their understanding and situational awareness. In contrast, substitutive AI generates actionable results devoid of human intervention, like clinical trial orders or drug prescriptions. However, true substitutive AI, completely obviating human involvement, does not yet exist.

Another aspect to consider is the degree of AI's autonomy and problem-solving abilities in the way AI operates. Some AI types make decisions based on pre-programmed rules, inferring or evaluating data according to these rules. For example, a chess program evaluates moves and selects the best one using a scoring formula.[43] In healthcare, some good examples of this type of AI are Clinical Decision Support Systems (CDSS) that use predefined rules based on medical guidelines, best practices, and clinical knowledge to assist healthcare professionals in making informed decisions; or a triage system to assess patient symptoms and determine the urgency of their condition following predefined rules. Alternatively, more flexible AI programs operate without predefined rules, formulating their own as they learn from data (adaptive AI).[44] This is the case with Natural Language Processing (NLP) for clinical text analysis, which is a system that learn from textual data to comprehend medical concepts, extract insights, and generate knowledge without explicit rule-based programming. Another example refers to healthcare chatbots/virtual assistants that learn from interactions to improve responses without being explicitly programmed with rigid rules.

40 Harwood, Maltby and Mukaetova-Ladinska, 2019.

41 Pieper and Gleasure, 2025.

42 Froomkin, Kerr and Pineau, 2019: 45.

43 Gualdi and Cordella, 2021: 2300.

44 Gill, Xu, Ottaviani, et al., 2022: 100514.

1.9 *Black, Grey and White Boxes*

Regarding the level of explainability, three different types are usually identified.[45] The kind of AI, that is considered explainable, also known as 'white boxes', refers to systems that provide users with clear insights into the underlying mechanisms governing behaviour, predictions, and influential variables. While these systems offer perceptibility in their operations, their capabilities are often confined to tasks such as prediction and pattern matching. Positioned in the middle ground, semi-explainable AI systems, referred to as 'grey boxes', exhibit a level of explainability that falls between full clarity and opacity. In contrast, opaque AI systems, labelled as 'black boxes', conceal the internal processes responsible for generating outputs, providing high accuracy but sacrificing accountability in the process (see Chapter 3, Section 3.4.8).

There are two forms of the black box.[46] In the stringent scenario, the decision-making procedures are entirely inaccessible to human understanding. There is no means of discerning the AI's rationale behind a decision or prediction, identifying which information significantly influences the AI's outcome, or obtaining a hierarchical classification of the variables processed by AI based on their importance. The black box in this form is resistant to post-analysis through the reverse engineering of AI outputs. By contrast, in the case of weak black boxes, the decision-making process remains opaque to humans. Nonetheless, unlike fully opaque systems, it is feasible to infer the importance of the variables considered by AI through reverse engineering. This, albeit to a limited extent, may enable the prediction of how the model will formulate its decisions.

Most AI used in healthcare is a black box.[47] The challenge posed by the 'Black Box Problem' arises from the inherent difficulty in comprehending the decision-making mechanism of AI, leading to an inability to anticipate its decisions or results. The lack of visibility into the developer's instructions or design hinders the ability to discern the intent of the creator or user, rendering intent tests unattainable. This, consequently, gives rise to complexities in evaluating fault in legal liability cases, whether civil or criminal.[48]

45 Wadden, 2021: 94; Raposo, 2024a: 1049–1072.
46 Bathaee, 2018: 906.
47 London, 2019: 15–21.
48 Raposo, 2024a: 1049–1072; Raposo, 2025.

CHAPTER 2

Uses of AI in Healthcare Delivery

1 Possible Uses of AI in Healthcare Provision

1.1 *General Overview*

Considering the increasing gap between the supply and demand for staff in the healthcare sector, AI might not only be a possibility but an inevitability.[1] On a global scale, the world is expected to face a shortage of 18 million healthcare professionals, with a significant deficit of 5 million doctors.[2] This imbalance in healthcare workforce availability poses a significant challenge, particularly in developing countries. AI might mitigate this issue, as demonstrated by the many studies that confirm AI benefits in terms of improving the quality of care and access to healthcare.[3]

Even if that were not the case and there was enough supply of health professionals, AI might still be essential to achieve the higher goals that we envisage in healthcare delivery: enhancing population health, improving the patients' care experience, boosting caregiver satisfaction, and mitigating the escalating cost of care.[4] These intelligent machines outperform their human counterparts in terms of time efficiency and cost-effectiveness, offering substantial incentives for supporting doctors in making diagnostic decisions. By reducing time, cost, and medical errors, these systems enhance healthcare accessibility and quality.[5] As a consequence, the utilisation of these tools effectively diminishes the occurrence of medical errors.[6]

As AI uses in medicine are increasing by the day[7] – giving rise to what Nicholson Price named 'black box medicine'[8] – we should explore their most common and/or promising uses.

---

1 AI can be used, not only to provide healthcare but also to perform different administrative tasks. The use of NLP technology in AI systems presents opportunities for automation and thus have the potential to automate administrative responsibilities, such as documenting patient visits in electronic health records, streamlining clinical workflow, and allowing doctors to dedicate more time to patient care. Cf. Bajwa, Munir, Nori, 2021: e188–e194.

2 World Health Organization, 2016.

3 Naik, Hameed, Shetty, et al., 2022: 4; Raposo, 2015: 3; Meskó, Hetényi and Győrffy, 2018: 545.

4 Bajwa, Munir, Nori, et al., 2021: e188–e194.

5 Froomkin, Kerr and Pineau, 2019: 100; Sion, 2019:55–60.

6 Astromskė, Peičius, and Astromskis, 2021: 509.

7 Nuffield Council on Bioethics, 2018.

8 Price II, 2015: 421.

 | HTTPS://DOI.ORG/10.1163/9789004543980-004

1.2 *Triage*

The concept of triage typically conjures images of swift decisions related to urgent interventions for critically ill patients or accident victims. However, this concept extends beyond emergency situations and finds application in remote patient assistance, particularly within regions with limited access to medical resources and personnel, as seen in telehealth applications.

One possible use of AI in triage concerns AI-based patient triage systems that use language analysis along with searches of large databases to correlate symptoms and medical conditions. Subsequently, based on the results obtained, these systems make recommendations regarding the best medical referral.[9] Frequently, triage is done using conversational agents, which encompass ML and NLP, with the capacity to comprehend and scrutinise spoken and written language. They employ these technologies to engage with individuals through speech, text, or other forms of input and output on platforms ranging from mobile applications to web-based and audio-based. Many of these agents are specially designed to incorporate NLP, allowing users to communicate with them as if they were interacting with a human being. Subsequently, the agent can analyse the input and respond in a conversational manner.[10]

When dealing with triage decisions in emergency care contexts, where such choices can determine life or death, there is a natural hesitation to delegate such crucial judgments to machines. Nonetheless, AI, functioning as an outcome predictor, could serve as an aid to emergency technicians or doctors.[11] A study conducted in 2017 focused on emergency room triage for patients with acute abdominal pain, revealing an 'acceptable level of accuracy'.[12]

1.3 *Management of Medical/Clinical Information*

A major value of AI in the context of information management is related to the recording of information. Through NLP and digital assistants, an efficient way of 'taking notes' is achieved, thus reducing the weight of one of the main

9 Heaven, 2018. One could foresee, for instance, that the GPT-based natural language application might serve as a foundation for conversational agents directly engaging with individuals, perhaps acting as an initial interaction point or, more debatably, for assessing non-critical patients during triage. Cf. Mittelstadt, 2021: 30.

10 Milne-Ives, De Cock, Lim, et al., 2020: e20346.

11 The Mayo Clinic introduced an AI-driven 'digital triage platform'. This platform, provided by Diagnostic, assumes the role of conducting clinical intake for patients and suggesting diagnoses and hospital risk assessments (McGrail, 2020). These solutions are now accessible through websites or smartphone apps, having evolved from decision trees initially crafted by medical professionals to incorporate AI elements.

12 Farahmand, Shabestari, Pakrah et al., 2017: e5.

obstacles to the use of EHRS: the demand for massive amounts of written information.[13]

This is especially relevant because information has become so vast that it is difficult to analyse. CDSSS that operate through AI are being used to interpret large amounts of data contained in EHRS. Such data include, for example, lab results, imaging studies, radiology reports, electrocardiograms, genetic testing, family history, medications taken, and hospital admissions. Computerised alerts provided by these CDSSS offer healthcare providers targeted and timely information capable of reducing clinical error.[14]

Moreover, AI can enhance patient safety by generating computerised alerts that notify doctors of critical information, such as potential drug interactions, abnormal test results, or deviations from standard care protocols. By delivering this targeted and timely information, AI not only supports clinical decision-making but also plays a crucial role in reducing medical errors and improving patient outcomes.[15]

1.4 *Analysis of Exams and Diagnosis*

The human eye and the human brain have their limits. After many hours of looking at exams, it becomes practically impossible to find something, leading to human mistakes. AI, in turn, can avoid many such mistakes, and thus it plays a significant role in various aspects of radiology, especially image interpretation.[16] It can interpret exam results with a minimal margin of error.[17] For example, it can classify echocardiograms to determine whether the patient has any heart disease, and it does so with an accuracy (92%) that surpasses

13 Terry, 2019: 141.

14 Gianfrancesco, Tamang, Yazdany, et al., 2018: 1544–1547.

15 Liu, Wright, Patterson, et al., 2023: 1237–1245.

16 A radiologist typically views 4,000 images in a computed tomography scan of various parts of the body (pan scan) in polytrauma patients. Searching for an occult fracture in this context is a very meticulous job, which often leads to visual fatigue and encourages errors (Challen, Denny, Pitt et al., 2019: 231–237), many of which are likely to be detected by the AI system.

17 AI is becoming so dominant in the analysis and interpretation of medical exams (Tang, 2019: 20190031) that it has been claimed that 'the role of the radiologist will be obsolete in five years' (Khosla, 2017). Still, some studies tend to demonstrate otherwise. In the examination of contemporary research encompassing AI's role in mammography screening, a systematic review was conducted, shedding light on the accuracy of diagnostic tests. The findings indicate that radiologists continue to exhibit superior performance compared to AI in the realm of breast cancer detection (Freeman, Geppert, Stinton, et al., 2021: n1872).

that of humans (79%).[18] It is also useful in pathology. For example, to diagnose breast cancer, ML performs better than a panel of 11 pathologists.[19]

Based on the success attained by AI in the diagnosis field, some hospitals and healthcare systems are integrating CDSS platforms to leverage the power of ML in aiding diagnostic decisions and predicting treatment outcomes. These CDSS platforms function by constantly monitoring the data entered by doctors into the EHR. As information is logged, the CDSS performs real-time analyses, considering not only the EHR data but also relevant clinical information from various unrelated sources. These additional sources encompass test results from pathology laboratories, radiological departments, genetics departments, ambulatory settings, and research findings stored in biobanks, clinical trials, and genome sequence databanks. Drawing on this comprehensive pool of information, the CDSS generates diagnostic recommendations using algorithms typically developed with rules derived from established clinical guidelines and published medical research reviews.

1.5 *Prediction and Risk Calculation*

AI can predict which patients are more likely to suffer a heart attack within ten years, through an analysis of their medical data, namely risk factors such as high blood pressure, cholesterol, smoking, and age.[20] Studies point out that AI can successfully (percentage of accuracy: almost 80%) predict the likelihood of a heart attack from 30 minutes to 24 hours before.[21] AI algorithms thus outperformed humans in predicting this incident.[22] The same trend applies to the prediction of specific heart diseases. Moreover, ML techniques have significantly enhanced the accuracy of cardiovascular risk prediction.[23] Within surgical practices, AI possesses the capability to compile a wide array of information, ranging from patient-specific risk factors and anatomical data to the natural progression of diseases, patient preferences, and cost considerations. This amalgamation enables AI to prognosticate the potential outcomes associated with surgical decisions.[24] This improvement enables the identification of more patients who could benefit from preventive treatment, while also reducing unnecessary treatment for others, and highlights the potential of AI

18 Madani, Arnaout, Mofrad, et al., 2018: 6.
19 Griffin, 2021: 76.
20 Weng, Reps, Kai, et al., 2017: e0174944.
21 Alamgir, Mousa and Shah, 2021: e30798.
22 Ramsey, 2023.
23 Pal, Parija, Panda, et al., 2022: 1100–1113.
24 OECD, 2020: 8.

systems to enhance medical outcomes and optimise patient care in specific domains.[25]

A particularly useful tool is the use of CDSS for risk calculation, as patient outcomes can be predicted. AI can construct patient profiles by combining data from EHRS and incorporating factors like genetic traits, lifestyle choices, and socio-demographics. Public health systems can then use these profiles to forecast which patients might necessitate re-hospitalisation, allowing them to allocate resources efficiently. On the other hand, private health insurers can associate these profiles with risk levels to determine insurance premiums and offer tailored packages for disease groups not covered by existing plans. This data-driven approach enhances decision-making for insurers and health systems, improving patient care and risk management.[26]

1.6 *Therapeutic Options*

The potential impact of AI in treatment planning has gained significant credibility.[27] One of the great 'heroes' in this domain is the supercomputer Watson. Watson 'interprets' data (clinical information) from cancer patients, based on which it suggests treatment options. Of course, human doctors can also decide which treatment is most appropriate, but Watson has a competitive advantage: it gets more accurate results in a matter of seconds.[28]

Among the possible areas of application, the treatment of sepsis stands out as a crucial focus.[29] Sepsis, being a leading cause of death globally and the most common cause of mortality in hospitals,[30] requires precise management of intravenous fluids and vasopressors. Suboptimal decision-making in sepsis treatment often leads to unfavourable outcomes. The results were astounding, as the AI doctor consistently outperformed human doctors in selecting the most effective treatment options. This promising development fuels hope that computational models like this could significantly enhance clinical decision-making, ultimately leading to improved patient outcomes in the future.

---

25 Combs, 2019. Another striking example takes place at Stanford Hospital, where AI is being used for mortality prediction, that is to say, to anticipate which patients are more likely to die in a 3–12 month period and thus would greatly benefit from palliative care (Avati, Jung, Harman, et al., 2018).

26 Lysaght, Lim, Xafis, et al., 2019: 305.

27 Komorowski, Celi, Badawi, et al., 2018: 1716–1720.

28 Ferrucci, Levas, Bagchi, et al., 2013: 95–96.

29 Lauritsen, Kalør, Kongsgaard, et al., 2020: 101820.

30 World Health Organisation, 'Sepsis', 3 May 2024, https://www.who.int/news-room/fact-sheets/detail/sepsis.

1.7 *Personalised Medicine*

Personalised medicine brings with it the opportunity to tailor healthcare interventions for individual patients, based on their distinct disease profiles, diagnostic or prognostic insights, and responses to treatment.[31] This tailor-made treatment potential considers genomic variations as well as influential factors in medical care, including age, gender, geographical location, race, family medical history, immune composition, metabolic makeup, microbiome composition, and susceptibility to environmental factors.[32] This purpose involves collecting diverse data from individuals, such as genetic information, physiological monitoring data, or electronic medical records, and utilising these data to customise treatment approaches guided by advanced models.

The use of expansive datasets alongside predictive models represents a challenge for human doctors if not assisted by AI models. The fundamental objective of personalised medicine is to anchor all stages of a patient's medical journey in their individual biology rather than relying solely on population-based considerations.[33]

1.8 *Robotic Surgeries*

1.8.1 Types of Robotic Surgeries

In the field of medicine, robot technology has emerged as a promising advancement. There are three primary applications of robotic systems in cyber surgery: providing external support to the surgeon's actions, enabling off-site surgeons to participate in procedures, and facilitating the manipulation of complex surgical instruments.

Based on these different aims, a triad of robotic roles emerges within the field of surgical applications:[34] (a) Active systems that engage in predefined tasks while remaining within the purview of the operating surgeon's control; (b) Semi-active systems that afford surgeons the ability to enhance the pre-programmed actions of the system; (c) Master-slave systems which exclusively rely on a surgeon's involvement, devoid of any autonomous elements. Up until now, we do not have fully automated (and thus autonomous) surgical interventions. Robotic interventions are either passive (the robot merely correctly places the instruments, but it does not manipulate them; the doctor

31 In this study the concept of 'personalised medicine' is used as synonym of 'precision medicine'. For the controversy around their equivalence see Delpierre and Lefèvre, 2023.

32 See, for instance, the deep learning method called AINU that uses nanoscale nuclear features to identify cellular heterogeneity, with applications in regenerative medicine, virology, and cancer biology. Cf. Carnevali, Zhong, González-Almela, et al., 2024.

33 Bohr and Memarzadeh, 2020: 29.

34 Denecke and Baudoin, 2022: 795957.

does) or, at most, semi-active (both the surgeon and the robot control the instruments).[35] While certain surgical robots rely on continuous input from surgeons, such as the Da Vinci and Zeus robotic surgical systems that employ robotic arms controlled by surgeons,[36] none of them uses AI in the way we typically understand it today (e.g., ML or autonomous decision-making). They are robotic-assisted surgical systems that enhance a surgeon's precision and control, but all actions are directly controlled by the surgeon. Still, AI-driven features are either in development or available in separate systems. There is a growing trend towards increased autonomy of robots in other medical domains. Fogel and Kvedar even suggest a future scenario where robotic surgery might largely supplant conventional human surgical procedures by the late 2050s.[37]

1.8.2 Advantages of Robotic Surgeries

By offering enhanced precision and smaller instruments, robots contribute to less invasive surgeries.[38] An ongoing transformation is reshaping surgical technology, with a pronounced emphasis on diminishing the invasiveness of procedures and medical interventions that were previously deemed infeasible can now be performed, leading to reduced recovery times. This involves minimising incisions, reducing open surgeries, and incorporating flexible instruments and cameras to augment surgical manoeuvres. This approach of minimally invasive surgery is hailed as the future, although it is still in its early stages, necessitating further refinements to render it less burdensome for patients and to economise time and costs.

Navigation systems offer surgeons valuable assistance by pinpointing spatial positions and providing insights into anatomical orientation queries. Contemporary surgical navigation systems capitalise on real-time tracking of markers. This is achieved through a stereoscopic camera emitting infrared light, effectively ascertaining the 3D coordinates of prominent structures.[39]

Distinct motor skills are requisite for minimally invasive surgery as compared to traditional surgery due to the diminished tactile feedback inherent in

35 Griffin, 2021: 71.

36 The da Vinci and Zeus systems comprise a pair of robotic arms intricately interconnected to a control panel via a fibreoptic cable. These robotic arms exhibit remarkable dexterity, enabling them to execute exceedingly delicate motor movements with unparalleled precision, surpassing the level of accuracy achievable through conventional open surgical procedures (Adhamov, 2023: 196–19).

37 Fogel and Kvedar, 2018: 5.

38 Reddy, Gharde, Tayade et al., 2023.

39 Wang, Cao, Chen et al., 2021: 904–922.

tool-based approaches. To address this, sensors designed to deliver finer tactile stimuli are under development. These sensors leverage tactile data processing to translate sensor inputs into discernible data or stimuli for surgeons.[40] Artificial tactile sensing confers several advantages over direct physical touch, including an extensive repository for comparative sensations, standardisation in evaluating quantitative attributes among surgeons, continual enhancement, and uniform training levels. This innovative avenue holds the potential to revolutionise surgical procedures by enhancing precision, expanding knowledge sharing, and shaping a more standardised approach across the surgical community. Additionally, this technological advancement extends the horizons of surgical possibilities, allowing interventions in anatomical regions that would otherwise remain inaccessible to human hands.[41]

These features offer distinctive benefits in terms of reducing human error due to precise pre-programmed mechanical movements.[42] It allows surgeons to perform minimally invasive surgeries because surgical robots help surgeons navigate around vital structures with minimal surgical dissection. By delineating the so-called hepatic boundaries, the system works in such a way that the surgeon is alerted when he/she is leaving the designated 'safe zone'.[43] Another important safety function associated with surgical robots is the possibility of automatically turning off potentially dangerous surgical equipment whenever the surgeon approaches the patient's vital structures.[44] Another 'safe' feature is the ability to detect the exact location of the surgical instrument at all times, without the need for real visual observation of the surgery.[45]

One of the greatest achievements of robotic surgeries is to allow surgeons to perform a surgical intervention at a distance, that is, when the doctor and the patient do not share the same physical space.[46] The system allows the surgeon

40 Naeini, AlAli, Al-Husari, et al., 2019: 1881–1893.

41 OECD, 2020: 8.

42 To better understand the potential of robotic surgeries, compare robotic and laparoscopic surgeries. Unlike the latter, which employs a basic tube with limited functions, a robotic-assisted surgical device grants the surgeon increased dexterity akin to a second wrist or an additional finger. Primarily embraced in urology, these robotic surgical devices found early success due to their suitability for delicate procedures involving inaccessible pelvic bowl tissues. The device's range of motion and tissue illumination surpass what human hands could achieve independently (Pasquale, 2022: 6).

43 Griffin, 2021: 70.

44 Griffin, 2021: 70.

45 Griffin, 2021: 70.

46 The first telerobotic surgery – a milestone in surgical innovation – was performed on a patient by Professor Jacques Marescaux, on September 7, 2001, and became known as the 'Lindbergh operation'. It was a 54-minute telesurgical cholecystectomy, where the doctor,

to operate seated at a console at a remote distance, if needed, guided by a 3D image of the surgical field. The surgeon can use master control by holding their fingers below the screen.[47] Telerobotics is an integral part of the broader fields of telemedicine[48] and cybermedicine,[49] as a modality that aims to provide health services at a distance (telemedicine) using the resources of the digital world (cybermedicine). Telesurgery systems are robotic surgery systems capable of receiving and transforming surgical data in real-time. They operate through robots, which work through AI, without the surgeon having any physical contact with the patient. It is, however, a rather complex procedure. It is necessary to take into account several conditioning factors, from those related to hardware to those related to functionality and technical data streaming.[50]

Amidst the occurrence of severe acute respiratory syndrome (SARS) between 2002 and 2003, healthcare practitioners resorted to tracheal interventions, such as intubations, suctions, and resuscitations, in their tireless efforts to provide care.[51] However, these very procedures became conduits for the transmission of the pernicious disease, leading to the contamination and infection of respiratory doctors. Nebulisers and high-flow mechanisms, intended to aid in the treatment process, inadvertently contributed to the spread of the respiratory syndrome. This scenario highlights a compelling case for the integration of ML- driven autonomous robotic systems or robotic-assisted endoscopic procedures, particularly advantageous for patients carrying highly contagious pathogens. By incorporating autonomous capabilities, these procedures could augment teleoperated interventions, allowing surgeons to operate from outside the confines of the surgical room.[52]

Some fatalities were linked to robotic surgeries (it remains to be seen what the percentage is when compared to non-robotic surgeries). In 2002, a tragic incident occurred in Florida wherein a man lost his life following a flawed kidney operation conducted using the da Vinci system. Subsequently, the family of the deceased filed a lawsuit in state court, alleging negligence on the part

---

based in New York, conducted an operation on a patient situated in France. Cf. Anvari, McKinley and Stein, 2005: 460–464.

47 For instance, the Da Vinci system provides three or four robotic arms, an endoscope arm, and two or three instrument arms that carry out the surgeon's commands.

48 Raposo, 2016a.

49 Raposo, 2017a: 29–42.

50 Currently, the Intuitive Surgical's Da Vinci Surgical System is the leading company in the field of telesurgery, supplying over 210 devices. The company gained FDA approval in July 2000 to perform advanced surgical techniques.

51 Zemmar, Lozano, and Nelson, 2020.

52 O'Sullivan, Nevejans, Allen et al., 2019: 4.

of the hospital due to the surgeon's inadequate training and lack of experience in operating the system. Eventually, the case was resolved through an out-of-court settlement.[53] On the surface, the prospect of a robot equipped with diminutive, agile, and even laparoscopic probes surpassing the capabilities of an average surgeon – and possibly, down the line, even the most accomplished ones – appears glaringly evident. Yet, the majority of automated systems lack adaptability, rendering them fragile.[54]

1.9 *Caregiver Robots*

The so-called 'caregiver robots'[55] – for children, the elderly, people with reduced mobility, and people with mental illnesses – make it possible to replace a wide category of family and professional caregivers. These AI-powered robots can handle a variety of tasks, from reminding the patient to take their medication to giving them the medication. This technology can take many different forms: 'robotic' cribs designed to help the baby sleep better,[56] voice systems that become companions for the elderly (Palro)[57] and that can even train their memory through conversations and quizzes (ElliQ),[58] or even robots like the RIBA that can lift and carry a person.[59]

In their interactions with patients, caregiver robots demonstrate a range of traditional communication skills that are essential for effective and compassionate care.[60] These skills include the ability to understand and process natural language, which allows them to engage in meaningful conversations and respond appropriately to the patient's needs and concerns. Beyond mere language comprehension, caregiver robots are adept at expressing emotions, which is crucial in building trust and rapport with patients. By conveying empathy, concern, and encouragement through their tone, facial expressions,

53 Cf. Allain, 2013: 1056.

54 For instance, contemporary automobile assembly lines use industrial robots to affix windshields onto new vehicles. Curiously, when it comes to aftermarket windshield replacements, human technicians are favoured over robots. The rationale behind this preference lies in the intricacy of the tasks involved, such as removing a shattered windshield, preparing the frame for replacement and fitting the new windshield. These activities demand a level of real-time adaptiveness that contemporary robots, as they stand, are far from achieving (Pasquale, 2022: 6–7).

55 Matthias, 2015: 169–192; Boch, Ryan, Kriebitz, et al., 2023: 110.

56 Kelly, 2017.

57 Inoue, Sasaki and Nakamura, 2015: 980–983.

58 Tech Desck, 2019.

59 Soffar, 2025.

60 Yuan, Coghlan, Lederman, et al., 2023: 1637–1654.

and body language, caregivers can create a supportive and comforting environment for patients, even if it is merely an emulation of feelings unique to humans.

1.10 *Non-Medical Tasks*

Medical practice is not only about providing patient care or practising medicine itself; rather, it involves paperwork and routine duties that do little to benefit patients but severely contribute to doctor burnout. The burden of data searching and data entry weighs heavily on providers, with one study revealing that almost half of a doctor's time is spent on desk tasks, leaving only a quarter for direct patient interaction.[61] AI has the potential to transform healthcare delivery significantly, particularly by reducing the workload on doctors and preventing burnout. Through the implementation of AI-driven speech-recognition software, it becomes possible to automatically transcribe patient conversations and provider notes, streamlining administrative tasks like form filling.

This advancement not only relieves doctors of routine burdens but also has the potential to enhance the doctor-patient relationship. With AI handling administrative duties, doctors can devote more time and attention to their patients, potentially improving the overall quality of healthcare interactions. Currently, during patient visits, doctors spend only around half of their time engaging with patients, with approximately a third dedicated to managing EHRs and desk work.[62]

The automation of routine tasks has the capacity to enable certain processes to occur more efficiently at the outset of healthcare procedures.[63] For instance, administrative functions like scheduling, billing, coding, and payment can be automated. AI has found a significant role in streamlining and enhancing these routine, repetitive tasks, ensuring greater efficiency, accuracy, and impartiality. Errors in these areas generally lead to administrative or financial repercussions.[64] Assuming AI's eventual accuracy, such support can lead to a reduction in transcriptional errors and may even enhance privacy, as details are processed by machines rather than human scribes.

In addition, AI can automatically collect the most relevant medical information from patient medical records, which in turn reduces the time spent

---

61 Sinsky, Colligan, Li, et al., 2016: 753–760.

62 For a study in this regard see Pinevich, Clark, Harrison et al., 2021: 788–799.

63 Price II, 2019: 71–72.

64 OECD, 2020: 10.

searching for specific patient information. It can also be very useful in the identification of general medical information from medical literature by using natural-language processing.[65]

Discharge planning AI tools can estimate discharge dates for patients, the steps needed to get them discharged, and flag those who are clinically (almost) ready to be released from hospital, along with a list of steps to be taken before discharge. Some systems can even schedule the necessary follow-up appointments.[66] NLP systems can be used for the automation of routine or labour-intensive tasks, such as searching and navigating EHRS or automating preparation of medical documentation and orders.[67]

2 Does AI Surpass Humans in Healthcare Delivery?

2.1 *The Added Value of AI in Healthcare*

In some medical specialities, AI's success is stunning. Many claim that AI equates to, or even surpasses, human doctors.[68] For instance, it is claimed that AI surpasses human pathologists in detecting breast cancer.[69] Likewise, in radiology, AI is sometimes considered to be the co-author of the diagnosis together with the medical staff.[70]

It is undeniable that AI is becoming particularly accurate in its results.[71] In scenarios characterised by a condensed realm of possibilities, AI systems

65 Ahuja, 2019: e7702.
66 Mittelstadt, 2021: 33.
67 Mittelstadt, 2021: 33.
68 This claim can be found, for instance, in a 2017 study carried out by the Massachusetts General Hospital and MIT (Conner-Simons, 2017). See also University of California (2025).
69 Bejnordi, Veta, Johannes van Diest, et al., 2017: 2199–2210.
70 Kudina, 2021: 6–7. This also raises the question of whether AI would replace radiologists. See Huang, Wittbrodt, Teague and al., 2025 (underscoring AI's role as a collaborative instrument and not a substitute for human expertise).
71 In 2023, a group of researchers appraised the capabilities of ChatGPT on the United States Medical Licensing Exam (USMLE), encompassing the Step 1, Step 2 CK, and Step 3 assessments. Remarkably, ChatGPT achieved scores around or above the passing benchmark for all three exams, all without any dedicated training or reinforcement efforts. Furthermore, the model exhibited notable coherence and profound insights in its explanations. These outcomes imply that Large Language Models (LLMs) have great potential in supporting medical education and conceivably, aiding clinical decision processes (Kung, Cheatham, Medenilla et al., 2023: e0000198).

Around the same time, another team assessed the capabilities of another LLM, known as Flan-PaLM, on the United States Medical Licensing Examination (USMLE), unveiling

might even conceive optimal solutions instead of settling for mere satisfaction. If the criterion for adopting AI is human performance, AI has gained its spot in healthcare: 'AI systems will be introduced when they make fewer errors than healthcare providers, not when they are perfect'.[72]

2.2 *The Drawbacks of AI in Healthcare*

However, it is not all about AI benefits. There are also significant failures, and the fact is that AI makes mistakes that no human being would.[73] Blind acceptance of AI suggestions without exercising clinical judgment can expose doctors and their patients to avoidable errors.[74] Doctors must know when AI information is helpful and when it might be harmful, and must guide their consultation accordingly.

Thus, the apparent absolute success of AI technology in healthcare must be treated with caution. This leads to the conclusion that despite the advancements in AI, the role of human doctors is unlikely to disappear in the near future.

Take the case of radiologists to illustrate this statement. Unlike AI models, which perform single tasks, the responsibilities of radiologists go beyond merely reading and interpreting images. Currently, only a small fraction of these tasks can be handled by AI. Moreover, radiologists play a crucial role in

---

intriguing insights. Flan-PaLM employs a diverse set of prompting strategies to attain exceptional precision across various MultiMedQA multiple-choice datasets. This proficiency is exemplified by its remarkable 67.6% accuracy on MedQA (US Medical Licensing Exam-style questions), a significant advancement exceeding the previous benchmark by over 17%. However, an assessment involving human evaluators unveiled notable gaps in its performance. To bridge these gaps, the researchers introduced instruction prompt tuning, a technique for aligning LLMs with new domains by employing a limited number of exemplars. The outcome, Med-PaLM, demonstrates promising results, albeit trailing behind clinical experts. The study concluded that comprehension, knowledge retention, and logical reasoning are enhanced with the enlargement of the model scale and the implementation of instruction prompt tuning, which suggests the prospective value of LLMs in medical contexts (Singhal, Azizi, Tu, et al., 2023: 172–180).

72 Holm, Stanton, Bartlett, 2021: 175.

73 Nguyen, Yosinski and Clune, 2015: 427.

74 Watson for Oncology evaluated information from a patient's medical record and displayed potential treatment options ranked by confidence level, based on evidence-backed training. Watson was deployed at UB Songdo Hospital in Mongolia, where it was reported that its suggestions were followed at a rate of nearly 100% by doctors who had little or no training in cancer care. Apparently, the system recommended a drug that was clearly contraindicated for a patient with that clinical history. This error was detected by an oncology specialist, but it is conceivable that a non-specialist doctor could miss the error (IBM, n.d.).

collaborating with other healthcare professionals to discuss diagnoses and treatment plans. They also provide direct patient care, including administering local ablative therapies, conducting image-guided medical procedures like cancer biopsies and vascular stents, and tailoring imaging examinations based on the patient's condition. Additionally, they integrate imaging findings with other medical records and test results, ensuring comprehensive and accurate diagnoses. Furthermore, radiologists possess expertise in defining the technical parameters for imaging procedures, taking into account the specific needs of individual patients. Their ability to interpret images within the broader context of a patient's medical history and collaborate with colleagues makes them indispensable in providing holistic patient care.[75]

2.3 *Medical AI Capacities versus Human Doctors' Capacities*

A fundamental difference emerges in the decision-making mechanisms of humans and contemporary AI systems, and these disparities can propel AI to propose solutions that defy human expectations. Human decision-making is confined by the cognitive constraints inherent in the human brain. Faced with temporal limitations, humans are incapable of meticulously evaluating the entirety, or even a significant portion, of the available information.[76] In contrast, AI programs can explore a multitude of possibilities within a designated timeframe. This ability allows AI to explore solutions that humans might not have considered. These solutions can include options that people might overlook or dismiss in favour of more familiar or intuitive choices.[77] In complex situations, as seen in cases like computational pathology, an AI's approach can be vastly different from what a human mind would produce, sometimes in ways that even the creators of the AI could not predict. One drawback is that it also means that humans may lack the ability to scrutinise potential solutions that may have eluded human comprehension.[78] Despite the vast capabilities of AI systems, attempts to align their learning approaches with human

75 Also in this sense, see Davenport and Kalakota, 2019: 97.

76 Scherer, 2016: 364–365.

77 Scherer, 2016: 365.

78 For instance, back in 2012, the medical AI system Watson successfully completed the U.S. doctor licensing examination. Despite its capability to provide answers to queries, Watson falls short of autonomously and comprehensively assessing medical issues. It lacks the capacity for independent conceptualisation and for searching for novel, pertinent information due to its deficiency in advanced cognitive abilities needed for critical thinking and for diverse solution generation based on intrinsic value. See Khan and Alamäki, 2022.

practices have revealed a fundamental gap.[79] The core issue lies in the difference between how humans derive actionable insights and how AI processes information statistically. While AI systems are adept at gathering and analysing statistical data, replicating the nuanced decision-making process inherent in human expertise has proven elusive.

Additionally, the integration of AI into real-world medical settings can face significant challenges. Complexities in handling vast and varied medical data, including inconsistencies in medical records, abbreviations, human errors, and diverse documentation styles, have proven difficult for even the most advanced AI systems to manage effectively. Logistical issues such as system compatibility and cost overruns have further complicated the successful implementation of AI in healthcare. These challenges highlight the ongoing struggle to bridge the gap between AI's potential and its practical application in medical practice.[80]

2.4 *The Added Value of Combining AI and Humans in Healthcare Delivery*

The advantages of combining AI with humans in healthcare delivery are patent in a study from the Harvard Medical School that compares the accuracy of ML systems to human pathologists in detecting breast cancer.[81] The ML system had a 92% success rate; however, human doctors achieved 96% accuracy. Harvard then combined the human pathologists' findings with those of the ML systems, and the accuracy increased to 99.5%. That is, combining both reduces errors by almost an order of magnitude (from 40 per thousand to just 5 per thousand) and would represent 56,000 fewer misinterpreted breast exams per year in the US alone, where the study was conducted. This study highlights the significant added value of integrating AI with human expertise in healthcare. By leveraging the strengths of both AI systems and human professionals, we can achieve unprecedented accuracy and reliability in medical diagnostics. This synergy not only enhances patient outcomes but also paves the way for more efficient and effective healthcare delivery.

---

79 Strickland, 2019.

80 Take again the case of Watson. Despite all its knowledge and significant outputs, it suffered a severe setback when the University of Texas Cancer Center cancelled its collaboration with Watson because of incompatibilities with the hospital records system, as well as alleged violations of hospital procurement regulations. In the end, the project seemed to fall apart due to cost overruns and the extraordinarily complex work of structuring and preparing huge amounts of data to be ingested by Watson's ML systems (Birmingham, 2017).

81 Goh, Gallo, Hom, et al., 2024: e2440969.

By incorporating human feedback and human-AI collaboration strategies, we can create a partnership where AI provides insights,[82] while humans apply their expertise for final decision-making.[83] This collaboration ensures oversight, quality control, and validation of AI predictions, minimising errors and biases. It also fosters continuous improvement, as both AI and humans learn from each other, which in turn builds trust and acceptance. Human feedback further enhances AI's adaptability, allowing it to manage complex cases that go beyond its training data.

82 Tschider, 2023: 332.
83 Sezgin, 2023.

CHAPTER 3

Medical AI and Patient Safety

1 AI and Patient Safety

1.1 *Medical Errors and Patient Safety*

Human error is one of the main causes – if not the main cause – of harm in healthcare,[1] as paradigmatically underlined by the 2000 study by the US Institute of Medicine.[2] The list of human factors leading to errors encompasses the multitude of elements influencing work performance and interpersonal dynamics within teams, alongside the lack of expertise in using new technologies. In the healthcare sector, collaborating with diverse professionals on a routine basis often gives rise to safety concerns and hindrances in effective communication (including verbal, non-verbal, and written errors), which elevate the likelihood of medical errors and consequent harm to patients.[3]

Therefore, new processes and mechanisms are being envisaged to promote patient safety and reduce medical errors[4] (and consequently reduce the likelihood of medical liability).[5] Such paths involve, for instance, patient engagement and education,[6] simulation training,[7] and medication reconciliation programs,[8] among others.

However, it is impossible for doctors to keep up with new advancements in medicine.[9] The constant influx of new medical knowledge is so overwhelming that it leaves little room to assess the information's relevance or implement it practically in medical practice.

In recent years, AI has taken centre stage, combining immense datasets to gather problem-solving acumen and expertise.[10] This process unfolds at a

1 Sameera, Bindra and Rath, 2021: 328–335.
2 Institute of Medicine (US) Committee on Quality of Health Care in America, Kohn, Corrigan, Donaldson, 2000: 2.
3 Brennan and Oeppen, 2022: 30–33.
4 Rodziewicz, Houseman and Hipskind, 2023.
5 In this book, 'medical liability' refers to civil liability for medical malpractice, excluding criminal liability unless expressly stated otherwise.
6 Kendir, Fujisawa, Fernandes, et al., 2023.
7 Aggarwal, Mytton, Derbrew, et al., 2010: i34–i43.
8 Barnsteiner, 2008.
9 Doximity, 2022.
10 About the historical development of AI uses in healthcare see Davahli, Karwowski, Fiok, et al., 2021.

 | HTTPS://DOI.ORG/10.1163/9789004543980-005

velocity and scale that surpasses human capacity. AI harnesses the potency to attain remarkable precision in predictive actions like diagnosis, prognosis, and therapeutic interventions. In doing so, healthcare outcomes stand to be markedly enhanced. Accompanied by ever-advancing computational capabilities, AI systems possess the remarkable ability to analyse personal characteristics, medical records, extensive literature, and other medical data swiftly and accurately.

1.2 *The Added Value of AI for Patient Safety*

AI systems possess several advantages over humans, which frequently lead to superior cognitive outcomes. One of their key strengths lies in their capacity to process enormous volumes of data at a speed far beyond human capability.[11] Additionally, these algorithms perform calculations and analyse information without being influenced by personal biases or self-interest, ensuring that every step of the decision-making process is fully considered.[12] Moreover, unlike humans, AI is not hindered by physical or mental constraints such as fatigue, stress, or emotional involvement, enabling it to operate with unwavering consistency and efficiency.[13]

The focus of safety assurance revolves around establishing a sense of certainty in the safety of a system. Demonstrating this safety assurance is often accomplished through a safety case, a document prepared by the technology or service developers. This document presents a logical argument that is substantiated by a collection of evidence. Within the safety case, the rationale is provided for why the system is considered sufficiently safe to function as intended within a specific setting.[14]

Safety-wise, the advantages of AI have been widely discussed in the medical literature.[15] AI can use sophisticated algorithms to learn, which it does by drawing on massive amounts of healthcare data and then using the insights gained to aid clinical practice. In its more sophisticated iterations, such as ML and DL, it can also possess learning and self-correction capabilities, enhancing its accuracy through continuous improvement guided by the feedback it receives. The AI system can help reduce errors in diagnoses and therapeutic decisions, both of which are inevitable in human clinical practice.[16] It can extract useful

11 Aldoseri, Al-Khalifa and Hamouda 2024.

12 Starke, Baleis, Keller, et al., 2022. Still, AI is far from unbiased, as we shall see.

13 Starke, Baleis, Keller, et al., 2022.

14 Habli, Lawton and Porter, 2020: 252.

15 On the benefits and risks of using AI in healthcare, Powell, 2025. See also Angus, 2020: 1043–1045.

16 Kumar, Koul, Singla, et al., 2023: 8459–8486.

information from large patient populations to help make real-time inferences regarding health risk alerts, as well as predict outcomes, which allows for predicting the patient's prognosis more accurately than doctors.[17] Through these inherent 'capabilities', the AI system empowers healthcare practitioners to offer personalised and top-notch patient care by providing them with the latest medical insights derived from recent literature and clinical practices.[18] One goal of AI is the so-called 'democratisation of experience' by allowing the average doctor to reach the specialist's level of sophistication in diagnosis or treatment.[19] Moreover, AI is not susceptible to fatigue and can also be useful as an adjunct diagnosis tool in situations where doctors may be working under time pressure outside of normal working hours.[20] Last but not least, it can relieve health professionals of time-consuming administrative tasks, thus releasing doctors for relevant medical tasks, and increasing their time for care practices.[21]

2 Legal Regulations Aimed to Improve the Safety of Medical AI

2.1 *Medical IT Regulation*

There is a disparity in regulatory oversight between medical practice and the Information Technology (IT) industry. While medicine operates under strict regulations, the world of IT has remained relatively unregulated.

Despite the limited regulatory landscape, the IT industry has embraced ethical guidelines for professionals. For instance, the Association for Computing Machinery, the world's largest computing society, issued a comprehensive Code of Ethics and Professional Conduct in 2018.[22] The presence of these ethical codes establishes a framework for technologists to follow in their professional pursuits.[23]

Unlike the tightly controlled medical field, adherence to these ethical codes in IT is often voluntary, resulting in diverse practices across the industry. As technology continues to play an increasingly significant role in our lives,

17 Yu, 2018: 722.
18 Bohr and Memarzadeh, 2020: 25–60.
19 Cohen, 2020: 1446.
20 Doctor performance worsens with sleep deprivation, and for example fractures are more likely to be misdiagnosed at night than during the day. See Kuo, Harrison and Jones, 2021: 106133.
21 Mennella, Maniscalco, De Pietro, et al., 2024.
22 Association for Computing Machinery, 2018.
23 Smith, 2021: 540.

discussions surrounding the need for more robust regulations and oversight persist. Striking a balance between innovation and ethical responsibility remains an ongoing challenge for the industry and society as a whole.

2.2 *Product Safety Norms*

The General Product Safety Directive,[24] supplemented by sectoral legislation, had long been the set of norms that ensured the safety of all consumer products, even those not covered by specific EU regulations. However, from December 2024, it was replaced by the General Product Safety Regulation.[25] The EU product safety legislation works to guarantee that products on the EU market meet stringent health, safety, and environmental standards, allowing them to circulate freely within the EU.

Many of the EU's product safety regulations were established before the advent of digital innovations like AI, the IoT, or robotics. Consequently, these regulations may not always include specific provisions addressing the novel challenges and risks posed by these technologies. Nonetheless, the existing product safety framework is designed to be technology-neutral, meaning it can still be applied to products that incorporate these advanced technologies. For instance, within the EU product safety legislation, software updates may be likened to safety maintenance activities, provided they do not bring about significant alterations to a product already available on the market and do not introduce unforeseen risks not initially assessed. Nevertheless, if a software update substantially transforms the product it is applied to, it could trigger a re-evaluation of the entire product, potentially categorising it as a new product. In such cases, adherence to the pertinent safety product regulations must be re-examined when the modification occurs.[26]

Subsequent legislative measures within this framework, particularly in sectors like medical devices and automobiles, have already taken into account certain aspects of the emergence of digital technologies. This includes considerations such as automated decision-making, treating software as a distinct

---

24 Directive 2001/95/EC the European Parliament and of the Council of 3 December 2001 on general product safety, https://eur-lex.europa.eu/legal-content/EN/ALL/?uri=celex%3A32001L0095.

25 Regulation (EU) 2023/988 of the European Parliament and of the Council of 10 May 2023 on general product safety, amending Regulation (EU) No 1025/2012 of the European Parliament and of the Council and Directive (EU) 2020/1828 of the European Parliament and the Council, and repealing Directive 2001/95/EC of the European Parliament and of the Council and Council Directive 87/357/EEC.

26 A comment to this legal Framework in European Commission, 2020.

product, and facilitating connectivity. However, they are not always able to keep pace with sudden technological advances.

2.3 *The Medical Device Regulation*

2.3.1 The Concept of Medical Device

The Medical Device Regulation (MDR) represents the most significant set of norms applicable to medical products.[27] To fall under the jurisdiction of the MDR and receive the designation of a medical device, a product must align with the predefined categories of use stipulated within the regulation.[28] The classification as a medical device is of utmost importance because oftentimes it leads to the qualification as a high-risk AI system under the AI Act (AIA).[29]

The MDR applies to products designed for the well-being of humans, serving one or multiple objectives such as diagnosing, preventing, monitoring, predicting, treating, or alleviating diseases, injuries, or disabilities.[30] Additionally, it encompasses products involved in investigating, altering, or substituting anatomical features, physiological processes, or pathological states and items that provide information through in vitro examination of specimens sourced from the human body, including organs, blood, and tissues (Article 2(1) MDR).

Sometimes, AI is embedded in medical hardware and therefore forms an integral part of a medical device (e.g., an AI-enabled ventilator with predictive control). In other cases, software for medical devices can stand alone as independent software, possessing its distinct medical purpose and qualifying

---

27 Regulation (EU) 2017/745 of the European Parliament and of the Council of 5 April 2017 on medical devices, amending Directive 2001/83/EC, Regulation (EC) No 178/2002 and Regulation (EC) No 1223/2009 and repealing Council Directives 90/385/EEC and 93/42/EEC.

Its sister regulation is the Regulation (EU) 2017/746 of the European Parliament and of the Council of 5 April 2017 on in vitro diagnostic medical devices and repealing Directive 98/79/EC and Commission Decision 2010/227/EU, which applies solely to in vitro medical devices. To simplify, and considering the similarities between both, this book will only analyse the MDR.

28 See Hauglid and Mahler, 2023: 1–23; Palmieri, Walraet and Goffin, 2021: 344–345.

29 Regulation (EU) 2024/1689 of the European Parliament and of the Council of 13 June 2024 laying down harmonised rules on artificial intelligence and amending Regulations (EC) No 300/2008, (EU) No 167/2013, (EU) No 168/2013, (EU) 2018/858, (EU) 2018/1139 and (EU) 2019/2144 and Directives 2014/90/EU, (EU) 2016/797 and (EU) 2020/1828 (Artificial Intelligence Act).

This is not an absolute and automatic label. To be considered a high-risk AI system it is necessary, not only to be a medical device but also to be subject to a third-party conformity assessment.

30 The manufacturer, as outlined in Article 2(12) of the MDR, holds the responsibility of determining the intended purpose of the device.

as a medical device in its own right (software as a medical device).[31] A prime example is smartwatch applications designed to alert users and healthcare practitioners about irregular heartbeats, identifying cardiac arrhythmia. An alternative case involves software that guides or influences a hardware medical device, while also possessing a medical purpose, which is also classified as a medical device software. For instance, the software responsible for measuring blood glucose levels, calculating insulin dosages and controlling insulin pumps falls into this category.

By contrast, software aimed at guiding or impacting the use of a medical device's hardware but not aligning with the purposes outlined in Article 2(1) of the MDR, falls outside the scope of medical device software if it does not generate information for any of the medical purposes specified in the MDR definition. Such software does not qualify as medical device software. Nevertheless, if the manufacturer intends this software to be used in line with the functions specified in Article 2(2) of the MDR, it might be classified as an accessory,[32] thereby still being governed by the MDR framework.[33]

Determining the device's intended purpose holds dual significance, serving not only to establish its identity as a medical device but also crucially impacting its categorisation. Recital 19 emphasises the importance of identifying whether the software is intended for general purposes by the manufacturer. Instances of such software include those supporting administrative tasks, promoting a healthy lifestyle, managing electronic patient records, offering various forms of clinical decision support, or handling data formats. This principle extends to devices that monitor overall well-being and physical activity, such as sports monitoring devices like Fitbit, are explicitly exempt from this regulation (Recital 19 MDR).[34]

Consequently, AI systems designed for tasks like analysing extensive patient data to uncover patterns, extracting new insights, or facilitating strictly administrative decision-making processes do not fall under the classification of medical devices. For instance, AI may play a role in supporting decisions regarding the high-level management of resources, whether they be human or otherwise, or in serving as the foundation for changes in how healthcare institutions are organised or how health services are delivered. These applications may all be

31 Gerke, Babic, Evgeniou, et al., 2020: 1–4.

32 Accessories are products meant to complement medical devices and aid in their functioning. While not classified as medical devices themselves, accessories are subject to the same legal standards and regulations.

33 Medical Device Coordination Group Document MDCG 2021–24, 2021. See also Ludvigsen, Nagaraja and Daly, 2022: 78–93.

34 Palmieri, Walraet and Goffin, 2021: 347–348.

collectively referred to as 'medical AI', but they do not meet the criteria for classification as 'medical devices' according to the MDR and they will only become subject to the norms applicable to 'high-risk' AI systems under the AIA if they are included in the lists outlined in Annexes I or III of the AIA.[35]

The manufacturer's specification of the device's intended use allows for the assessment of potential hazards it might present to users and the subsequent classification of a risk category following regulatory guidelines. As devices may carry varying levels of risk, the manufacturer's adherence to increasingly rigorous obligations becomes imperative to ensure the device's safe utilisation across its complete lifecycle.

2.3.2 The Certification Procedure

To determine the risk level of medical devices, they are classified into four classes: class I (low risk), class IIa (moderate risk), class IIb (medium risk), and class III (high risk).[36] Class I devices are considered low risk due to their minimal potential for harm. On the other hand, class III devices, like implantable pacemakers and breast implants, pose higher risks to patients' lives and well-being, making them subject to more stringent regulations.

The MDR establishes a system for regulating software-based medical devices. The classification of software like AI within this system depends on its intended use, as defined by the manufacturer, and the associated risk level. This means AI software could be categorised differently depending on how it is intended to be used and the potential risks it presents. MDR Rule 11, which pertains to medical device software, designates it as the least demanding risk class, class I, necessitating less stringent safety criteria.[37] However, exceptions within this rule are established, contingent on the software's intended function.[38] The MDR specifically identifies software types designed to aid diagnostic or therapeutic decisions, categorising them as class IIa. Further clauses assign the AI device to other classes as the level of risk increases. Decision-support systems, which are part of AI-based medical devices, generally fall under class IIa, unless they significantly impact a patient's health, in which case they may be categorised as class IIb or class III.

This provision highlights the increasing demands tied to higher software risks and emphasises the crucial importance of intended use. This notion echoes the CJEU's ruling C-329/16, issued on December 7, 2017, interpreting the

35 Hauglid and Mahler, 2023: 9–10.

36 Medical Device Coordination Group Document MDCG 2021–24, 2021.

37 Palmieri, Walraet and Goffin, 2021: 349.

38 For the different levels of risk see Fraser, Biasin, Bijnens, et al., 2023: 467–491.

former Medical Device Directive.[39] In this ruling, the CJEU tackled whether stand-alone software facilitating a medical-social support system to determine a prescription drug qualified as a medical device. The CJEU's response reaffirmed the manufacturer's designated intended use as a cornerstone of the legislation.[40]

Under the MDR, the certification process for medical devices is critical for ensuring their safety and efficacy in healthcare settings.[41] Before a medical device in a higher risk class (class IIa, IIb, III for medical devices) can be introduced to the market, both the product and the manufacturer's documentation must undergo evaluation by a Notified Body.[42] This requirement extends to class I devices that are sold in a sterile condition (class Is), have a measuring function (class Im), or are reusable surgical instruments (class Ir). In contrast, products in the lowest risk class (class I for medical devices) are self-assessed for conformity by the manufacturer without needing the involvement of a Notified Body.[43]

For effective safety oversight, the regulation mandates that Member States undertake the process of appointing a Notified Body, under the procedure defined in Article 42 of the MDR. These designated bodies are tasked with conducting a thorough assessment of conformity to ascertain the device's alignment with the stipulated MDR prerequisites.[44] Successful conformity attests to the device's safety during usage and qualifies it to receive the CE label, a declaration of its alignment with European standards.[45]

---

39 Judgement of 7 December 2017, Snitem and Philips France, C-329/16, ECLI:EU:C:2017:947. The former Medical Device Directive was Council Directive 93/42/EEC of 14 June 1993 concerning medical devices.

40 'Article 1(1) and Article 1(2)(a) of Council Directive 93/42/EEC of 14 June 1993 concerning medical devices, as amended by Directive 2007/47/EC of the European Parliament and of the Council of 5 September 2007, must be interpreted as meaning that software, of which at least one of the functions makes it possible to use patient-specific data for the purposes, inter alia, of detecting contraindications, drug interactions and excessive doses, is, in respect of that function, a medical device within the meaning of those provisions, even if that software does not act directly in or on the human body'.

41 The European Conformity (CE) mark is a key indicator that a medical device meets the necessary product safety and performance requirements. Cf. Antich-Isern, Caro-Barri, and Aparicio-Blanco, 2021: 120992.

42 Notified Bodies are required to remain independent of the manufacturers they control. However, complete independence is difficult to guarantee, as these bodies are often private companies responsible for issuing certifications.

43 Medical Device Coordination Group Document, MDCG 2021–24, 2021.

44 French-Mowat and Burnett, 2012: S22–S28.

45 Schröttner and Baumgartner, 2023: 39–61.

2.3.3 The Risk Assessment

One of the challenges in assessing the risk level of AI-based medical devices is the uncertainty of their impact on patient outcomes.[46] Unlike physical devices with direct interactions, AI systems' risks often arise from the way their recommendations are used by healthcare professionals in decision-making. Hence, the accuracy of the data provided and the intended impact on a doctor's clinical decision-making are essential parameters in evaluating the risk level.

The level of automation in AI systems also plays a crucial role in risk assessment.[47] Highly automated AI CDSS with advanced cognitive functions, such as information acquisition and analysis, carry higher risks.[48]

Explainability is another concern with AI-based medical devices.[49] AI's inability to explain the reasoning behind its recommendations raises questions about its reliability and trustworthiness. Lack of explainability in decision-making may lead to hesitancy among healthcare professionals in accepting AI-generated suggestions, thereby affecting patient care.

Moreover, technological failures, misdiagnoses, or incorrect treatment scenarios are potential risks associated with AI-based medical devices. Addressing these risks is essential to ensure patient safety and the ethical use of AI in healthcare delivery.

2.4 *The AI Regulation*

2.4.1 General Structure of the AI Regulation

In April 2021, the European Commission released its proposal for a legal framework for AI in the EU – the AIA.[50] Later on, two other proposals came to the public arena: one from the Council[51] and another from the European

46 Cf. Lagioia and Contissa, 2020: 258.

47 Cf. Lagioia and Contissa, 2020: 258–259.

48 For example, systems like Watson, which suggest specific clinical decisions, present greater risk compared to moderately automated devices.

49 Cf. Lagioia and Contissa, 2020: 260.

50 European Commission, Proposal for a Regulation of the European Parliament and of the Council Laying Down Harmonised Rules on Artificial Intelligence (Artificial Intelligence Act) and Amending Certain Union Legislative Acts, https://eur-lex.europa.eu/legal-content/EN/TXT/?uri=celex%3A52021PC0206.

51 Council of Europe, Proposal for a Regulation of the European Parliament and of the Council laying down harmonised rules on artificial intelligence (Artificial Intelligence Act) and amending certain Union legislative acts – General approach, 25 November 2022, https://data.consilium.europa.eu/doc/document/ST-14954-2022-INIT/en/pdf.

Parliament.[52] At the moment this book is published, the AIA has already been published and has come into force.[53]

The AIA defines a tiered system of risk classes, each with increasingly strict requirements.[54] It introduces a pyramid-shaped classification that categorises AI into four risk classes based on the associated usage risks (even though the later introduction of a legal framework for General Purpose AI (GPAI) disrupted this classification, as GPAI does not quite fit into the original structure). The provider's compliance obligations escalate depending on the assigned risk class, ensuring the product's secure utilisation across its lifecycle.[55]

At its pinnacle are AI systems linked to unacceptably high risks in their usage. Progressing through the pyramid, AI systems with elevated risks are identified – called high-risk – which require a conformity assessment. At the foundation of the pyramid lie AI systems with negligible or no risks, necessitating fewer or even no stipulated requirements. AI medical devices are, by default, considered high-risk AI systems if they are subject to the third-party conformity assessment (as indeed happens with most AI medical devices).[56]

The AIA's approach to identifying high-risk AI systems is rooted in referencing the list of acts outlined in Annex I and Annex III. By doing so, the regulation designates all AI configured as products covered by the regulations listed in Annex I as high-risk AI. Notably, the MDR[57] is one of the legal acts mentioned in Annex I, contributing to the intricate landscape for AI-based medical devices approval process.

2.4.2 Legal Requirements Imposed on High-Risk AI Systems

For high-risk AI systems to obtain the CE marking of conformity and be allowed to enter the EU market, the AIA sets out a series of compliance obligations

52 European Parliament, Artificial Intelligence Act, Amendments adopted by the European Parliament on 14 June 2023 on the proposal for a regulation of the European Parliament and of the Council on laying down harmonised rules on artificial intelligence (Artificial Intelligence Act) and amending certain Union legislative acts (COM(2021)0206-C9-0146/2021-2021/0106(COD)), https://www.europarl.europa.eu/doceo/document/TA-9-2023-0236_EN.pdf.

53 However, its enforceability will come in stages.

54 On the application of such requirements to AI systems used in healthcare see Van Kolfschooten and Van Oirschot, 2024: 105152.

55 Drabu, McAdams, Bonser et al., 2021.

56 Palmieri and Goffin, 2023: 420–421.

57 Regulation (EU) 2017/745 of the European Parliament and of the Council of 5 April 2017 on medical devices, amending Directive 2001/83/EC, Regulation (EC) No 178/2002 and Regulation (EC) No 1223/2009 and repealing Council Directives 90/385/EEC and 93/42/EEC.

(Article 8 AIA et seq.), with risk management becoming central (Article 9 AIA). It ties closely with the MDR's risk management framework to pinpoint, assess, and mitigate 'reasonably foreseeable risks' that high-risk AI systems might present to health, safety, or fundamental rights, such as privacy and data protection. Data governance (Article 10 AIA), an essential component, aligns with the data protection framework but goes further with AI-specific mandates for training, validation, and testing datasets (see Chapter 3, Section 3.1). This includes an evaluation of the 'availability, quantity, and suitability of the data sets that are needed' (Article 10(2)(e) AIA), an examination of potential biases that could impact health and safety, and a consideration of 'specific geographical, contextual, behavioural or functional setting within which the high-risk AI system is intended to be used' (Article 10(4) AIA).[58]

The AIA outlines specific duties for high-risk AI system providers, who bear responsibility for ensuring their systems meet all compliance requirements (Article 16 AIA). To achieve this, providers must establish an AI quality management system, as required under Article 17 of the AIA. This quality management system forms the core infrastructure for ensuring ongoing quality and compliance (Article 17 AIA), integrating key elements like AI documentation (Article 18 AIA), and auto-generated logs (Article 19 AIA).[59] Additionally, providers based outside the EU must appoint an authorised EU representative (Article 22 AIA).

Within the AIA, Chapter IX introduces an obligation for providers to create a monitoring plan and conduct post-market monitoring specifically for high-risk AI systems. This includes post-market monitoring (Article 72 AIA), reporting serious incidents (Article 73 AIA), implementing corrective actions (Article 20 AIA), and working with competent authorities (Article 21 AIA) to maintain regulatory adherence.[60] This mandate aligns with the existing provisions of the MDR. This strategic alignment ensures a harmonised approach to monitoring and safeguards across both the AIA and the MDR. These post-market monitoring obligations focus on ensuring that providers gather, document, and analyse data from deployers or other sources throughout the AI system's lifespan to assess ongoing compliance.

---

58 Aboy, Minssen and Vayena, 2024.
59 Aboy, Minssen and Vayena, 2024.
60 Aboy, Minssen and Vayena, 2024.

A key concern with this approach is its heavy reliance on deployers for oversight and reporting (Article 26 AIA).[61] In healthcare delivery, the hospital will be the deployer. However, it is not excluded that in some situations the doctors will also be considered as deployers.

2.4.3 Intersection between the AI Regulation and the Medical Device Regulation

This intersection between the MDR and the AIA presents practical challenges. Although both regulations employ risk-based classification, they follow distinct identification processes. The AIA's risk categorisation does not necessarily alter the classification under the MDR.[62] This means that particular AI-based software used for medical diagnosis can be classified as a medium-risk medical device under the MDR while simultaneously being considered a high-risk AI system under the AIA (see Chapter 3, Section 2.4), due to the AIA referencing the MDR in Annex I. A noteworthy consequence is that operators of medical devices using AI will be presumed to be employing AI in a high-risk capacity. Consequently, manufacturers and providers will need to adhere to the requirements of both the AIA and the MDR according to the relevant medical devices' risk classification.

High-risk AI systems bound by EU product safety regulations must follow the prescribed conformity assessment procedures.[63] Compliance verification

61 The AIA defines deployers as any natural or legal person, public authority, agency, or entity that uses an AI system under its control. Notably, patients are excluded from this definition. It is assumed that if an AI system is provided to a patient by a hospital, the hospital and its healthcare professionals would be considered the deployers responsible for compliance under the regulation (Solaiman and Malik, 2024).

62 Health Action International, 2023.

63 When the urgency for technological advancement escalates, implementing changes becomes notably challenging and time-consuming. This scenario is recognized as the 'dilemma of technology control', also known as the 'Collingridge dilemma' (Collingridge, 1980). As we come to understand the repercussions of technology, our ability to manage it effectively diminishes substantially. This is due to the technology having already gained substantial momentum and embarked on its own trajectory. The Collingridge dilemma posits two main issues: when a technology is new, its impacts are difficult to predict and thus hard to regulate effectively; but once the technology is well understood and its impacts are clearer, it may be too entrenched to easily control or modify. In the context of AI, this dilemma is evident. Early-stage AI systems can be unpredictable, and the full extent of their societal, ethical, and economic impacts may not be immediately apparent. Regulating AI at this stage requires foresight and flexibility, yet it is challenging to create regulations that anticipate all potential issues without stifling innovation. Conversely, as AI systems become more integrated into various sectors, their impacts become more

involves affixing the 'CE' mark of conformity to the product, signifying adherence to European standards. Notified Body assessment, mandated for high-risk AI, aims for enhanced rigour and impartiality compared to self-assessment.

3 Reduction of the Quality of Healthcare Delivery and Threats to Patient Safety

3.1 *Data*

3.1.1 The Imperative Need for Data

The process of 'training' the AI system corresponds to making the computer learn how to select a good output for any new input it might observe. AI training methods vary, but they all depend on access to a body of large and accurate training data.[64] Because AI relies heavily on the data an algorithm is trained on,[65] the main problems are based on flawed and insufficient training data.[66] Informally, this problem is known as 'garbage in, garbage out'.[67]

In healthcare settings, this involves, for instance, clinical data, that is, data associated with clinical cases. These encompass a range of information that transpires during interactions between doctors and patients, patients' subjective articulations of symptoms and the trajectory of their ailment (many of which are data collected in patients' health records),[68] intertwined with

---

apparent and the necessity for regulation becomes clearer. However, by this point, the technology may be so deeply embedded that implementing effective controls is difficult without significant disruption.

Thus, the Collingridge dilemma underscores the balancing act required in AIA: policymakers must strive to create adaptive regulatory frameworks that can evolve alongside the technology, anticipating risks without hindering beneficial advancements. This approach seeks to address the inherent unpredictability of early-stage AI while remaining robust enough to manage the complexities of more mature AI systems.

64 Bathaee, 2020: 155–158.

65 '[A]ny ML system or human-trained algorithm will only be as trustworthy, effective, and fair as the data that it is trained with' (Gerke, Minssen and Cohen, 2020: 303).

66 In their paper, Zou and Schiebinger (2018: 324–326) discuss the ground-breaking work of Esteva and colleagues (Esteva, Kuprel, Novoa, et al., 2017: 115–18), who used ML to detect skin cancer. It is emphasised that less than 5 percent of the images this model was trained on featured individuals with dark skin.

67 The expression 'garbage in, garbage out' intends to express the idea that when AI is 'fed' with bad data (incomplete, erroneous, biased) the output will be as bad as the input (Geiger, Cope, Ip, et al., 2021: 795–827).

68 A substantial part of the data used to train AI comes from EHRS, which contain very detailed information about the patient, including clinical notes and laboratory values, allowing the application of Large Language Models (LLMS) to extract coded vocabularies.

pertinent clinical particulars like case data, laboratory findings, imaging assessments, and the clinical insights provided by the healthcare practitioner. Even after the system reaches the market and is used in real medical practice, it is advantageous to provide the system with additional data to guarantee better predictions.[69]

The volume of data is essential. To understand a disease and the functioning of the human body is more complicated than other uses of AI and thus to rely on enough data, from multiple sources, is essential.[70] Data are required, not only for training but also for testing the AI system, that is, for its validation. This is typically achieved by setting aside a portion of the data, known as the 'validation set', subsequently used to evaluate the algorithm's performance against specific evaluation criteria. The validation set serves as a benchmark for measuring the algorithm's progress and ensuring its effectiveness throughout the training process.[71] A shortage of large datasets further complicates medical decision support.[72] Other times, the problem is not the absence of data, but the immensity of data. As mentioned, the performance of a network improves with increased depth (more layers) and more rounds of training. However, it is crucial to strike a balance, as excessively deep networks or excessive training iterations can also introduce complexity and potential performance degradation.[73]

The challenge relates not only to the quantity of data but also to the quality.[74] The main issue here is that frequently data come in an unstructured and incomplete form. Uncertain and complex datasets, as well as probabilistic, unknown, incomplete and imbalanced datasets present significant challenges for AI training.

3.1.2 The Source of the Data

Gathering data for the training of medical AI involves many different methods, ranging from physical sensors, encompassing both wearable and non-wearable

---

Still, this might not be the most accurate data source. Vide Hoffman and Podgurski, 2013: 515–21.

69 When the AI model is operating with the data provided by the doctors, the latter must make sure that AI is 'fed' with reliable, accurate and up-to-date data. For instance, when the AI system is embedded in an EHR, the doctors cannot copy-paste data from another patient or from a past appointment with that patient, no matter how similar the situation might appear. Cf. Paterick, Patel, Ngo, et al. 2018: 558–561.

70 Golden, Popescu, Israel, et al., 2024.

71 Froomkin, Kerr and Pineau, 2019: 47.

72 Holzinger, Langs, Denk, et al., 2019.

73 Xu, Coen-Pirani, and Jiang, 2023.

74 Jogan, Kurada, Vasisht, et al., 2024.

configurations, to the incorporation of biosensors. Moreover, information reservoirs like EHRs contribute to this data landscape.

A major obstacle to the use of good data relates to privacy and data protection concerns,[75] especially in light of the General Data Protection Regulation (GDPR).[76] The GDPR applies to any processing of personal data, an extremely broad concept.[77] Personal data under the GDPR encompasses any information that relates to an identified or identifiable natural person, known as a 'data subject'.[78] This includes direct identifiers like names, identification numbers, and location data, as well as indirect identifiers such as online identifiers or factors specific to the physical, physiological, genetic, mental, economic, cultural, or social identity of that person. The term 'processing' is equally wide-ranging, covering any operation or set of operations performed on personal data, whether or not by automated means. This includes activities such as collecting, recording, organising, structuring, storing, adapting, or altering data. It also covers more complex operations like retrieving, consulting, using, disclosing, disseminating, aligning, combining, restricting, erasing or destroying personal data.

The first requirement for data processing under the GDPR is to find a legal ground for establishing the lawfulness of the personal data processing. Without a valid legal basis from Article 6(1) of the GDPR, the processing of personal data will be considered unlawful and can result in penalties: consent (Article 6(1)(a) GDPR); performance of a contract (Article 6(1)(b) GDPR); compliance with a legal obligation (Article 6(1)(c) GDPR); protection of vital interests (Article 6(1)(d) GDPR); public interest or official authority (Article 6(1)(e) GDPR); and fulfilment of legitimate interests (Article 6(1)(f) GDPR).[79]

Medical AI requires health and genetic data, which are a special category of personal data (Article 9(1) GDPR), whose processing is forbidden unless an exception applies. This exceptional case involves first the finding of an additional legal ground in Article 9(2) of the GDPR: explicit consent (Article 9(2)(a) GDPR); matters of employment, social security, or social protection law (Article 9(2)(b) GDPR); protection of vital interests (Article 9(2)(c) GDPR); use of data by not-for-profit bodies (Article 9(2)(d) GDPR); data made

75 Topol, 2019: 52.

76 Regulation (EU) 2016/679 of the European Parliament and of the Council of 27 April 2016 on the protection of natural persons with regard to the processing of personal data and on the free movement of such data, and repealing Directive 95/46/EC (General Data Protection Regulation).

77 Purtova, 2018: 40–81; Finck and Pallas, 2020: 11–36.

78 Karthik, 2023.

79 Kotschy, 2020: 321–344.

public by the data subject (Article 9(2)(e) GDPR); use of data for legal claims (Article 9(2)(f) GDPR); substantial public interest (Article 9(2)(g) GDPR); use of data within healthcare and social care (Article 9(2)(h) GDPR); public health issues (Article 9(2)(i) GDPR); archiving, research, and statistical purposes (Article 9(2)(j) GDPR).[80]

To avoid GDPR issues, some AI developers resort to synthetic data.[81] However, the use of synthetic data comes with its own set of challenges. One of the primary concerns is the accuracy and reliability of AI models trained on synthetic data compared to those trained on real personal data. Real data often contains nuanced patterns, outliers, and correlations that are difficult to replicate in synthetic data.[82] As a result, models trained on synthetic data may not perform as well in real-world scenarios, where such subtleties can significantly impact outcomes.[83]

Despite the protective legal regime set forth by the GDPR, there are still several loopholes in data protection. The problem arises because the foundation of data protection remedies centres on individual rights, which stem from the system's alignment with a human rights framework. Conversely, algorithmic harms commonly emerge from the categorisation or stigmatisation of groups within systems, leading to a gap in protection that these frameworks struggle to address effectively.[84]

80 Georgieva and Kuner, 2020: 365–384.

81 This move is based on the idea that synthetic data are a type of anonymised data, which is not certain yet (Anagnostaki, 2023) and also that anonymised data are a legal shield against the GDPR, which is not guaranteed either, namely because this type of data can be reidentified using AI reverse engineering (El Mestari, Lenzini and Demirci, 2024; Levine, Philpot, Nightingale, et al., 2024: 116, 118).

82 Rafiei, Rad, Sikora et al., 2024. In the opposite sense, Savage, 2023.

83 For instance, take the case of IBM Watson Health's cancer AI algorithm (Watson for Oncology). The system was trained with synthetic data, that is, artificial data from non-existent patients created by AI itself. Synthetic data are seen as a promise for AI training, but at least in the case of this AI medical system, the results are not promising. Concerns have been raised regarding the accuracy and safety of its recommendations, as it has been reported to provide 'unsafe and incorrect' treatment suggestions (Ross and Swetlitz, 2018). Because the software was trained using artificial data, it lacked the necessary sophistication to decipher uncertain, intricate, or untidy patient EHRs. Consequently, this shattered the trust of medical professionals and healthcare facilities (Arnold, 2021: 127).

84 Edwards and Veale, 2017: 22.

3.1.3 Big Data

3.1.3.1 *The Concept of Big Data*

Doctors – and humans in general – have access now to an overwhelming volume of information, but they cannot get all the benefits that theoretically could result from that information because its volume exceeds the human capacity for effective utilisation.[85] Computers, in contrast, have the ability to handle an immense volume of health-related information, marking a significant leap in capacity compared to the past.[86]

Big data is a concept that refers to extremely large and complex datasets that, due to these features, cannot be easily managed, processed, or analysed using traditional data processing tools and methods. These datasets are characterised not only by their massive size but also by their variety (big data encompasses structured, semi-structured, and unstructured data from various sources such as sensors, social media, digital transactions, multimedia content, and more),[87] velocity,[88] and volume.[89]

The cutting-edge technologies of AI, ML, IoT, neural networks, and quantum computing have now come into play. These powerful tools enable companies to process vast amounts of data more efficiently than ever before, unlocking valuable insights and driving innovation at unprecedented levels to uncover patterns and insights.[90] Embracing these advanced technologies empowers

---

85 Senger and O'Leary, 2018: 289–290.

86 About big data in healthcare, Batko and Ślęzak, 2022.

87 A variety of data types and formats fall under the umbrella of big data, including structured data like traditional databases, semi-structured data such as XML files and JSON, and unstructured data like text, images, and videos. Coping with this diversity necessitates adaptable tools and methodologies. More details in Sakr, 2020.

88 Data are generated and collected at an unprecedented speed in today's digital age. For instance, social media interactions, online transactions, and sensor data continuously contribute to data flow. Real-time processing and analysis are often necessary to extract meaningful insights from this rapidly generated data (Oussous, Benjelloun, Lahcen, et al., 2018: 431–448).

89 Big data involves datasets that are too large to be managed and processed by traditional databases or software. This immense volume of data requires specialised tools and technologies to store, process, and analyse data. Cf. Batko and Ślęzak, 2022.

90 An illustrative and widely known anecdote involves the American supermarket Target, which, as it seems, was employing big data customer profiling to identify potential customers for targeted offers. The story gained prominence due to its implications for privacy and the potential insights that big data can provide.

In this case, Target utilised its vast amount of customer data, including purchase history and browsing patterns, to identify features (that is, patterns) indicating that a certain customer was pregnant. What happened is that Target's data scientists discovered that certain products, such as unscented lotions and supplements like calcium and magnesium, were often purchased by pregnant women in the early stages of pregnancy.

businesses to stay ahead in the data-driven era and deliver exceptional value to their customers and users. While in the past decision-support systems relied on expert knowledge to achieve high-level decision performance in structured and limited problem areas, AI research and development are increasingly embracing data-driven methods, which are gradually replacing the traditional knowledge-based approach in specific domains.

3.1.3.2 *Factors Leading to the Rise of Big Data in Healthcare*

One factor driving the rise of big data in healthcare is the increasing accessibility of genomic data.[91] Nowadays, whole-genome sequencing is widely available, and advanced sequencing technologies ensure swift, extensive, and accurate genetic data retrieval. Consequently, individual health records are progressively incorporating personal genomic information. These breakthroughs empower healthcare professionals with unprecedented insights into diseases, especially in fields like cancer genomics and pathogenomics. The rise of genomic testing also facilitates the availability of population-level data, against which a patient's genetic profile can be assessed.

Another contributing factor to the data torrent arises from the development of highly accurate and portable sensors and the emergence of the Quantified Self-Movement.[92] Today, our surroundings proliferate with products that continuously collect information about both us and our environment. Nearly everyone who carries a smartphone automatically accumulates activity and location data, and those employing smart devices like watches or scales may amass even more comprehensive data.[93]

Using this information, Target developed a pregnancy prediction model that assigned a 'pregnancy prediction score' to each customer based on their shopping habits. With this model, Target could send targeted advertisements and promotions to customers who were likely to be expecting a baby, such as offering them discounts on baby-related products. The case gained attention when a father angrily confronted a Target store manager after his teenage daughter received pregnancy-related coupons in the post. It turned out that Target's data analysis had correctly predicted the daughter's pregnancy based on her shopping habits, making the store aware of that fact even before her father was aware of the situation. This incident sparked discussions about the ethics of data collection, privacy, and the power of predictive analytics. Cf. Mai, 2016: 192.

91 O'Connor and McVeigh, 2025.

92 The Quantified Self Movement entails the self-monitoring, measurement, and quantification of every facet of daily life through the use of technology, including applications and smart wearable devices like activity trackers or sleep monitors. See Swan, 2013: 85–99.

93 In addition to healthcare professionals, patients are provided with the chance to employ patient-facing applications. This domain encompasses various technologies such as voice assistants, text-based chatbots, smartphone apps with wearable sensors, and other devices (Cerrato and Halamka, 2019). These applications empower individuals to

Furthermore, the digital transformation and standardisation of medical records play a crucial role in overcoming historical information deficits that have hindered the development of CDSS.[94] EHRs serve as a convenient, centralised source from which CDSS can draw patient-specific information. This means that multiple streams of data – ranging from records held by primary care doctors and specialists to information gathered in emergency rooms, urgent care clinics, or during visits to various healthcare providers throughout a patient's lifetime, including family medical history – can now coexist in a single accessible location.

The abundance of data available to doctors in the modern era is staggering, and ongoing technological advancements suggest that doctors may soon gain access to more than just a patient's extensive medical history. This expanded access could encompass not only records from prior medical consultations but also personal genomic data, extensive behavioural data collected from various sources, wearable devices, environmental data, and the vast repository of global medical knowledge, all in real-time.

3.1.3.3 *Structured Data*

Accurate and effective AI systems require well-structured data inputs.[95] In healthcare, the flexibility of AI models is limited by the constraints of the data they were originally trained on. Consequently, before using all available data, a meticulous process for organising and preparing the data is necessary. This process involves grouping patients based on specific criteria, segmenting the data to focus on regions of interest relevant to AI interpretation, and applying filters to assess the quality of image acquisition. The need for structuration is present even in DL systems, known for their capacity to automatically extract pertinent mathematical representations from datasets.

There is, however, the risk that relying too heavily on structured data may narrow the breadth and diversity of information that can inform AI-driven healthcare processes. The fact is that because these methods are trained to accommodate specific input types (e.g., medical image data), they are unable to operate when confronted with alternative input modalities like patient

independently assess symptoms, seek treatment, and self-manage conditions like respiratory illnesses, diabetes, and epilepsy. The integration of AI within these systems enhances their functionality by enabling personalised tracking of a patient's condition and providing tailored guidance (Asan, Bayrak and Choudhury, 2020). However, this book will only deal with AI systems integrated into healthcare delivery in hospital settings.

94 Castaneda, Nalley, Mannion, et al., 2015: 1–16.

95 Arogundade, 2023: 31–39.

statements.[96] Despite this risk, structured data are essential for developing reliable AI systems.

3.1.3.4 *Labelled Data*

In order to train AI algorithms effectively, it is essential to use properly labelled data.[97] Data labelling is the process of annotating or tagging data with specific labels or information to make it understandable for AI models. This step is crucial in training AI, particularly for supervised learning, where the model learns from labelled examples to make predictions or classifications on new, unseen data.

For instance, for tasks like tumour detection, this dataset would consist of images (the inputs) paired with annotations provided by expert radiologists indicating the desired output, such as tumour/no-tumour classification or detailed tumour-contour segmentation. The presence of human-annotated labels in the image dataset is crucial for the algorithm's training process.[98]

Also, take the case of Traditional Chinese Medicine (TCM), where unstructured clinical cases accumulated over time pose data conversion challenges. Bridging the gap between human-generated clinical cases and machine-understandable language relies on core technologies like NLP and medical text recognition. The hurdle here is the intricate and non-standardised nature of TCM expressions, leading to a costly and convoluted process of manual annotation, which significantly impacts data quality.[99]

The accuracy of data annotations is crucial for the quality of the dataset.[100] However, annotating the dataset is a time-consuming and labour-intensive task that requires robust validation of the ground truth diagnosis. Even when data are both accurate and representative, their value can be diminished due to issues that arise during the annotation process. For instance, the gold standard for diagnosing pulmonary nodules is through pathological biopsy, but not all cases allow for such an invasive procedure. The annotation of AI data, particularly in the context of CT scans for lung nodules, requires the expertise of experienced doctors.[101]

One challenge arises when dealing with rare findings, because in those cases it is extremely difficult to obtain a sufficient number of examples to train

96 Asan, Bayrak and Choudhury, 2020.
97 Taye, 2023.
98 Gunasekara, Kaldera and Dissanayake, 2021.
99 Zhang and Zhang, 2023.
100 European Society of Radiology, 2019: 4.
101 Zhang and Zhang, 2023.

an algorithm in such a way that it can confidently identify these conditions. Additionally, the presence of variation in the data poses a risk of inadvertent overfitting, where the algorithm interprets random noise as an abnormality.[102]

3.1.3.5 *Overfitting*

Overfitting occurs when an ML model becomes excessively fine-tuned to the dataset it was trained on, causing it to lose its generalisability to other datasets. A model is deemed overfitted when its characteristics are closely aligned with the initial dataset to the extent that using it for future data would yield flawed or inaccurate results, ultimately leading to suboptimal decision-making.[103] The model is excessively focused on irrelevant noise rather than effectively learning how to perform when faced with new case scenarios, leading to poor performance on new datasets. In a sense, it is similar to memorising a maths textbook and then facing entirely new problems during an exam. Rote memorisation without understanding will not yield much success.[104] The same can happen with overfitting. It can result in interpretation errors that hinder the network's ability to generalise its findings to unseen data.

In healthcare, a major consequence is the inability of the AI system to consistently operate across diverse patient groups. As these methods are intricately tied to specific tasks and trained using specific datasets, they often struggle to perform effectively when confronted with input data that deviates even slightly from their training data. In instances where new medical trends or unanticipated diseases emerge, the reliability and validity of AI-derived results may be compromised.

An illustrative example of overfitting in the healthcare domain involves a predictive model developed to identify patients at risk of a certain medical condition based on a specific dataset. Let's say the model is trained on a particular population with a unique set of attributes, such as age, gender, and medical history. If the model becomes overly tailored to these specific features during training, it might perform exceptionally well when tested on the same dataset but struggle to generalise its predictions to a different population. As a consequence, when applied to a broader and more diverse patient population, the model might produce inaccurate predictions due to its overreliance on the nuances of the training data.

To address overfitting, obtaining more training data is often the most effective solution. Additionally, conducting multiple rounds of training and testing on diverse datasets progressively enhances the network's performance and

102 European Society of Radiology (ESR), 2019.
103 Gadepally and Lakhani, 2022.
104 Bashir, Montañez, Sehra, et al., 2020: 347–358.

enables the assessment of accuracy and generalisability before the algorithm is deployed for widespread use. Another approach to mitigate overfitting is through 'data augmentation'. This involves introducing variability into the training data by modifying certain aspects, such as adding artefacts or imperfections like the ones the model may encounter in real-world scenarios.[105]

3.1.3.6 *Underfitting*

Overfitting contrasts with underfitting, where the model's grasp of patterns is minimal, leading to crude predictions.[106] Imagine a model predicting running times solely based on the runners' heights; it might work at the height extremes but falters for similar heights in the middle.

In healthcare, underfitting occurs when a model is too simplistic and fails to capture the complexities of the data, resulting in overly generalised and inaccurate predictions. For instance, consider a model designed to predict patients' recovery times after surgery. If the model only considers a single factor, such as the patient's age, it may make some accurate predictions for very young or very old patients, as these age groups often have more predictable recovery patterns due to well-known physiological factors. However, for patients who fall in between these age extremes, the model is likely to struggle. This is because recovery times can be influenced by a wide range of other factors, such as the patient's overall health, the presence of comorbidities, the type of surgery performed, lifestyle factors like diet and exercise, and even social support systems. By relying solely on age, the model misses these crucial variables, leading to crude predictions that do not reflect the real-world diversity of patient recovery experiences. As a result, the model might predict similar recovery times for a 40-year-old marathon runner and a 40-year-old smoker with diabetes, simply because they share the same age. In reality, these two patients would likely have very different recovery trajectories, but the underfitting model cannot account for these differences, leading to potentially inaccurate and unhelpful predictions.

This highlights the importance of including a variety of relevant features in the model to improve its accuracy and reliability. This shortcoming reminds us that models must adapt to training data patterns, and if those patterns do not hold in new data, predictions will suffer.

---

105 For instance, when training a network to recognise cystic lesions on ultrasound, the training dataset should include examples with internal hyperechoic spots caused by artefacts, teaching the network to identify 'non-perfect cysts' and to generalises better to real clinical images (European Society of Radiology (ESR), 2019: 3).

106 Bashir, Montañez, Sehra, et al., 2020: 347–358.

3.2 *Bias*

3.2.1 AI Biases in Healthcare

AI biases refer to systematic errors or prejudices in the outputs generated by AI systems, often reflecting the biases present in the data used to train these systems or the design of the algorithms themselves. In certain cases, attributes like ethnic group might statistically correlate with outcome variables such as the likelihood of property theft convictions, fraudulent tax claims, response to payday loan ads, or job advancement. Despite statistical validity, societal and political preferences often wish to mitigate such correlations.[107]

Healthcare is not immune to these drawbacks. AI models – for instance, AI-CDSS – can unintentionally perpetuate or even amplify biases, particularly when creating profiles based on collective data rather than individual characteristics. When AI systems create profiles by comparing individuals within a dataset, they often draw on patterns and correlations between various attributes like gender, ethnicity, socioeconomic background, and more. These profiles are not just reflective of individual traits but are influenced by how different groups are represented and treated within the data.[108]

Biases are especially at issue in matters of race[109] and gender,[110] latent in practically all aspects of our society. For example, in healthcare, a CDSS might be designed to predict patient outcomes by identifying links between diagnoses and social determinants of health such as race, income level, or education. However, if the underlying data contain biases – such as historical disparities in healthcare access or outcomes for marginalised groups – the AI system may learn and reinforce these biases. This can lead to outcome predictions that disadvantage these groups, exacerbating existing inequalities rather than mitigating them.

3.2.2 The Causes of Biases

Algorithmic bias encompasses two facets: bias instigated by human agency and bias derived from data sources.

Human-induced bias is introduced by developers, whether deliberately or unintentionally, due to the constant influence of personal morals and relevant incentives, which affect data training.[111] Examples include labelling data, handling inaccurate data, setting variables, and prioritising attributes, which can

107 Edwards and Veale, 2017: 28.

108 Lysaght, Lim, Xafis, et al., 2019: 304.

109 Obermeyer, Powers, Vogeli, et al. 2019: 447–453.

110 Cirillo, Catuara-Solarz, Morey, et al., 2020.

111 Howard and Borenstein, 2018: 1521–1536.

unintentionally reflect the developer's biases or limitations. The concern is not just about explicit developer bias but rather the unintentional and indirect embedding of discriminatory elements within these systems.[112]

Dataset-induced bias occurs when the underlying datasets reflect existing biases against vulnerable groups prevalent in society, as algorithms will inadvertently adopt and reproduce them in their outputs.[113] Reservations arise when it comes to decision-making based on historical data because the data's underlying structures often include unwanted correlations that most of us would rather not reinforce.[114] These correlations frequently pertain to 'protected characteristics', attributes like race, gender, pregnancy status, religion, sexuality, and disability, which many jurisdictions prohibit from influencing decision-making, whether directly or indirectly.[115] Moreover, medical AI may inherit bias from the flawed and prejudiced decisions made by human doctors, which also serve as AI's training data.

3.2.3 Examples of Bias in Healthcare

3.2.3.1 *Ethnic Group Biases*

A well-known case of ethnic biases in medical AI refers to the diagnosis of skin cancer. Suppose there is an AI-based CDSS that helps doctors find the best treatment for patients with skin cancer, but that was predominantly trained on Caucasian patients. In such a case, the AI software is likely to give less precise or even imprecise recommendations for subpopulations for which training data were missing, such as African Americans.[116]

---

112 It has even been said that algorithmic biases exceed human biases. In a test with seven deep learning algorithms, the latter showed greater discrimination than a panel of eleven pathologists (cf. Ehteshami, Veta, Van Diest, et al., 2017: 2199–2210). See also Miller, 2018.

113 About biases in AI used in healthcare, Kerasidou, 2021: 612–614; McCradden, Joshi, Mazwi, et al., 2020: e221–e223.

114 For instance, an algorithmic CV filtering system that solely relies on past applicant success rates might unwittingly replicate biases from manual CV filtering processes, neglecting women and/or ethnic minorities. Likewise, in healthcare delivery, an example could be an AI system designed to prioritise patients for treatment based on historical outcomes. If the algorithm relies heavily on past treatment success rates without accounting for socio-economic factors, it might unintentionally replicate existing biases. For instance, it could prioritise patients from wealthier backgrounds who have historically had better access to healthcare, while neglecting underserved populations, such as low-income individuals or ethnic minorities, who may have faced barriers to receiving timely and effective care in the past. This could perpetuate disparities in healthcare access and outcomes, rather than addressing them.

115 Edwards and Veale, 2017: 28.

116 Gerke, Minssen and Cohen, 2020: 304.

Another example relates to clopidogrel, a drug used to prevent heart attacks and strokes. The effectiveness of clopidogrel is linked to the CYP2C19 gene, and specific alleles affect its efficacy. However, clinical studies primarily involved participants of European ancestry, resulting in the erroneous conclusion that the drug's efficacy extended broadly. In reality, the drug's effectiveness varies among different ethnicities, potentially leading to inadequate treatment for certain patient groups.[117]

As another illustration, let's consider the case of hypertrophic cardiomyopathy, a condition where the heart's wall thickens abnormally, posing a risk of abnormal rhythms and sudden death, particularly among young athletes. Genetic tests are used to identify the disorder, but a study revealed that initial data inadequately represented Black Americans, leading to misdiagnoses and improper risk assessment for this population.[118]

Biases can have consequences not only on the therapeutic outcome, but also on the allocation of resources. For instance, an algorithm employed in the US for allocating healthcare resources was found to exhibit significant discrimination against African Americans. Thus, even when equally ill, African Americans were much less likely to be referred for treatment compared to White individuals.[119] The reason was that the algorithm's prediction model, designed to anticipate patients' future health needs from recent health expenditures, failed to accurately identify the needs of Black patients compared to their White counterparts. This discrepancy results from the fact that the algorithm's designers neglected to consider that historically, healthcare spending for Black Americans tends to be lower than for White Americans, despite similar health conditions. This difference arises due to various unrelated factors, such as unequal access to healthcare, substandard medical services, or lack of insurance. The algorithm employed insurance claims data to assess patients' health requirements, a seemingly logical approach, but it overlooked the inherent bias in healthcare spending by race. For White patients, the algorithm's recommendations were valid, as lower spending indicated fewer health conditions. However, this very correlation accentuated racial disparities in healthcare for Black patients. This leads to a flawed inference that individuals of Black ethnicity are healthier, due to the lower healthcare expenditures on them, despite them being notably more ill according to the datasets.[120] Inadvertently, this model intensified the existing biases in medical

117 Examples suggested by Price II, 2019: 92.

118 See Price II, 2019: 91–92.

119 Obermeyer, Powers, Vogeli, et al., 2019: 447–453.

120 Obermeyer, Powers, Vogeli, et al., 2019: 447–453.

care, perpetuating the unequal treatment of Black patients by recommending interventions that were suitable for White patients but not for those who faced barriers to adequate healthcare access.[121]

Similarly, studies have found that among women with breast cancer, Black women were less likely to be tested for high-risk germline mutations compared to White women, despite having a similar risk of such mutations. Consequently, an AI algorithm that relies on genetic test results is more prone to inaccurately assessing the risk of breast cancer for Black female patients in comparison to White female patients.[122]

These examples indicate that patients belonging to marginalised groups, particularly Black patients, might receive subpar medical care compared to White patients. Such cases heighten the potential for unintended consequences that propagate harm to individuals and society at large, thereby perpetuating existing health disparities.

Astonishingly, a recent study revealed that healthcare algorithms possess the capability to infer a patient's racial background based on medical images like chest and hand X-rays and mammograms.[123] This feature can lead to even more 'disguised biases', where the algorithm's predictions or recommendations are influenced by racial factors, even when they are not explicitly intended to be. Such biases can lead to unequal treatment and reinforce disparities in healthcare, as the underlying assumptions or patterns embedded in these algorithms might inadvertently perpetuate existing inequalities.

3.2.3.2 *Gender Biases*

Not only ethnic group biases but also gender biases are at stake here. The convergence of these biases within AI systems inadvertently perpetuates gender-based distinctions and inequalities, ultimately leading to unfavourable health outcomes for women. An alarming implication is the potential for algorithms to misinterpret symptoms – for instance, when determing what constitutes 'typical' indicators of a cardiac issue. Consequently, many women may face misdiagnoses.

Gender disparities extend to medication testing, with several drugs, ranging from anaesthetics to cardiovascular treatments, predominantly studied in male subjects.[124] This discrepancy fails to acknowledge the nuanced physio-

121 Obermeyer, Powers, Vogeli, et al., 2019: 447–453.

122 Parikh, Teeple and Navathe, 2019: 2377.

123 Gichoya, Banerjee, Bhimireddy, et al., 2022: e406–e414.

124 Historically, women have frequently found themselves excluded from clinical trials, spanning various health conditions and treatment assessments. This omission is particularly

logical differences between genders.[125] The need to rectify these gender-based discrepancies within healthcare and AI systems is a critical pursuit to ensure equitable and effective medical practices.

3.2.3.3 *Age Biases*

Age can also be a problematic factor. Let's take the example of an algorithm trained on a predominantly young, healthy population that uses vital signs and laboratory readings to recommend treatment for hypertension. Suppose further that current guidelines suggest that older adults may also benefit from personalised treatment, provided that higher blood pressure thresholds apply for hypertension than those used to indicate interventions for younger adults.[126] A health system that serves a predominantly elderly population may be responsible for damage resulting from its use, as it is likely to provide recommendations that are not in line with the current standard of care for this age group.

Parikh and colleagues highlight how doctors may overlook the diagnosis of myocardial infarction in older women due to atypical symptoms. Consequently, an AI algorithm trained on historical data (for instance, EHR data and existing practices) might fail to recommend testing for cardiac ischaemia in older women, potentially delaying life-saving treatment.[127]

3.2.3.4 *Contextual Biases*

3.2.3.4.1 What Are Contextual Biases

Typically, medical AI is trained in well-resourced environments, such as academic medical centres and state-of-the-art hospitals, which possess ample data and experienced practitioners, making it an intuitive choice for training. Thus, the data that feed the AI often refer to the places where this technology is used, which, given its current cost, is still limited to hospitals in more affluent communities.

Two primary factors contribute to this bias: variations in patient populations and treatment rankings influenced by available medical resources.[128] These factors can impact both the accuracy of diagnoses and the appropriateness of treatment options, leading to 'contextual bias'.[129]

---

pronounced in instances of Parkinson's disease, where diagnoses lean disproportionately toward men (Vaswani, Tropea, and Dahodwala, 2020: 1724–1735).

125 Marotta, 2022: 1311.

126 Example taken from Maliha, Gerke, Cohen, et al., 2021: 629–647.

127 Parikh, Teeple and Navathe, 2019: 2377.

128 Price II, 2019: 91.

129 Price II, 2019: 65–116; Price II and Cohen, 2024: 345–346.

Algorithms developed in high-resource hospitals may encounter significant challenges when applied in lower-resource settings, leading to suboptimal diagnoses and treatment recommendations.[130] Because the collected data refer to patients who have access to better healthcare and better nutrition, the outcomes obtained may not be adequate for poorer populations,[131] leading to ill-suited solutions for these vulnerable groups. These AI systems replicate the biases of the original system.[132] That is, because the system is trained with a certain type of data (usually, from privileged classes) it tends to be applied most to individuals belonging to that group. This is where the system reaches better results; but, in turn, the AI system keeps learning with data from these particular individuals, perpetuating its initial biases.[133] In sum, the adequacy of various options depends on the available resources, which further influence the algorithms' recommendations. In scenarios where treatment options are presented, the most suitable choice in a resource-abundant environment may differ significantly from one in a resource-constrained setting. Challenges also arise when certain recommended treatments are simply inaccessible in low-resource settings due to factors like expensive machinery, specific transportation needs, or storage conditions, particularly for drugs requiring refrigeration. Moreover, some treatments may yield excellent results when conducted by skilled experts with robust support systems, but their efficacy diminishes when attempted without such expertise. In sum, algorithms trained in well-endowed environments may favour treatments that are optimal only within those settings. When applied in low-resource settings, the quality of care may decline, albeit inconspicuously.

The discrepancy between high-resource training and low-resource deployment environments presents a significant challenge, resulting in a decline in the quality of AI recommendations for care. This limitation jeopardises the actual democratisation of excellence that medical AI promised.

Contextual biases also have consequences in terms of costs. In certain cases, AI may recommend the more expensive option simply because it is commonly favoured in high-resource settings, even though a lower-cost option would be more suitable for the context with fewer resources. This tendency can lead to anti-frugal effects, where cost-saving measures are undermined. While medical AI can potentially reduce some expenses since the software is typically

130 Price II, 2019: 104.

131 Minssen, Gerke, Aboy, et al., 2020.

132 The majority of the data employed to train these models originate from populations that are Western, educated, industrialised, rich, and democratic, often referred to as 'WEIRD' populations. Cf. OECD, 2020: 12.

133 Schönberger, 2019: 176.

cheaper to operate than hiring an additional diagnostician, it might also raise other costs by systematically aligning care patterns with those of high-resource and more expensive healthcare settings.[134]

3.2.3.4.2 Examples of Contextual Biases

For instance, a powerful chemotherapeutic cocktail might be appropriate in a top cancer centre where specialists closely monitor patients. However, it could pose undue risks in a hospital located in a poor community and lacking such safeguards, potentially leading to adverse outcomes.[135]

Likewise, in a resource-rich environment, a powerful chemotherapy drug might be most effective against cancer but could come with high risks of severe side effects necessitating skilful monitoring.[136] In the same vein, opting for surgery may be the right decision in a well-resourced setting but can be unfavourable in places lacking such resources. Performing the surgery to remove the cancer requires greater surgical skill, knowledge of liver and gallbladder anatomy, and the use of specialised instruments. However, the risks are significant, as mistakes during the procedure can lead to severe consequences. By contrast, stopping the surgery and transferring the patient to a more skilled surgeon with better equipment carries minimal immediate risk but no immediate benefit. Consider cholecystitis, a common inflammation of the gallbladder. In such instances, the surgeon must be vigilant during the laparoscopic surgery to detect signs of cancer. They must then decide whether to continue with the regular surgery or halt it and refer the patient to a better-equipped hospital. When a patient's symptoms appear to be cholecystitis but are actually due to gallbladder cancer, continuing the procedure as a routine laparoscopic cholecystectomy would be inappropriate, as it risks tumour spread and may compromise subsequent oncological management.

3.2.3.4.3 Contextual Biases as an Obstacle to Better Healthcare

Currently, there are vast disparities in the standard of care that patients receive based on the context in which they seek treatment.[137] These differences encompass various aspects, from the availability of specialised practitioners to the qualifications and credentials of the healthcare providers involved.

The democratisation of medical expertise is a significant promise of medical AI, which aims to enhance medical expertise by 'levelling up' the quality of

134 Price II, 2019: 98.

135 Example provided by Price II, 2019: 67.

136 Price II, 2019: 95–97.

137 Shepherd, Willis-Esqueda, Paradies, et al., 2018.

care provided.[138] It has the potential to enable a broader range of healthcare providers to deliver care at the level of excellent specialists. This levelling up could significantly reduce the variations in care experienced by patients, ultimately leading to improved medical outcomes and patient satisfaction, and truly democratising medical expertise, benefiting patients worldwide.[139] Its scalable nature makes it a powerful tool for expanding access to high-quality healthcare and levelling up the capabilities of healthcare providers across diverse contexts.

However, the true test lies in deploying this technology in low-resource settings like community hospitals, rural health centres, and practitioners' offices, which face unique challenges due to different patient populations and limited resources. In the context of medical AI, addressing such variations and contextual differences is crucial to ensure that the algorithms deliver accurate and equitable care across diverse patient populations. Failure to account for these gaps may not only compromise patient outcomes but also expose healthcare providers and manufacturers to legal challenges and claims of false or misleading marketing.

3.2.4 Biases Mitigation and Accountability

Some steps are available to identify AI biases and mitigate their impact. This is a crucial concern, as in cases where the model's output is erroneous, the result for the patient might fall significantly short of the medical standard of care.[140]

The first step is to verify what biases exist through testing and training. The subsequent consideration involves investigating whether bias within the data could account for the AI's accuracy decline in real-world scenarios. Unless the deployment of medical AI is integrated into a new data infrastructure that feeds back valuable information to the algorithm's development, any contextual issues are likely to persist unresolved as the algorithm is employed in novel settings.[141]

One approach to rectify this is to construct fair or non-discriminatory AI systems that exclude explicit input of these characteristics, even if they hold predictive value. However, this might not ensure fairness, as the excluded variables could still be interconnected with the included variables.[142] That is, it might be that such associations could be avoided by simply excluding variables

138 Price II, 2019: 73.
139 Bajwa, Munir, Nori, et al. 2021.
140 Bathaee, 2020: 161–162.
141 Price II, 2019: 100.
142 Edwards and Veale 2017: 29.

such as ethnic group, age or gender from the model, but non-sensitive attributes can easily act as so-called 'proxies'.[143]

When neglecting to consider AI biases that could have been detectable, liability might arise.[144] On the other hand, constraining AI based on the acknowledgement of biases does not necessarily alleviate liability, as it could instead indicate the user's recognition of a specific risk, and that it is a conscious decision to keep using that AI system in a hazardous way. Constraints imposed on AI might diminish liability if they effectively minimise biased outcomes, but if they are insufficient, they might demonstrate the user's awareness of the potential for discrimination should they rely on the AI's judgment, which, once again, is a potential indicator of liability.[145]

3.3 *Lack of Transparency*

3.3.1 Transparency versus Explainability

Transparency and explainability are concepts that cannot be confused.[146]

Transparency is about describing the different steps used by AI in its decision-making process. Ultimately, it might involve the disclosure of the algorithms, the very complex techniques employed, and various kinds of information that most likely laypeople are not able to understand. In certain cases, disclosing the mere use of AI may be considered an essential aspect of transparency.[147] It is often naively treated as a panacea for ethical issues arising from new technologies.[148]

Transparency primarily involves revealing details about the AI system, encompassing its operational mechanisms, including the underlying rationale. The latter aspect pertains to explainability, which can be viewed as a subset of transparency in a broad sense. Since explainability holds its distinct significance, it is frequently regarded as an independent concept.[149]

---

143 Schönberger, 2019: 176. For example, a patient's postal code might seem like a neutral piece of information, but it can act as a proxy for socioeconomic status, ethnic background, or access to healthcare facilities, all of which are correlated with health outcomes.

144 Raub, 2018: 529–570.

145 In this line of reasoning, Bathaee (2020: 159) suggests that an alternative strategy for the AI user (or the AI creator, I shall add) might involve refraining from imposing any restrictions. They could then argue that any harm stemming from the AI's opinions was unforeseen and unintended.

146 Raposo, 2024b: 1–12.

147 However, as AI becomes more prevalent, this information may become less significant unless the AI system plays a critical role in the given context.

148 Mittelstadt, 2021: 16.

149 Distinguishing between the two see, amongst many others, Matulionyte, Nolan, Magrabi, et al. 2022.

If we define the lack of transparency as the absence of detailed information about an AI system (such as its goals, the data used for training, the algorithms at play, performance metrics, bias identification and mitigation, human oversight, and compliance with legal standards),[150] then it follows that explainability is a component of transparency.[151] However, even with this understanding, it is important to acknowledge that transparency and explainability are distinct concepts. Explainability is a specific aspect of transparency, focusing on the clarity of how and why an AI system produces certain outcomes. While many aspects of transparency can be addressed through technical and legal measures, explainability often encounters challenges due to current technological limitations, making it difficult to provide clear explanations.

In most cases, explainability is far more useful for doctors and patients, but in some scenarios (such as auditing) transparency will be the one taking the lead. Transparency builds trust among stakeholders, particularly doctors and patients, which is key to the successful implementation of AI in clinical practice,[152] but transparent AI does not solve the problem of unexplainable AI. In sum, explainability and transparency are not the same, but they are certainly interrelated.

3.3.2 Lack of Transparency in Medical AI

Not long ago, an AI algorithm developed by Google for breast cancer screening received attention for its promising performance: it has been shown to improve the speed and robustness of breast cancer screening, to generalise well to populations in multiple countries beyond those used for training, even to outperform radiologists in specific situations.[153] However, this work also received some criticism in the media and AI community, as it was presented with almost no details on how the algorithm was built or on its main technical descriptions, which highlights the need for more transparency in AI.[154] This is a good example of the lack of transparency in medical AI.

Lack of transparency can endanger patient safety and pose significant security issues. The feasibility of inspecting systems and understanding how they work is crucial. 'A simple but key aspect of determining the safety of any healthcare application depends upon the ability to inspect the application – to literally disassemble and examine a physical application to determine how the

150 Matulionyte, Nolan, Magrabi, et al. 2022: 158–159.
151 Matulionyte, Nolan, Magrabi, et al. 2022: 158–159.
152 Hardt and Chin, 2020: 39; Shah, 2020: 45.
153 McKinney, Sieniek, Godbole, et al., 2020: 89–94.
154 Char, Abramoff and Feudtner, 2020: 11.

parts work together, to see the mechanisms at work, and thus better understand how the application might fail'.[155]

Information about algorithm functionality is often intentionally hidden (lack of transparency), and even when accessible it is difficult to comprehend (transparency but without explainability). The reasons why this happens are mostly related to the strategic choices of AI developers. One primary motive for AI developers to avoid complying with transparency requirements is to safeguard their competitive advantage and IP rights by concealing their algorithms and decision-making processes from competitors, regulatory authorities, and the public.[156] Additionally, concerns about non-compliance with complex legal frameworks, such as those in the EU, can also contribute to the reluctance to disclose information.[157]

3.3.3 AI Transparency in EU Law

The GDPR places significant emphasis on transparency in AI systems, ensuring that individuals are informed about the processing of their personal data, including when AI is involved. Requirements for transparency are outlined in Articles 13 and 14 of the GDPR, which involve informing individuals about automated decision-making, including profiling, and providing meaningful information about the logic, significance, and potential consequences of such processing.[158]

The AIA further underscores the importance of transparency (see Chapter 3, section 2.4.2). Article 13 specifies transparency requirements for high-risk AI systems, detailing the information to be provided, the intended recipients, and the responsibility of AI providers. Additional provisions, such as Article 12, address the logging and monitoring of high-risk AI systems, a vital element to ensure transparency. Annex IV outlines technical documentation requirements, including the integration of software components and computational resources used in developing AI systems.

While the GDPR and the AIA underscore the importance of transparency in AI systems, ensuring that individuals are informed about the use of their personal data, the functioning of AI, and the potential impact on their rights and freedoms, one might question whether these measures are sufficient. Could AI companies, in some cases, calculate that the benefits of concealing certain information outweigh the potential costs of fines? This raises concerns that, despite regulatory frameworks, the incentive to maximise profits might lead

155 Char, Abramoff, and Feudtner, 2020: 11.
156 Kokoulina, 2023: 28–56.
157 Cheong, 2024.
158 Wulf and Seizov, 2024: 235–256.

some companies to prioritise financial gain over full transparency, particularly if the penalties for non-compliance are perceived as manageable or insignificant compared to the advantages of withholding information.

3.3.4 Improving AI's Transparency in Healthcare

There are several avenues available to improve the transparency of AI technologies in healthcare.

Traceability, for instance, is considered a fundamental requirement for credible AI and refers to transparently documenting the entire AI development process, including tracking how the AI model works in real-world practice after deployment.[159] More specifically, traceability requires maintaining a complete account of (i) model details (intended use, type of algorithm or neural network, hyperparameters, as well as pre- and post-processing steps), (ii) training and validation (collection process, data composition, data acquisition and labelling protocols) and (iii) monitoring of AI tools (performance metrics, failures, periodic evaluations).

To enhance the transparency of AI technologies in healthcare, one promising avenue is the concept of an AI passport, which involves standardised documentation or certification that provides stakeholders with comprehensive information about an AI system's capabilities, performance, and ethical considerations.[160] Although the specific form and implementation of the AI passport are still under development, the idea has gained traction as a potential tool for promoting transparency and accountability in the development and deployment of AI systems. It is worth noting that the AI passport is not yet mentioned in existing EU regulations regarding transparency in AI.

3.4 *Lack of Explainability and the Black Box Problem*

3.4.1 Lack of Explainability

In the past, traditional AI systems aimed to mimic an idealised model of clinical decision-making, relying on Boolean logic with explicit rules.[161] These systems were designed with branching rules, as in the following example: 'If the patient is at least 40 years old and has hypertension, then provide treatment X'. Due to their explicit rules based on ordinary medical language, these systems were fully explainable, and their decision-making process was understandable.

However, as time passed, it became apparent that building AI systems with this architecture posed challenges in competing with healthcare professionals

159 Mora-Cantallops, Sánchez-Alonso, García-Barriocanal, et al., 2021.
160 European Parliament, 2022: 48–50.
161 Holm, Stanton, and Bartlett, 2021: 178.

in terms of performance.[162] Despite their explainability, traditional AI systems struggled to match the decision-making capabilities of human healthcare providers.[163]

The development of more efficient AI systems brought with it a collateral problem: systems that are opaque and which humans struggle to understand,[164] because it is not possible to present the underlying reasons for decisions.[165] As referred to by Cohen: 'in some but certainly not all instances, the rules used by algorithms used in medical AI may be not just not explained but not possible to explain'.[166]

These systems are opaque in the sense that, at the receiving end, we have no concrete sense of how an algorithm arrived at the specific classification of any given input.[167] Often, not even the designers of these systems are able to provide a human-understandable representation of them.[168] The latter, despite comprehending the processes through which the systems generate classification models, often find themselves perplexed by the models themselves.

For instance, trained pathologists possess the ability to clarify why they have deemed specific areas of a scan clinically concerning. However, the rationale behind an ML image-analysis algorithm's attribution of significance to a particular finding or image may remain unclear. This happens because 'being able to predict an outcome is not the same as understanding what actually causes it'.[169]

An explanation that clearly reflects how a decision was made is usually more valuable than just stating the decision itself. Sometimes, the topic is so important or complicated that people would expect some effort to be put into explaining the reasoning behind the opinion before it is given. This helps people judge how accurate the opinion is and how well it matches the situation. The reason for this approach is to allow the recipient to evaluate the opinion's validity and assess the accuracy of its underlying perspective.[170]

The existence of explanations should not be confused with a real understanding of the AI system. Some authors make a distinction in their research

---

162 Tozzo, Angiola, Gabbin, et al. 2021: 527.
163 Holm, Stanton, and Bartlett, 2021: 179.
164 Burrell, 2016.
165 Price, Gerke and Cohen, 2019: 1765.
166 Cohen, 2020: 1431.
167 Mittelstadt, Allo, Taddeo, et al., 2016: 6.
168 Schönberger, 2019: 177; Mittelstadt, Allo, Taddeo, et al., 2016: 6.
169 Sgaier, Huang, and Charles, 2020: 52.
170 Bathaee, 2020: 166.

between interpretation (interpretability) and explanation (explainability). Montavon et al., for instance, define interpretation as the process of translating an abstract concept into a domain that can be perceived and understood by human experts. On the other hand, they define explanation as a set of features within the interpretable domain that have played a role in generating a particular decision for a given example.[171] For Kiseleva et al., the core of explainability is rooted in the ability to clarify information for people, emphasising the importance of their capacity to create and convey explanations; interpretability, in turn, relates to the ability to understand and make meaning of something.[172]

3.4.2 White Boxes and Black Boxes

When discussing AI explainability, three distinct types of AI are often mentioned. The first type is Explainable AI, also known as 'white box', which allows users to clearly understand how the AI behaves. While they offer intrinsic explainability,[173] their performance may be relatively lower.[174]

The second type, semi-transparent AI or 'grey boxes', lies between opaque and fully transparent systems. They provide some insights into their processes, but still withhold complete access to their inner workings.

Lastly, there are opaque AI systems, famously referred to as 'black boxes', where users and developers lack access to the internal mechanisms that drive the system's outputs. Opacity, defined as the potential concealment of an AI system's internal processes, renders such systems impervious to reverse engineering. These black box systems are often highly accurate but lack accountability and responsibility.[175] Opacity is not deliberately created.[176] Instead, it is often inevitable, and it happens because of the very mechanisms used to develop AI systems.[177]

---

171 Montavon, Samek and Müller, 2017: 1–15.
172 Kiseleva, Kotzinos, and De Hert, 2022.
173 Ghassemi, Oakden-Rayner, and Beam, 2021: e745.
174 Yang, Ye and Xia, 2022: 31–32.
175 Raposo, 2024a: 1049–1072.
176 Although some secrecy on the part of the developers may exacerbate this to protect intellectual property and keep the inner workings of an AI system secret and shielded from reverse engineering. See, Scherer, 2016.
177 Scherer, 2016.
Burrell explains opacity in ML algorithms as a product of their 'high dimensionality and complexity (Burrell, 2016: 1–12).

This overall lack of explainability, often referred to as 'black-box AI',[178] poses a significant challenge that could undermine the efficacy of AI and subsequently reduce its adoption within the healthcare domain. The term 'black box medicine' was created to characterise this emerging paradigm of limited transparency and understanding.[179]

3.4.3 Examples of Unexplainable Medical AI

Cases of unexplained medical AI have been the object of several studies. Ross and Swetlitz, for instance, investigated the application of IBM Watson for Oncology.[180] At the UB Songdo Hospital, Mongolia, this AI system is (or was) employed to offer guidance to generalist doctors with limited expertise in oncology. Watson works by analysing a patient's medical record, identifying potential treatment options from a list, and assigning a percentage score to each option based on its suitability for the patient. These recommendations are then presented to the doctor as a ranked list. The challenge arises because Watson operates opaquely, providing no explanations for the treatment scores it assigns. Despite this opacity, suggestions from Watson are followed at UB Songdo Hospital almost 100% of the time.[181]

Other concerning aspects arose when Ross and Swetlitz described an oncologist's experience using Watson at a South Korean hospital. The AI recommended a chemotherapy drug called taxane for a patient whose cancer had not spread to the lymph nodes, even though taxane is typically used for cases where cancer has spread to the lymph nodes. Watson supported its recommendation with a study showcasing the drug's effectiveness in patients with lymph node spread, leaving the oncologist puzzled and unable to discern the rationale behind the recommendation. Deploying the same technology in areas with less clinical expertise exposes patients to potential harm, as there is a lack of clinical safeguarding. In such situations, patients could receive inappropriate treatments due to the opacity of the AI system and the absence of expert oversight.[182]

To address epistemic opacity, the concept of 'computational reliabilism' was proposed.[183] This concept acknowledges our cognitive limitations in understanding algorithms fully and focuses on instilling trust in the algorithm and its outcomes. Under this conception, the belief in AI results is justified due to

178 Examples of black boxes in healthcare in Poon and Sung, 2021: 582 (in the domains of gastroenterology and hepatology).

179 Price II, 2015: 421.

180 Ross and Swetlitz, 2017.

181 Ross and Swetlitz, 2017.

182 Ross and Swetlitz, 2017.

183 Durán and Jongsma, 2021: 329–335; Durán and Formanek, 2018: 645–66.

the existence of a reliable algorithm that consistently produces trustworthy outcomes.[184] Still, its effective added benefit is yet to be demonstrated.

In the medical sphere, opacity is especially problematic. The intricate nature of mathematical formulas encapsulated within bits and bytes renders them incomprehensible to the average doctor or patient, rendering them essentially useless without specialised training in AI technologies.[185] To bridge this gap, doctors and patients require an explainability model that is accessible and understandable to lay persons.

3.4.4 The Legal Imposition of Explainability in the EU

Under EU law, explainability is the target for AI systems (health-related and non-health-related), as imposed by several norms.

Article 15(1)(h) of the GDPR requires data controllers to explain how decisions are taken in automated decision-making and profiling[186] ('meaningful information about the logic involved'). Recital 71 of the GDPR provides additional context to Article 15(1)(h), elaborating on the need for safeguards in automated decision-making processes.

Similar impositions can be found in the text of the AIA. Article 13(1) states that high-risk AI systems 'shall be designed and developed in such a way as to ensure that their operation is sufficiently transparent to enable deployers to interpret a system's output and use it appropriately'.[187] While the term 'transparent' is used, it is clear that the focus is on enabling users to interpret the system's outputs, meaning that explainability comes into play.[188]

3.4.5 Consequences of the Lack of Explanations in the Medical Field

The lack of explainability in the medical field has been severely criticised.[189] Some critiques are well-founded, while others are less so.

A major issue relates to patient informed consent (see Chapter 6, Section 4.3). The introduction of opaque AI algorithms and cumbersome informed

184 Durán and Jongsma, 2021: 332.

185 Pande, 2018.

186 Automated decision-making refers to decisions made solely by automated means without human intervention, including profiling, which involves the use of personal data to evaluate certain aspects of an individual, such as their performance at work, economic situation, health, preferences, or behaviour (Bygrave, 2020).

187 About this legal imposition, European Commission – Directorate-General for Justice and Consumers, 2019.

188 Raposo, 2024b; 1–12; Raposo, 2025.

189 The WHO, for instance, has recognised the critical importance of combating opacity through provisions to ensure transparency, 'explainability' and intelligibility in the design and use of AI in healthcare. See Italian Committee for Bioethics, 2020: xiii; World Health Organization, 2021.

consent forms limit the level of autonomy and shared decision-making power between doctor and patient.[190] Due to these opaque AI systems, it has become increasingly difficult for patients to understand the decision-making process and the different ways in which their data can be processed.[191]

This, in turn, undermines trust in healthcare providers. According to the assertions made by Grote and Berens, the absence of explainability in medical decision-making has the potential to erode the 'epistemic authority' of doctors.[192]

Besides, opacity leads to a general reduction of trust in technology. Rigorous regulatory approval and validation processes contribute to doctors' and patients' trust in technology, particularly for high-risk technologies.[193] The lack of trust affects not only the individuals impacted by AI decisions, such as patients, but also the medical professionals employing these systems, as well as the AI developers themselves.[194]

AI opacity also affects accountability, as it becomes difficult to identify the source of AI errors and to define responsibilities when they go wrong.[195] The more opaque the AI algorithm, the harder it becomes to pinpoint responsibility for errors in patient care or medical decisions. Consequently, the burden of responsibility may fall more on the doctors using a non-explainable medical AI tool, who, in turn, will be unable to explain their medical decisions or errors. In fact, the use of black-box medicine might increase the standard of care expected from medical practitioners, particularly in high-risk practices such as surgery or administering risky drugs.[196]

The lack of explainability in black-box AI has been used as an argument to exclude medical liability when AI is involved.[197] This argument is subject to debate.[198] Doctors often employ various technologies and equipment without fully understanding their inner workings, yet they are held accountable for their operation and expected to address any errors or mishaps that may occur.[199] This practice is generally accepted because drugs and medical devices are considered reliable, meaning they are sufficiently safe and effective

190 Vyas, Eisenstein and Jones, 2020: 874–882.
191 McKeown, Mourby, Harrison, et al. 2021: 1–21.
192 Grote and Berens, 2020: 205–211.
193 Quinn, Jacobs, Senadeera, et al., 2022.
194 Burrell, 2016; Ferretti, Schneider and Blasimme, 2018: 328.
195 Wachter, Mittelstadt and Floridi, 2017a.
196 Price II, 2018: 301.
197 Matthias, 2004: 175–183.
198 Explaining this argument in detail, Raposo, 2025.
199 Also in this sense, see Durán and Jongsma, 2021: 333.

for use in human patients. I acknowledge that the same cannot yet be said for AI medical devices, as most still lack adequate evidence to demonstrate their safety and accuracy.[200] However, in the future, clinical trials and rigorous testing are likely to provide this necessary evidence. As a result, we can expect and should advocate for more flexible requirements for explainability in medical AI, once their safety and precision are thoroughly validated.[201]

3.4.6 Types of Explanations

To enhance understanding of how the AI system operates, four types of explanations can be employed: contrastive, selective, causal, and social explanations.[202] While contrastive explanations delve into the specific inputs that led to a decision, selective explanations highlight the most relevant factors based on human judgments. For those lacking expertise in the specific domain, like patients, causal explanations prove more valuable, focusing on the underlying causes rather than mere statistical correlations. Likewise, medical experts could greatly benefit from causal explanations, as they offer the rationales behind AI decisions, facilitating further investigations. Equipped with such explanations, doctors can assess the coherence of the system's arguments in light of medical literature, clinical practice, past cases, and individual patients. The explanations empower doctors to discern the significance of specific inputs in generating the output and evaluate their reliance on AI CDSS recommendations. Social explanations emphasise the value of adopting an interactive and conversational approach, where information is customised to align with the recipient's beliefs and understanding. For example, doctors might need to monitor the progress of their explanations by noting what has already been communicated to the patient and by deducing what the patient has understood on their own.

When approaching the matter from a medical standpoint, the need to differentiate between two tiers of explainability becomes apparent.[203] At the forefront is the first level of explainability, which grants the recipient comprehension of how the system reaches its general conclusions. Drawing a parallel to laboratory testing, where we possess knowledge of the biological and biochemical reactions yielding outcomes, in this type of explainability, we have the capacity to furnish rankings that unveil the significance of individual features within AI-powered CDSSs. The subsequent stratum, designated as second-level

200 Cf. Raposo, 2025.

201 In detail about this thesis see Raposo, 2025.

202 I follow the classification of Lagioia and Contissa, 2020: 275–276.

203 Amann, Blasimme, Vayena, et al., 2020.

explainability, delves into identifying the key features for individual predictions. This dimension enables meticulous scrutiny of individual forecasts, thus allowing potentially erroneous predictions to be detected. While instances may arise where first-level explanations are deemed adequate, other contexts will require second-level explanations to ensure patient well-being and safety.

From a different perspective, now based on the scope of the explanation, it is possible to distinguish the ones that encompass all patterns and variables considered by the system (model-centric explanation) and the ones that are focused solely on those that are relevant to the individual patient's case (subject-centric explanation).[204] A model-centric explanation aims to provide a comprehensive overview of how the AI model functions as a whole. It includes details about the data inputs, the algorithms used, the decision-making processes, and the overall patterns detected by the system. Model-centric explanations are valuable for understanding the general behaviour and performance of the AI model, ensuring that it operates within expected parameters and adheres to regulatory standards. They are particularly useful for developers, regulators, and researchers who need to validate and improve the AI system's reliability and safety.[205] On the other hand, a subject-centric explanation aims to provide personalised insights into why the AI system made a specific recommendation or diagnosis for that particular patient. It highlights the key factors and data points that influenced the AI's decision, making it easier for healthcare providers and patients to understand the rationale behind the AI's conclusions. This personalised approach enhances trust in the AI system by demonstrating transparency and relevance to the patient's unique medical context.[206]

3.4.7 The Possible AI Explanations

3.4.7.1 *Explainable AI*

To address the need for explainability, the field of Explainable AI (XAI) has emerged. XAI focuses on developing methods and techniques to make AI systems and their decisions more understandable to humans. Various models have been developed in this regard, such as Local Interpretable Model-Agnostic Explanations (LIME) and Shapley Additive exPlanations (SHAP). LIME generates explanations at a local level, highlighting the impact of individual features in the algorithm's prediction.[207] By manipulating these features, end-users can

204 Lagioia and Contissa, 2020: 272.
205 Jakubik, Vössing, Kühl, et al., 2024.
206 Guerrero Quiñones, 2024.
207 Ribeiro, Singh and Guestrin, 2016; Slack, Hilgard, Jia, et al., 2020: 180–186.

observe how they affect the prediction outcomes, providing insights into the 'weight' of each feature. Similarly, the SHAP technique quantifies the contribution of each feature to the algorithm's prediction, offering a deeper understanding of the prediction for a given data sample.[208]

3.4.7.2 *Analogies with Human Explanations*

One strategy commonly used by doctors to explain information or reasoning to patients is the explanation by example: a doctor may give a certain diagnosis because the patient's case (their medical history, for instance) is similar to paradigmatic cases that the doctor has encountered before. An analogous technique can be used in AI, which uses case-based learning methods and thus allows the identification of cases most similar to the case that needs to be explained.[209] By using these methods, doctors can ensure that the machine is not only making the same decisions a human doctor would make but that it is doing so based on the same reasons as a human doctor.[210]

Moreover, in the field of machine vision models, other techniques allow the AI system to justify its decision. One such technique involves a heat map on the medical image, which shows that the machine vision model is looking at the relevant parts of the image when reaching its decision.[211] In both scenarios, the goal is to provide insight into the reasoning process behind a diagnosis or decision. For the doctor, this involves guiding the patient through the scan, helping them understand which parts are of concern and why. Similarly, in machine vision, the heat map serves as a visual explanation, showing that the AI model is focusing on the correct and clinically relevant regions of the image when arriving at its conclusion.

In addition to pointing to the scan, there are other ways a medical decision can be rationalised. For instance, a radiologist might highlight the most pertinent features of the patient's clinical process during a consultation. This

---

208 Rasheed, Qayyum, Ghaly, et al., 2022; Slack, Hilgard, Jia, Singh et al., 2020: 180–186.

209 Harned, Lungren and Rajpurkar, 2019: 5.

210 Harned, Lungren and Rajpurkar, 2019: 5.

211 Harned, Lungren and Rajpurkar, 2019: 5; Savage, 2022. Techniques like Grad-CAM (Gradient-weighted Class Activation Mapping) are commonly used to create heat maps. Grad-CAM is a powerful tool for visualizing and understanding the inner workings of convolutional neural networks. By highlighting the regions of an image that are most influential in a model's decision-making process, Grad-CAM helps bridge the gap between AI predictions and human interpretability, making AI systems more transparent and trustworthy. It works by visualizing the gradient of the model's output with respect to the activations of the final convolutional layers. The result is a heat map that highlights regions of the image that have the most influence on the final decision. Cf. Kumar, Abdelhamid and Tarek, 2023: 34–4.

could involve referencing previous medical history or discussing key diagnostic findings that have influenced the decision-making process. Similarly, an AI system used in medical decision-making can offer a structured and transparent explanation by listing the key factors that contributed to its decision.[212] The AI system can identify and present the most relevant diagnoses that align with the patient's symptoms and imaging results. By ranking or highlighting these diagnoses, the system can show which conditions it considered most likely based on the available data. The AI can also list previous medical procedures that may have influenced its analysis. For example, a prior surgery or treatment might alter the expected appearance of tissues in an imaging scan, which the AI can take into account.

3.4.7.3 *The Limitations of the Analogy*

However, the current state of technology poses challenges in offering comprehensive explanations for many AI models.[213] For instance, while heat maps are visually informative tools that can highlight areas of interest in a medical image, they do not inherently provide a detailed explanation of the underlying AI algorithms or the decision-making processes involved. Heat maps are primarily designed to show the distribution or concentration of data, i.e., they indicate where the AI model is 'looking' within an image when it makes a prediction. However, they do not offer insight into the specific logic or reasoning that led the AI to focus on these areas, nor do they explain how the model arrived at its final decision. Moreover, the map alone does not provide context regarding why these areas are significant, what specific features the AI detected, or how those features correlate with particular medical conditions. Ultimately, heat maps do not reveal the inner workings of the AI model, such as the weights and biases in a neural network or the specific patterns the model learned during training. As a result, while heat maps can enhance the interpretability of AI decisions to some extent, they fall short of offering a comprehensive explanation that fully rationalises the AI's decision-making process.[214]

To achieve a deeper understanding, it is necessary to complement heat maps with other techniques that can elucidate the AI's reasoning. For example, combining heat maps with feature attribution methods can help pinpoint

212 Harned, Lungren and Rajpurkar, 2019: 5.

213 Ali, Abuhmed, El-Sappagh, et al. 2023.

214 Ghassemi, Oakden-Rayner and Beam, 2021): e746.

specific features, like shape, texture, or intensity, in the highlighted areas that influenced the model's decision the most.[215]

Ultimately, the level of explainability depends on how complex the AI system is.[216] When AI diagnoses a medical condition using just a few parameters from a patient's EHR, explanations are not difficult to provide. For instance, if age is one of the factors, you can alter the input age to see if the outcome changes. Yet, as the number of parameters increases, the potential input combinations skyrocket. When an AI considers thousands of parameters, it would take countless lifetimes to fully comprehend the decision-making boundary it establishes. Age might significantly affect outcomes when combined with a few other factors, but when it is one among thousands of parameters, its significance could depend on specific circumstances. Even then, it may not decisively influence outcomes in many instances. Consequently, determining the decisive input parameters for the AI model becomes an insurmountable task.[217]

3.4.7.4 *The Limitations of Explainable AI*

The aforementioned solutions represent technical approaches to enhance the explainability of AI systems. However, translating complex computer and mathematical-level explanations into understandable medical terms still poses a challenge for medical AI. This complexity arises because AI primarily engages in statistical correlation analysis, differing from medical causation analysis.[218] On the other hand, providing explanations for AI extends beyond technical considerations and also involves legal and ethical dimensions.[219] The development of comprehensive and legally compliant frameworks for AI explainability remains an ongoing challenge.

To improve the explainability of AI algorithms, AI developers need to involve clinical end users from the beginning of the development process to select the best explainability approach for each application and ensure that

---

215 Feature attributions are methods used in AI to determine the importance or contribution of individual features (inputs) to a model's prediction (Zacharias, Von Zahn, Chen et al., 2022: 2161). Suppose the AI system analyses several features and predicts that a particular patient has a high risk of developing diabetes. To understand why the AI made this prediction, feature attributions are used to assess the importance of each input feature in the decision-making process. The analysis might reveal that blood sugar levels were the most influential factor, contributing 40% to the prediction, body mass index might be the second most important feature, contributing 30%; family history of diabetes might account for 20% of the decision, and age and physical activity levels could each contribute 5%.

216 Borrego-Díaz and Galán Páez, 2022: 1579–1601.

217 Bathaee, 2020: 149–150.

218 Zhang and Zhang, 2023.

219 Osasona, Amoo, Atadoga, et al., 2024: 322–335.

the chosen explanations are useful and well accepted in clinical practice.[220] Also, regulatory bodies can play an important role when considering the traceability and explainability of AI tools as prerequisites for certification.[221]

Following this path, we might reach a hypothetical future where AI systems offer understandable evidence and explanations. Even in that scenario, doctors will not become mere intermediaries due to two crucial reasons. Medical experts are the ones who possess the necessary domain expertise to interpret and evaluate the significance and accuracy of the evidence and explanations[222] (assuming AI's capability to provide such evidence). The ability to explain forms the cornerstone of the interaction and trust between doctors and patients throughout the entire treatment process, fostering cooperation in devising effective treatment plans, and for that, human practitioners are essential.[223]

3.4.8 Explainability versus Accuracy

The pursuit of explainability in XAI can potentially compromise the accuracy of the technology.[224] Simpler AI approaches are easier to explain but yield less accuracy, while complex DL systems offer greater accuracy but are challenging to explain comprehensively. This raises a choice between simpler, more explainable systems with reduced results,[225] and more accurate systems with less explainability.[226] It is not an easy choice, as less accurate technologies can lead to incorrect clinical decisions and unfavourable patient outcomes. The key question might be whether the AI's decision-making process is more important than the correctness of the decision.

In various non-medical scenarios, consumers often prioritise results over explainability. Friedrich and colleagues provide the following example: When we visit a mechanic, we are not necessarily concerned with understanding the specific engine issue; we simply want the problem fixed so that the car functions properly. Once the problem is diagnosed and resolved, the mechanic does not need to explain the details to the customer. For many consumers, the key factors of interest are the cost and restoration of functionality. Similarly, in numerous cases, patients prioritise outcomes over the intricacies of the

220 Jesso, Kelliher, Sanghavi, et al., 2022.
221 Balasubramaniam, Kauppinen, Rannisto, et al., 2023.
222 Davenport and Kalakota, 2019: 95.
223 Lagioia and Contissa, 2020: 276.
224 Quinn, Jacobs, Senadeera, et al., 2022.
225 Athey, 2017: 483–85.
226 Matulionyte, Nolan, Magrabi, 2022: 151–80.

process.[227] Millard and Kerr put it brilliantly: 'Watson's expertise would not be primarily based on its ability to provide such explanations; its rate of success at performing particular tasks would seem to do the heavy lifting where expertise is concerned'.[228]

3.4.9 How Decisive Is Explainability?

3.4.9.1 *The Controversy Around Explainability*

Scholars are divided regarding the importance of explainability, from those who cannot live without it[229] to those who are more focused on the accuracy of the outputs than on their explainability.[230] Despite efforts to advocate for a more nuanced approach to this duality, the trade-off still puzzles experts.[231]

In 2017, in its influential AI Now report, the AI Now Institute, established at New York University, advocated against the utilisation of black box AI in critical domains such as criminal justice, welfare, education and… healthcare.[232] The institute recommended that these high-stakes areas should not rely on opaque AI systems, emphasising the need for transparency and explainability. The medical field operates in high-stakes situations and places a significant emphasis on trust, which raises concerns about the lack of explainability in black-box AI systems.[233]

However, this should not be a deterrent to the use of black-box AI in medicine. Explainability is important but not essential.

3.4.9.2 *Absence of Explainability in Other Domains*

Numerous instances exist where users lack comprehensive knowledge about how a product operates, yet they maintain their trust and usage of the product. For instance, one does not necessarily have to possess computer programming skills or automotive engineering expertise to place trust in the functionality of the technology.[234] Users must be assured that their devices are fulfilling their

---

227 Friedrich, Mason and Malone, 2022: 7.

228 Millar and Kerr, 2016, p. 113.

229 Grote and Berens, 2020: 205–211; McDougall, 2019: 156–160.

230 London, 2019: 15–21.

231 Crook, Schlüter and Speith, 2023: 316–324; Hulsen, 2023: 652–666.

232 Campolo, Sanfilippo, Whittaker, 2017.

233 Although eventually more problematic for doctors than for patients, according to Friedrich, Mason, and Malone, 2022.

234 For instance, consumers are not required to comprehend the intricate workings of their cell phones or vehicles, but they do require a level of confidence that these devices will reliably execute specified tasks.

intended functions and not engaging in undisclosed activities that might jeopardise their well-being, such as intrusive surveillance.

Other domains where unexplainable decisions are accepted are art and music (where critics are often subjective and based on personal taste and experience), sports coaching decisions (coaches make strategic decisions during games based on their knowledge of the sport, the players, and the situation; these decisions are often accepted by teams and fans even if the exact reasoning behind them is not fully articulated), or hiring decisions (employers often make hiring decisions based on interviews and subjective assessments of candidates' fit for the company culture; the exact reasons for choosing one candidate over another may not be fully explainable but are accepted as part of the hiring process), amongst others. On the other hand, causal explanations and the apparent transparency of knowing the causal relationships between the input and output do not necessarily translate to understanding the implications of using AI and its assumptions.[235]

In sum, we are continually surrounded by complex devices that we use daily without understanding the details of their design or function. Our interaction with these devices reflects a broader trust in the systems and expertise that guarantee their reliability and safety.[236] Thus, the challenges associated with explainability in black box AI should not be seen as a roadblock to its use.[237]

3.4.9.3 *Absence of Explainability in Medicine*

In the particular domain of medicine, there are many phenomena that doctors can't fully explain, yet still incorporate into their practice. Many things in medicine are opaque. Ultimately, even decisions taken by human doctors are opaque, as the human brain is the ultimate black box.[238] London refers to these situations where doctors navigate without hesitation in this way: 'Large parts of medical practice frequently reflect a mixture of empirical findings and inherited clinical culture'.[239] That is, despite lacking knowledge of the underlying causal system, doctors may 'simply' rely on their experience to make recommendations.

Take the case of a junior doctor seeking guidance from a more experienced senior. When the junior has solid reasons to trust the senior's judgment over their own, they are still working in the best interest of the patient by following

235 Lorenzini, Ossa, Shaw, et al., 2023: 427.
236 Gordijn and Ten Have, 2023: 284.
237 Shortliffe and Sepúlveda, 2018: 2199–2200; Amann, Blasimme, Vayena, et al., 2020.
238 Brożek, Furman, Jakubiec, et al., 2023.
239 London, 2019: 17.

the advice, even if they don't completely grasp the senior's reasoning. A comparable scenario can arise when a general practitioner relies on the expertise of a specialist to guide their decisions.[240]

A prime example is drug prescription. The human body's intricacies make it challenging to predict the effects of drugs. Doctors prescribe drugs, while it is accepted that patients will not understand their complexities.[241] Often, even doctors may not entirely comprehend how specific drug molecules function within a patient's body, as the precise mechanisms of action may not be fully elucidated.[242] Even regulatory approval does not guarantee a complete understanding of a drug's mechanisms.[243]

There is a distinction, though. In the case of drugs, the lack of complete understanding is commonly accepted because safety and efficacy are demonstrated through clinical trials.[244] Pharmaceutical interventions undergo extensive clinical validation, following transparent and regulated processes that take considerable time and require decades of evaluation for human use approval. In contrast, AI systems have not fully implemented this model yet.

Things might be changing, though. There are reports of clinical trials evaluating the performance of AI algorithms.[245] This can open the door for the demonstration of the safety and efficacy of AI healthcare applications, even those deemed black boxes. If this trend continues, it could make unexplainable AI more acceptable by disregarding explanations as long as the results are proven safe and effective through standardised clinical trials, similar to the approach taken with drugs.[246]

3.4.9.4 *The Relevance of Explainability for Patients*

What we might question is how relevant explainability is for patients. Even when doctors understand the inner workings of a device, drug or procedure,

---

240 Example taken from Cohen, 2020: 1459.

241 Raposo, 2025.

242 Pruski, 2024; London, 2019: 18.

243 A notable example refers to acetaminophen, whose exact mechanism of action remains only moderately understood despite its widespread use. Theories have emerged to explain its effects, but a comprehensive understanding is lacking. However, the safety and efficacy of acetaminophen as a pain medication have been established through rigorous validation in randomised controlled trials (RCTs). See Shaughnessy, 2022.

244 Singh and Loke, 2012.

245 For example, one evaluating an AI algorithm for the diagnosis of diabetic retinopathy (Huang, Wang, She, et al., 2022) and another one evaluating an AI system for the detection of breast cancer on mammograms (Cè, Caloro, Pellegrino, et al., 2022: 795–816).

246 Raposo, 2025.

the fact is that patients might not be that interested in the explanation.[247] A study revealed that when doctors inform patients about having melanoma, the patients seldom question the doctor's decision. Instead, they readily accept the biopsy and follow the recommended treatment.[248] This highlights the longstanding acceptance of medical decision-making opacity, where patients have usually accepted diagnoses and treatments without questioning the doctor's decision-making process.[249]

Still, it seems that AI brings a different mindset, and it might be the case that the importance of explanations to patients varies depending on whether it is a human doctor or an AI doctor. While people could tolerate opacity in doctors, the same leniency might not extend to the opacity of the AI systems they use. The difference is that patients can take advantage of their doctors' professional obligation to be accountable for their practice, as stipulated by several professional guidelines[250] while the same is not yet valid for AI systems.

While there is a possibility of improving the explanatory capabilities of black-box AI systems in medicine, a certain degree of inherent opacity will persist unless all AI becomes completely explainable, which may be an unlikely scenario.[251] We might have to accept fewer explanations when there is a well-defined risk-to-benefit ratio; i.e., when dealing with low-stakes decisions where the risks of mortality and morbidity are minimal, there is room for less explainable AI systems.[252] In contrast, when the level of risk increases, so shall the demand for explanations.[253] This balance ensures that AI is used to

247 Friedrich, Mason and Malone, 2022.

248 Sukel, 2017: 36–39.

249 Mann, Berdahl, Baker, et al. (2022), Farah, Murris, Borget, et al. (2023: 134) and Quinn, Jacobs, Senadeera, et al. (2022) argue that the analogy does not apply because 'most patients may not need to know how aspirin suppresses the production of prostaglandins in order to put trust in its medical use; but most patients probably do want to know "Why does the computer think I have cancer?" before they can have confidence in its judgment'.

Ratti and Graves also exclude the analogy because 'Clinicians use pharmaceuticals to directly treat patients, while AI tools are used as instruments to assist clinicians and s physicians' (Ratti and Graves, 2022: 803).

250 Smith, 2021: 535–545.

251 Taylor, 2024: 1695–1704.

252 AI tools can assist in preliminary health screenings, such as flagging potential issues based on basic patient data (e.g., temperature, heart rate) or reviewing medical imaging for common, non-life-threatening conditions. In these cases, the AI's suggestions are typically reviewed by a human practitioner, and the low stakes involved mean that a detailed explanation from the AI may not be necessary, especially when the system is known to have high accuracy.

253 In situations where AI is used to diagnose serious or life-threatening conditions, such as cancer or heart disease, the stakes are incredibly high. Patients and healthcare providers

enhance, rather than undermine, the quality and trustworthiness of patient care.

3.5 *Imprecision and Flawed Results*

3.5.1 AI Is Not Perfect

In healthcare, the application of AI has been fraught with missteps and setbacks, where lofty expectations have often clashed with real-world outcomes. Mistaken results can be caused by the previously mentioned AI's biases, but that is not the only cause. Inaccuracies can also occur when, although technically the prediction is correct, the model and result are inaccurate representations of real-world equivalents.[254]

This limitation arises from AI's reliance on structured data and the absence of contextual understanding that humans possess. Consequently, AI may fail to identify straightforward mistakes in data or make judgments that would be readily apparent to a human observer. Gathering and analysing information is not the same as having the wisdom and judgment that come from real-world experience and expertise, which only humans possess.[255]

The harsh truth is that AI can be very easily fooled. Even though AI has the same success rate as doctors in many specialities, and sometimes even higher, no technology is perfect, and tests of image classifiers show that systems still confuse cats with frogs.[256] Another ridiculous example involves a DL system confusing blue and orange wavy lines for starfish.[257]

---

need to trust that the AI's recommendations are based on sound reasoning. An AI system that recommends a particular treatment or surgical intervention must provide a clear and understandable rationale, allowing doctors to make informed decisions and discuss the options transparently with patients. Thus, explainability becomes more pressing (though not indispensable). See Ossa, Starke, Lorenzini, et al., 2022: 6.

254 Imagine a hospital uses an AI model to diagnose diseases based on patient symptoms and medical history. The model is trained on a large dataset of past patient records, including symptoms, diagnoses, and outcomes. Suppose, however, that a patient recently started a new medication that significantly alters some of their symptoms, a factor that was not present in the training data. The AI model does not account for this new medication and thus while its prediction is technically correct based on the available data, it inaccurately represents the patient's current health status. See Schönberger, 2019: 178; Crawford and Calo, 2016: 311–313.

255 Arnold, 2021: 124.

256 Wexler, 2017.

257 Matsuzaki, 2018: 267.

AI methods, lacking 'common sense', often struggle to detect simple errors or make decisions that would be obvious to humans.[258] Obviously, a similar misdiagnosis of cancer could have far more serious results.[259]

The potential for safety vulnerabilities within AI systems could place a larger number of patients in jeopardy compared to the likelihood of a single misdiagnosis by a human doctor, as automated systems tend to replicate a greater number of errors.[260] On the other hand, given the pervasive nature of AI compared to physical technologies like drugs, some argue that managing the consequences of an 'unsafe' algorithm presents greater challenges. While withdrawing a drug from the market can be difficult, it remains possible. However, AI systems, being less visible and interpretable, and often integrated into a hospital or health system's IT infrastructure, pose unique difficulties for their 'withdrawal'.[261]

'The inconvenient truth', as coined by Panch et al. points out a clear weakness in the operational effectiveness of AI algorithms.[262] AI advancements do not always align perfectly with the complexities of real-world medical procedures,[263] often requiring careful integration and human oversight to ensure that these technologies enhance patient care without introducing new risks or uncertainties.

3.5.2 Possible Reasons for Imprecise Outcomes

Oftentimes, the problem lies in poor testing. Recent research conducted a comprehensive review of numerous studies touting AI's superiority over

258 Shaw, 2019: 1–11.

259 Safety vulnerabilities can endanger patient safety. Back in 2015, a medical robot was being used in British hospitals for heart valve repair surgery. However, not only did the robot commit serious operational errors, but it also disrupted the correct actions of human doctors, tragically leading to a patient's death (Zhang and Zhang, 2023). Another instance occurred in 2019, when the U.S. Food and Drug Administration (FDA) urgently recalled Zimmer Biomet's ROSA Brain 3.0 robotic surgical system due to software glitches that caused the robot's arm to be positioned incorrectly (U.S. Food and Drug Administration, 2021).

260 Zhang and Zhang, 2023.

261 Marwaha, Landman, Brat, et al., 2022.

262 Panch, Mattie, and Celi, 2019.

263 Imagine a scenario, envisioned by Challen et al., where a diagnostic system developed to identify skin malignancies, primarily trained on clinical images of biopsied lesions, might exhibit reduced efficacy when employed to screen the wider population (Challen, Denny, Pitt, et al., 2019: 233). This outcome arises from the contrasting nature of lesion appearances and the distinct risk profiles of patients encountered in the general population.

radiologists, but it revealed that only a handful of these studies were validated in distinct populations from the ones used for algorithm development.[264]

Some errors are also caused by the inability of the algorithm to understand and connect the different concepts used. Within the healthcare sector, there is a diversity of terminologies employed by different agencies to characterise identical medical conditions. Different doctors frequently employ varying descriptors for similar clinical presentations, resulting in a lack of internal consistency. As a consequence, when AI algorithms attempt to interpret this input data, translational errors can arise.[265]

Mistakes frequently result from inaccurate data. When the mistaken outcome was based on mistaken input, we have a case of 'garbage in, garbage out'. In an illustrative case, Rich Caruana and his colleagues reported that a neural network, while demonstrating higher accuracy than alternative methods in diagnosing the probability of death from pneumonia, assigned asthmatic patients a lower probability compared to the general population. This finding was considered 'counterintuitive'. What happened is that patients with a history of asthma typically receive direct admission to the intensive care unit for intensive medical intervention, which in turn lowers their probability of death. Without such aggressive care, asthmatic patients face a higher risk of mortality from pneumonia. Consequently, the scoring system used in this case was misleading, as it failed to reflect the patients' underlying medical needs. This surprising result can be explained by the data used to train the system. The positive aspect is that the intensive care provided to asthmatic pneumonia patients has proven to be highly effective, reducing their risk of pneumonia-related mortality compared to the overall population. On the flip side, this carries a drawback: since the prognosis for these patients is more favourable than average, AI models trained on these data inaccurately infer that asthma mitigates the risk, even though individuals with asthma actually face a significantly higher risk if not hospitalised.[266]

In other situations, the AI algorithm was trained to cut costs and not for the patient's well-being.[267] This training could potentially direct users toward clin-

---

264 Reardon, 2019.

265 Lawton, 2012.

266 Caruana, Lou, Gehrke, et al., 2015: 1721–1730.

267 Suppose an AI system is implemented in a hospital to reduce the rate of readmissions, a common quality indicator that is often tied to financial penalties from insurance companies or government programs. The algorithm is trained to identify patients who are at high risk of readmission and recommend discharge plans that minimise the likelihood of them returning to the hospital within a certain period. However, if the AI is primarily focused on reducing costs associated with readmissions, it might recommend discharging

ical decisions aimed at enhancing quality indicators but not necessarily aimed at improving healthcare outcomes.[268] Ultimately, while controlling healthcare costs is important, AI algorithms must be designed with a balanced approach that equally values patient well-being.

In other cases, the problem is not AI per se, but the contextual conditions. Factors such as the hospital's access to a reliable internet connection and suitable lighting conditions for capturing images significantly determine the tool's efficacy.[269]

Another issue is that AI systems, especially those that are always learning, may update frequently. For example, a system that predicts outcomes that affect treatments may create a kind of self-fulfilling prophecy,[270] reinforcing its own predictions as it keeps learning.[271] This constant updating can lead to the reinforcement of erroneous patterns. For instance, if an AI system is tasked with predicting patient outcomes to guide treatment decisions, it might initially observe and learn from the outcomes of its recommendations. If the system predicts that a particular treatment will be ineffective for certain patients, and that prediction influences the decision to withhold or alter the treatment, the AI might observe that the outcome aligns with its prediction. Over time, the system could reinforce this pattern, increasingly predicting poor outcomes for similar patients, even if the initial prediction was based on flawed data or incorrect assumptions.

It should also be noted that in light of their probabilistic nature, AI models should be expected to fail in certain instances.[272] Unlike deterministic algorithms that produce the same output for a given input every time, AI models – particularly those based on ML – rely on probability and statistical patterns

---

patients as soon as they are medically stable rather than ensuring they are fully recovered. For example, the algorithm could suggest sending patients home with minimal follow-up care or without providing adequate support for managing their condition at home, simply because these options are cheaper.

268 Char, Shah, and Magnus, 2018: 982.

269 An investigation into the deployment of an AI-powered retinal scanning tool in Thailand aimed at detecting diabetic eye disease revealed the fundamental influence of socio-environmental variables; remarkably, instances arose where the AI system's use resulted in delays and disruptions due to inadequate conditions. Cf. Beede, Baylor, Hersch, et al., 2020: 1–12.

270 This phenomenon can create a feedback loop, where the AI's predictions start influencing real-world outcomes in a way that seems to confirm its own expectations. This is known as a self-fulfilling prophecy. In the context of healthcare, such a loop could have serious consequences. See King and Mertens, 2023: 127–152.

271 Challen, Denny, Pitt, et al., 2019: 235.

272 Schönberger, 2019: 177.

to make predictions or decisions. As a result, there is always a margin of error inherent in their functioning. AI models, especially those used in complex tasks like image recognition, NLP, or medical diagnostics, are trained on vast amounts of data. They learn to identify patterns and correlations within these data, which they then use to make predictions or decisions when faced with new, unseen data.[273] However, these models do not 'understand' in the human sense; instead, they calculate the likelihood of various outcomes based on the data they have been trained on.[274]

It is not only that AI might perform poorly, but also that its malfunctions might go under the radar.[275] The challenge of identifying random errors is amplified, often going unnoticed when the tool aligns with the doctor's judgment.[276] This scenario creates a situation where both the AI and the human expert might be wrong, yet the error goes undetected because the agreement between the two creates a false sense of confidence.[277] However, if both the AI and the doctor are wrong, the error is likely to go unnoticed.[278] The doctor may be less inclined to seek a second opinion or conduct further tests because the AI's agreement strengthens their confidence in the initial diagnosis. In some cases, the AI outcome is not necessarily bad, it is just not reliable enough to ground a decision.[279] Sometimes the algorithm provides conclusions that, though probable, cannot be attested with absolute guarantee, as they are based on statistical learning and computational learning theory. These are called inconclusive evidence and are not considered a solid base on which to ground a decision, obviously including a medical decision. Other types of 'unreliable' conclusions are the ones based on impenetrable evidence, that is, when the

273 Hendrycks, 2025: 64.

274 For example, an AI system used to diagnose diseases might determine that a certain set of symptoms has a high probability of indicating a particular condition. However, because this determination is based on probability, there will always be cases where the model's prediction is incorrect, either because the input data was atypical, or because the model's training data did not fully capture the variability of real-world cases.

275 In certain instances, AI algorithms have been designed with deceitful intentions. An example is Volkswagen's algorithm manipulation, which programmed the vehicles to deceitfully clear emissions tests by minimising nitrogen oxide emissions exclusively during testing scenarios (Char, Shah, and Magnus, 2018: 982).

276 Amann, Blasimme, Vayena, et al., 2020.

277 For example, imagine an AI system designed to diagnose pneumonia from chest X-rays. A doctor who already suspects pneumonia may feel validated when the AI confirms this diagnosis.

278 Amann, Blasimme, Vayena, et al., 2020.

279 Mittelstadt, Allo, Taddeo, et al., 2016: 4; Mittelstadt, 2021: 13–14.

relationship between the input and the output is not clear, thus preventing scrutiny of the decision.

Ultimately, many of the errors derive from the fact that the AI we currently have is 'narrow' or 'weak' AI, tailored to solve specific problems or tasks, but incapable of extending its capabilities limitlessly beyond the constraints of its training data and problem domain.[280] Even DL systems, which can autonomously derive appropriate mathematical representations from a specific dataset, are trained to handle a particular input type, such as medical image data, but they lack the ability to process alternative input types, such as statements from patients.[281] Researchers and technologists are making strides toward more advanced AI, but even the most sophisticated systems today are still considered weak AI.[282]

3.5.3 Replication of Errors

Another challenge relates to the replication of errors. One of the key advantages of medical AI is its scalability, which sets it apart from human expertise. Once an algorithm is developed and validated, it can be efficiently duplicated and applied in different settings.[283] This makes it much easier and more cost-effective to implement medical AI solutions across various healthcare environments, expanding access to specialised care beyond traditional limitations.

However, this feature also entails a drawback related to error replication. AI, like any other technology, has the infamous capacity to replicate and multiply errors.[284] When a doctor makes a mistake, the mistake affects only that particular patient. In contrast, when an 'AI doctor' makes a mistake, the implications can be far more widespread. AI systems, once trained and deployed, are often used across multiple settings – hospitals, clinics, and even different countries – affecting potentially millions of patients. The risk of widespread impact is further amplified by the diversity of healthcare environments in which AI systems

280 Even GPAI is still a form of weak AI.

281 Asan, Bayrak, and Choudhury, 2020.

282 'Emulation of intelligence can be achieved through various methods; when a set of methods with common theoretical underpinnings gains traction, it forms an approach to AI, such as the symbolic and connectionist approaches mentioned earlier on. However, the question of how to re-create intelligence to reach emergence of cognition and understanding in algorithms remains yet unanswered. Hence, we can claim that the current state of AI belongs to the weak type' (Vannuccini and Prytkova, 2023).

283 Oikonomou and Khera, 2024.

284 Topol, 2019: 51.

are deployed. Different countries and regions have varying levels of healthcare infrastructure, access to medical resources, and patient demographics. An AI system trained on data from one region might not perform as accurately in another, where the patient population differs significantly in terms of genetics, lifestyle, or prevalent health conditions.

3.5.4 Possible Ways to Mitigate Imprecise Outcomes

The key to solving some of these problems lies in regular retraining and evaluating performance, making use of technological improvements and new medical practices. This, however, leads to costly maintenance of datasets. The evaluation criteria for AI tasks should ideally be tailored to the complexity of the specific task. For instance, when assessing the accuracy of AI-enabled imaging applications for clinical use, validating them against existing quality standards for traditional imaging might be adequate. However, as AI applications progress to tasks involving prediction, diagnoses, and treatment, the standards for establishing proof should be substantially more stringent.[285]

A promising path is to build a robust establishment of the 'ground truth'[286] – a foundational concept in various AI applications – which, in turn, heavily relies on the careful selection, organisation, and preparation of real-world data to ensure its quality and relevance.[287] This involves verifying the machine's outputs during its learning phase as true positives or negatives. The ground truth serves as the standard for assessing the accuracy and reliability of the AI system. Ground truth refers to the precise, factual data or labels that represent the real-world outcomes or conditions the AI model aims to predict or analyse. For instance, in medical AI, ground truth might be the accurate diagnosis validated by medical professionals. However, in domains like healthcare, this process encounters challenges due to the inherent complexity and inaccuracy of data, starting with the fact that, in the clinical field, determinations of pathology often transcend binary distinctions and involve subjective viewpoints.[288]

---

285 Maddox, Rumsfeld and Payne, 2019: 31.

286 Read-Coop SCE, 2023.

287 Lebovitz, Levina, and Lifshitz-Assa, 2021: 1501–1525.

288 For instance, within oncology, consensus might not be reached regarding the definition of cancer necessitating medical intervention. Skilled pathologists can frequently find themselves at odds over histopathology and diagnoses, particularly in early-stage cases. When the objective of AI is to enhance early cancer detection, this divergence poses a dilemma in algorithm training. The resulting AI tool must avoid over-diagnosing or under-diagnosing tumours (OECD, 2020: 13).

Human control and vigilance to identify problems as they arise are imperative to guarantee AI's accuracy.[289] This, in turn, can increase costs and reduce the initial benefits of AI.[290] A judicious approach becomes imperative when contemplating the integration of AI into life-critical domains, not only because of the cost but also because of the potential mistakes introduced by humans. It requires a delicate balance between harnessing the potential benefits offered by AI and preserving the essential element of human oversight and expertise.

3.6 *The Standardisation of the Unusual*

AI can produce deep effects on the healthcare system, normalising practices and expectations that may not be universally accepted, cost-effective, or standard for those specific settings.[291] As these misfit effects persist and certain behaviours become systematised on a large scale, significant costs and harms can arise for organisations, health systems, patients, and healthcare providers.

An illustration of this involves an AI system trained on medico-administrative data in a context where doctors often modify their billing to claim the highest-paying codes for clinical procedures. As a result, the algorithm may incorrectly infer that these codes represent the usual or standard practices to be recommended.[292] This biased algorithmic inference arises because the billed procedure maximises the doctor's remuneration, even if it was not the one performed. This can lead to a series of not cost-effective recommendations, raising the risk of inappropriate treatment and care.

An additional hazard relates to potential negative outcomes in local healthcare administration practices. Let's consider a scenario inspired by the study of Bélisle-Pipon et al.: imagine that a payer, whether a public entity or a private insurer, mandates the adoption of prior AI recommendations as a prerequisite for reimbursing healthcare expenses. This policy shift could potentially lead to the exclusion of alternative forms of care from reimbursement.[293] The implications of such a decision would be substantial, significantly impacting the delivery of care. It would also limit the autonomy of patients, doctors, and administrators when making shared and appropriate decisions, particularly when human-recommended care differs from the new 'gold standard' established based on AI. In this context, AI has the capacity to reshape the

289 Beck and Burri, 2024: 104–130.

290 Panel for the Future of Science and Technology, 2022: 17.

291 Couture, Roy, Ganache, et al., 2021.

292 Example provided by Couture, Roy, Ganache, et al., 2021.

293 Bélisle-Pipon, Monteferrante, Roy, et al., 2023: 1514.

landscape of healthcare administration and decision-making, influencing the choices available to patients and doctors alike.

Moreover, it may introduce standards and practices that are external and not tailored to local organisations, perpetuating latent biases present in training data that may not align with specific health systems or care contexts.[294] Healthcare professionals who are accustomed to local practices and standards may be resistant to adopting AI tools that do not align with their experience or expectations. This can hinder the successful integration of AI into local health systems, reducing its potential benefits.

3.7 *Cybersecurity*

Within the domain of cybersecurity, patient data stands exposed and at risk of being compromised or stolen, a situation that could potentially enable malicious actors to exploit it for financial gain.[295] Instances of medical data breaches commonly encompass cyberattacks like unauthorised entry by hackers and internal data theft or loss by individuals with access to the data.[296] As a consequence, a prevailing atmosphere of mistrust has taken root.[297]

Take the case of a surgical robot. The security vulnerabilities inherent in robots engaging directly with patients can lead to malfunctions with potentially dire consequences. These malfunctions might manifest as direct harm inflicted by the robot or, more insidiously, through the provision of erroneous feedback to a surgeon. In instances of unanticipated robot behaviour, the user may grapple with uncertainty, unable to discern whether the robot is operating as intended or is being compromised by an external attack.[298] There

294 Suppose an AI system is developed and trained primarily in a healthcare environment that predominantly serves a population of Jehovah's Witnesses, where blood transfusions are not accepted due to religious beliefs. The AI system has been designed to recommend alternative treatments, such as the use of blood substitutes, medications to boost red blood cell production, and surgical techniques that minimise blood loss, as these align with the patients' religious practices.

When the AI system is introduced in a secular hospital, where the general patient population does not have religious objections to blood transfusions, it will continue to recommend alternatives to blood transfusions, even in situations where transfusions would be the most effective and immediate treatment option. These recommendations conflict with the established medical practices in this environment, where blood transfusions are routinely used and are considered the best standard of care.

295 Zhang and Zhang (2023) reported that in 2017, the healthcare sector accounted for 15% of worldwide data breaches, ranking second only to the financial sector.

296 An incident in 2020 involved a Chinese imaging AI being hacked, leading to the sale of its source code and training data on the dark web for 4 bitcoins. See Zhang and Zhang, 2023.

297 About the lack of public trust, Gómez-González and Gómez, 2020: 24.

298 Fosch-Villaronga and Mahler, 2021.

are examples of robotic surgeries being hacked, where an assailant possesses direct dominion over the robot's actions. The level of risk increases when the AI-based medical act is performed at a distance, using cybermedicine, such as might happen in telerobotic surgeries. In the latter case, a more visionary form of exploitation arises, rooted in the art of deceiving the robot's intricate algorithms.

A specific type of cyberattack, known as 'input attacks', manipulates the data fed into the AI system to produce incorrect results, thus undermining their accuracy and potentially leading to harmful consequences.[299] For example, an attacker could alter an MRI (Magnetic Resonance Imaging) scan's pixel values, causing the AI system to misclassify tissue as malignant instead of benign (or vice versa), potentially resulting in false diagnoses and harmful treatments.[300] What makes input attacks different from traditional cyberattacks is that they do not tamper with the AI system itself but only manipulate the input data. This makes them harder to detect, as they require subtle changes to medical images that remain invisible to the human eye.

Poisoning attacks differ in that they focus on the ML system's development phase. By altering the training data or process, the attacker ensures that the final deployed model is inherently defective.[301] Such attacks take place when the model's parameters and learning mechanisms are still being set.

Considering the nature of these risks, some argue that they do not fall under the obligation of disclosure as they are not inherent in the medical procedure itself.[302] However, this objection overlooks the fact that input attacks have become inseparable from certain AI-based medical procedures, making them 'inherent' risks that necessitate disclosure. While doctors are not expected to predict cyber-attacks, the reasonable person standard, as endorsed in the legal case Montgomery v Lanarkshire Health Board,[303] supports the need to disclose the risk of input attacks.

299 Polito and Pupillo, 2024: 10.
300 Example of Kiener, 2021: 707.
301 Polito and Pupillo, 2024: 10.
302 See Jones v. Papp, 782 S.W.2D 236.
303 Montgomery v Lanarkshire Health Board [2015] UKSC 11, [2015] 1 AC 1430.

3.8 AI 'Creativity' and Lack of Predictability

A fundamental aspect of AI that presents a formidable legal challenge revolves around the principle of predictability,[304] in which AI 'behaves' in a way humans cannot control or even predict.[305]

The erosion of control over AI can be categorised into two main types. The first is local control loss, signifying a state where the AI system slips beyond the control of the human or humans accountable for its operation and supervision. On the other hand, general control loss denotes a scenario where the AI system evades control by any human entity. The latter situation inherently magnifies the scale of public risk. Note that even when general control is lost, the ultimate scenario might not be hazardous. Significant public risk can be prevented if the AI system's goals align with the public interest.[306] However, we humans might not be fully prepared for AI to take the lead.

Scherer provides the example of C-Path, an ML system used in cancer pathology.[307] Pathologists speculated that examining the supportive tissue (stroma) encasing cancer cells, in conjunction with analysing the tumour cells themselves, could enhance cancer prognosis. However, in a comprehensive study, C-Path unveiled an unexpected revelation: the characteristics of the stroma held greater predictive value for breast cancer prognosis than the traits of the cancerous cells themselves.[308] This finding stood in stark contrast to both common sense and prevalent medical wisdom. Yet, Scherer argues that this apparent display of creativity is somewhat illusory, a by-product of the computational capabilities harnessed by these specialised AI programs.[309]

When it comes to AI that develops along the way, as it learns from the environment to which it is exposed (the so-called ML), additional difficulties arise, related to the lack of predictability as to how AI will behave. In this regard, Matthias talks about a 'responsibility gap'[310] to describe the impossibility of holding manufacturers morally liable due to the absence of culpability[311] (the

304 Regarding the application of this principle to AI, see Yampolskiy, 2024.

305 Instances of AI manifesting seemingly creative behaviour have surfaced, often echoing actions that, if undertaken by a human, would be deemed ingenious or indicative of unconventional thinking. This phenomenon is notably observed in computer chess programs, which execute moves contradicting the established tenets of human chess strategy (Scherer, 2016: 363).

306 Scherer, 2016: 367.

307 Scherer, 2016: 363–364.

308 Beck, Sangoi, Leung, et al., 2011.

309 Scherer, 2016: 364.

310 Matthias, 2004.

311 Santoni de Sio and Mecacci refer to up to four moral responsibility gaps (2021: 1057–1084).

same can be said about legal liability, even though the trend in the EU is to impose strict liability – that is, disregarding culpability – in many AI products). Some scholars challenge the concept of a 'responsibility gap' in AI systems, suggesting that there might never be a point where no one can be held responsible for outcomes.[312] The validity of these statements is, however, debatable.

Even if responsibility is not an issue, the lack of predictability affects the medical standard: how can we tell a doctor to behave in a certain way in a given scenario if we do not know what exactly the scenario is about, because we do not know, and cannot predict, how the AI system will behave/decide?

3.9 *Failures in Unpredictable Scenarios*

AI does not perform well in unpredictable scenarios. According to Belpaeme, ML (and the same can be said about any other form of AI) is better at 'dealing with a predictable world'. That is, the system works fine with 'anything inside a computer' (browsing history, playing chess, etc.) as these are predictable and stable environments.[313] There, AI excels because it works with well-defined and consistent parameters.[314] Data are often abundant and structured, allowing AI to learn and make predictions effectively,[315] and many digital tasks involve repetitive patterns that AI can learn from.[316]

However, when AI steps out of the digital world and is applied in the messy, complex reality of the real world, its performance often diminishes, revealing the challenges inherent in translating digital success into real-world effectiveness. Here, the environment is far less predictable, data are often unstructured or incomplete, and the variables AI must contend with are numerous and constantly changing.

---

312 Tigard is one of those scholars, advocating for a pluralistic approach to moral responsibility (Tigard, 2020). He contends that complex situations and unconventional entities often have responsibility attributed to them. In Tigard's view, there is no inherent technology-based gap of responsibility. To illustrate, consider the common practice of holding pet owners responsible for any damage caused by their pets, even when they have no direct control over the animal's actions and the unforeseeable harm it might cause.

313 Tony Belpaeme, Professor of Cognitive Systems and Robotics at Plymouth University, quoted in Sayburn, 2017: 89.

314 For example, AI algorithms in chess are provided with a finite set of moves and an unchanging board, enabling them to calculate optimal strategies with exceptional accuracy.

315 For instance, AI systems analysing browsing history can easily identify patterns because the data are logged and categorised systematically.

316 This is what happens when AI predicts user behaviour on a website or optimises search algorithms.

A surgical scenario is frequently referred to as one of those examples of a 'messy world'. In the operating room, the human body presents a highly complex and ever-changing environment. Unlike static images or predefined tasks, a surgery involves navigating through soft tissues, blood vessels, and organs, each with its own unique texture, movement, and variability. Surgeons must constantly adapt to what they see, feel, and anticipate as they operate, making real-time adjustments based on their extensive training and experience.[317] One of the most daunting challenges for AI in surgery is interpreting camera images of soft tissue. Soft tissues, such as muscles, fat, and connective tissues, do not have the clear, consistent structures found in hard tissues like bones. They can move in unpredictable ways and often look different depending on factors like lighting, angle, and even the pressure applied by surgical instruments.[318] For a human surgeon, interpreting these variations is already a complex task that requires years of training and experience. The human brain can process a multitude of visual cues simultaneously, drawing on a vast repository of knowledge about how tissues should behave and what abnormalities might look like. However, for AI, this task is exponentially harder.[319]

3.10 *Lack of Human Expertise*

As the nascent technology rapidly outpaces the training of practitioners, numerous providers find themselves unable to harness the full potential of the most advanced tools, leading to suboptimal utilisation or inadvertent errors in deploying AI systems for which they have not received adequate training.[320] Many doctors who are actively practising often find themselves facing digital health technologies in their workflow without the necessary training on how to safely integrate these technologies. This lack of preparation can lead to challenges when incorporating digital tools into clinical practice.[321]

With the healthcare landscape embracing AI systems, healthcare staff will need to acquire new skills and undergo training. They will be required to develop a comprehensive understanding of AI systems, including the technical knowledge to effectively operate and discern the limitations associated with these advanced technologies. Therefore, it is crucial for doctors to actively

317 Reddy, Gharde, Tayade, et al. 2023.
318 Sayburn, 2017: 89.
319 Lu, Li, Chen, et al., 2018: 368–375.
320 Marchant and Tournas, 2019: 34; UK Parliament, 2020: 3.
321 Rowland, Fitzgerald, Lungren, et al., 2022.

educate themselves about these systems to make informed decisions and uphold ethical standards in their medical practice.[322]

In a utopian way, this would involve acquiring proficiency in mathematical principles, AI essentials, data science, and related ethical and legal matters.[323] However, such an aim is difficult to achieve, as medical schools currently lack strong incentives for nurturing these areas of expertise, given the demanding nature of the existing medical curriculum and the scarcity of faculty members who possess AI expertise and can effectively instruct on integrating AI into clinical applications.

To make things more complicated, regulators may impose the testing of human factors and the provision of adequate training for human users of medical AI systems, recognising that the effectiveness of these systems often relies on how humans interact with them and interpret their outputs.[324] For instance, Article 14 of the AIA establishes that 'High-risk AI systems shall be designed and developed in such a way, including with appropriate human-machine interface tools, that they can be effectively overseen by natural persons during the period in which they are in use'; and Article 26(2) of the AIA states that 'Deployers shall assign human oversight to natural persons who have the necessary competence, training and authority, as well as the necessary support'.[325] This emphasis on human involvement and understanding highlights the importance of ensuring that users are well-prepared and equipped to effectively use AI systems in healthcare settings.[326]

3.11 *Lack of Empathy*

AI cannot simply be empathetic in a human way, and this is a harsh drawback in healthcare, because effective medical care depends on patient compliance with treatment, which, in turn, depends on a strong and trustworthy doctor-patient relationship.[327] An important predictor (or even the highest predictor) of trust is the patient's perception that the doctor is genuinely concerned with the patient,[328] and for that to happen the doctor must be

322 Char, Shah and Magnus, 2018: 983. Proper training should be included in standard medical education, as suggested by the American Medical Association, 2019.

323 Paranjape, Schinkel, Nannan Panday, et al. 2019.

324 Rajpurkar, Chen, Banerjee, et al., 2022: 35.

325 On the obligation of human oversight in the AIA, Enqvist, 2023: 508–535.

326 Gerke, Babic, Evgeniou, et al., 2020: 3.

327 Martin, Williams, Haskard, et al., 2005: 189–199. The biggest cause of poor outcomes in medicine is that approximately half of medical recommendations, including prescriptions, are not followed or taken as prescribed (Brown and Bussell, 2011: 304–314).

328 Cancer patients who felt that their doctor was sympathising with them when discussing the cancer diagnosis sought treatment and support groups more actively than patients

empathic. AI can be very good at cognitive skills; it can accurately represent emotions (although this would not be a trivial task) and properly relate situations to the desired results. Still, it lacks the capacity to show empathy, at least in the same way as a human.[329]AI tools today lack the emotional and contextual intelligence required to make decisions in situations that involve uncertainty, high risk, or emotional complexity. Certain decisions go beyond mere logic aimed at survival, requiring a deeper understanding of human emotions and the surrounding context. In such cases, AI falls short of handling the nuanced and emotionally charged aspects that are often essential for making well-rounded judgments.[330]

However, some changes are taking place in this domain, as I will discuss below.

3.12 *Lack of Trust*

3.12.1 Lack of Medical Trust

A comprehensive understanding of AI decisions is crucial for doctors, as they strive to know the reasons behind diagnoses and treatment recommendations.[331] This knowledge is essential for building trust and confidence in the decisions made.[332]

For AI systems to effectively assist doctors rather than make autonomous decisions, doctors must be able to trust them. Doctors must have confidence in the technology they rely on, which becomes challenging when they lack a deep understanding of how the AI system generates its outcomes.

However, this aim is not always achieved.[333] The opaque nature of many AI algorithms means that even the most experienced healthcare professionals

---

who did not feel supported and had longer periods of confusion and anxiety after receiving difficult information. In this regard see Dekker, Karchoud, Braamse, et al., 2020: 1399–1405.

329 Montemayor, Halpern and Fairweather, 2022: 1353–1359.

330 Lorenzini, Ossa, Shaw, et al., 2023: 427.

331 Juravle, Boudouraki, Terziyska, et al., 2020: 263–282.

332 Asan and colleagues propose the concept of 'optimal trust', a paradigm where both humans and AI systems embrace a certain degree of scepticism towards each other's decisions (Asan, Bayrak and Choudhury, 2020). This approach recognises that both humans and AI possess the potential to make errors and underscores the importance of mutual scepticism. By acknowledging the fallibility of both parties, this approach advocates for a balanced perspective on trust that fosters a constructive relationship between humans and AI.

Hengstler et al. discovered that perceived risk reduction relies heavily on trust, and this trust can be strengthened when the trustor has access to transparent algorithms. See Hengstler, Enkel and Duelli, 2016: 105–120.

333 For instance, a study conducted in the Netherlands revealed that IBM's Watson for Oncology faced difficulties in providing explanations for its diagnoses, leading human doctors to lose trust in the system (Durán, 2021).

may find it difficult to interpret or trust the system's recommendations, as doctors are understandably cautious about integrating technology that they do not fully comprehend into critical decision-making processes. As a result, the potential benefits of AI in improving patient care may not be fully realised if this trust gap is not addressed.

The lack of trust has been identified as a barrier to the widespread adoption of AI in the medical field by some experts.[334] This hesitation slows the progress of AI-driven innovations in medicine, highlighting the need for greater transparency and collaboration between AI developers and doctors to build trust and ensure safe, effective implementation.

3.12.2 Lack of Patients' Trust

In any relationship, trust is widely recognised as a fundamental element. In the context of healthcare, trust is particularly crucial due to the patient's vulnerability and dependency on the doctor, as well as the information asymmetry resulting from specialised medical knowledge. Patients cannot always predict their doctor's actions because doctors have a complex mix of knowledge and skills that go beyond simple patterns.

AI makes this knowledge gap even more complicated. While doctors who become proficient AI users can uphold the trust inherently linked with their role, the same cannot be said for most patients, whose comprehension of these intricate technologies remains limited. Instead, patients rely on trust, revealing their vulnerability as they build expectations around their doctor's expertise, values, and intentions. This lack of comprehension poses a significant barrier to placing trust in AI.[335]

The lack of explainability of many AI systems also fuels the lack of patients' trust. The healthcare relationship between professionals and patients is built on a unique bond that emphasises patient-centred care, respecting autonomy, and encouraging informed choices that align with patient values and preferences.[336] Information plays a critical role in infusing confidence[337] because the lack of it can unintentionally trigger a resurgence of paternalistic care ideals. This could marginalise patients, relegating them to roles of mere observers, detached from active participation in the medical decision-making panorama.[338]

334 Poon and Sung, 2021: 581.
335 LaRosa and Danks, 2018: 210–215.
336 Roter, 2000: 5–15.
337 Hatherley, 2020: 478–481; Holzinger, Carrington and Müller, 2020: 194.
338 Amann, Blasimme, Vayena, et al., 2020.

Another major factor contributing to the lack of trust in the doctor-patient relationship is the absence of empathy (see Section 3.11, Chapter 3). AI cannot simply replicate human empathy. An important predictor (or even the highest predictor) of trust is the patient's perception that the doctor is genuinely concerned with the patient,[339] and for that to happen, the doctor must be empathetic.

When trust is compromised, patients may hesitate to openly share all relevant clinical information.[340] Moreover, patients may feel uneasy, resulting in limited communication and potentially impacting the diagnostic capabilities of algorithms that depend on robust databases.

Statistical models offer higher accuracy compared to human intuition, suggesting people should prefer their advice. However, research indicates that individuals tend to favour human recommendations over computerised systems in various domains. Studies found people are more inclined to follow a doctor's recommendation rather than a computer's, possibly due to perceived differences in performance, including empathy.[341] Both consumers and professionals show this preference.[342]

When patients trust their doctors, they are more likely to follow medical advice, adhere to prescribed treatments, and engage in open communication about their health concerns.[343] This trust fosters a sense of partnership, where patients feel confident in their care and motivated to take an active role in their health management.

3.12.3 Excess of Trust

The transformation of medical authority from human doctors to algorithms presents a significant concern, often referred to as the 'collective medical mind' dilemma, as named by Danton et al.[344] The potential danger lies in the progression where AI systems, initially introduced as aids in decision-making, evolve into central points of medical judgments, in such a way that human

339 Cancer patients who felt that their doctor was sympathising with them when discussing the cancer diagnosis sought treatment and support groups more actively than patients who did not feel as supported and had longer periods of confusion and anxiety after receiving difficult information. In this regard see Dekker, Karchoud, Braamse, et al., 2020: 1399–1405.

340 Dugdale, Epstein and Pantilat, 1999: S34–S40.

341 Longoni, Bonezzi and Morewedge, 2019: 629–650; Torrent-Sellens, Jiménez-Zarco and Saigí-Rubió, 2021.

342 However, as Longoni and others point out, if consumer preferences are based on perceived performance, resistance to automated medical providers might not endure. See Longoni, Bonezzi and Morewedge, 2019: 630.

343 Pearson and Raeke, 2000: 510.

344 Char, Shah and Magnus, 2018: 981–982.

doctors feel compelled to adjust their diagnoses and recommendations to those issued by the machine.[345]

In such a context, it becomes uncertain whether the longstanding pillars of medical ethics – beneficence, non-maleficence, autonomy and justice[346] – can maintain their central position in the patient-doctor relationship, as they do now or as they are expected to.[347] If doctors feel compelled to accept what AI stands for, how can they promote these principles against AI judgments?[348]

The exceptionalism attributed to AI can largely be attributed to the notion that when an AI generates an outcome or recommendation, anyone who disagrees with it must provide a far more substantial justification supported by a vast amount of data, in comparison with the data the AI employed to reach that conclusion.[349] This belief is further reinforced by the technological imperative, where the mere existence of a sophisticated technological intervention exerts pressure to use it, given the perception of its superiority over conventional practices, despite potential risks. Presently, AI in medicine is experiencing its technological imperative moment.[350]

An additional challenge of this excess of trust is the slow erosion of what is commonly referred to as 'common sense' when operating with complex technologies. While AI algorithms excel in the field of processing vast quantities of information, their capacity for contextual comprehension and intuitive reasoning remains underdeveloped, emphasising the indispensable role of human judgment in critical decision-making contexts.[351] However, the absence of critical reasoning in the human players hampers their capacity to discern simple errors or deviations in data and decision-making processes that would be readily evident to human observers in a different scenario.

---

345 Kempt, Heilinger and Nagel, 2023: 1411.

346 Beauchamp and Childress' book (2013) is considered a classic for its detailed exposition of these four principles and their application, as well as its discussion of alternative approaches.

347 Blasimme and Vayena, 2020: 710.

348 In the already mentioned case of UB Songdo Hospital in Mongolia, the implementation of IBM's 'Watson for Oncology' was met with high adoption rates, with doctors following its suggestions at nearly 100%, even those with minimal cancer care training, a situation that obviously undermines their critical judgements and the capacity to overrule AI.

As previously noted, a concerning incident arose when Watson inappropriately recommended the drug taxane for a patient, despite the patient's medical history contraindicating its use. While an oncology specialist noticed this error, the possibility remains that a non-specialist doctor might overlook such a mistake. This incident is reported in Smith and Fotheringham, 2020: 133.

349 Jongsma and Sand, 2022: 230–231.

350 Couture, Roy, Ganache, et al., 2021.

351 Zhu, Gao, Fan, et al., 2020: 310–345.

The restriction of human judgment is another issue to consider. Overreliance on technology and disregard for human knowledge and human intuition are problems general to the introduction of any kind of technology, including AI.[352] As doctors increasingly rely on AI, there is a possibility that they may experience a gradual decline in their own skills or their ability to establish personal connections with patients, negatively impacting clinical practice and undermining the human aspect of healthcare.[353]

3.13 *Ethical AI*

3.13.1 Is AI an Ethical Entity?

Back in July 2018, Nick Bostrom, a philosopher from Oxford University, addressed the UK Parliament regarding AI and Future Generations. In his statement, he emphasised that intelligence is fundamentally rooted in computation,[354] not mathematics.[355] Bostrom pointed out that our tendency to confuse computation with mathematics may be the reason we mistakenly anticipate AI to possess qualities like pure ethics, omniscience, and eternity, while in reality, AI does not inherently possess these characteristics.

The fostering of the so-called 'ethical AI' is based on the concept of 'roboethics', which addresses the ethical conduct of humans in the process of designing, creating, and interacting with AI entities, such as robots[356] (although it is not restricted to robots, since many of the concerns it addresses are valid for other forms of AI).[357] Roboethics delves into the ethical dimensions surrounding the creation, configuration, and utilisation of robots. It also outlines the potential for both negative and positive impacts of artificially intelligent entities on humans. To achieve this objective, it is proposed that algorithms should incorporate moral components, leading to the development of artificial moral agents.[358]

To achieve this aim, two crucial aspects require consideration.[359] Firstly, the moral conduct of AI agents must be addressed to ensure their ethical behaviour towards human users and other artificial moral agents.[360] Constructing these agents in a manner that upholds moral and ethical principles is of utmost

352 Naik, Hameed, Shetty, et al., 2022.
353 Rajpurkar, Chen, Banerjee, et al., 2022: 35.
354 Computation involves the conversion of information through a physical process, consuming time, energy, and space.
355 Bryson, 2018.
356 Veruggio, 2006: 612–617.
357 Langman, Capicotto, Maddahi, et al., 2021.
358 Formosa and Ryan, 2021: 839–851.
359 Lupton, 2018: 5.
360 Martinho, Poulsen, Kroesen, et al., 2021: 481.

importance. Secondly, humans have a responsibility to adhere to ethical standards when designing, creating, and engaging with AI entities (Isaac Asimov's Three Laws of Robotics exemplify this ethical duty).[361]

3.13.2 Ethical AI in Healthcare

Within healthcare delivery, ethical AI becomes especially critical in some specific contexts.

One is mental healthcare. Following the conceptualisation of Lupton, AI systems can fall into two categories in this domain: implicit ethical agents, strictly following predetermined programming, or explicit ethical agents, capable of autonomously assessing ethical aspects by applying ethical principles to intricate scenarios, such as diagnosing depression in patients, prescribing treatment, and maintaining symptom monitoring.[362]

Another domain is resource management, as AI is not only being used to practice medicine but also to organise medicine. An ethical perspective is also essential in this domain. Making choices guided by AI, like prioritising patients for treatment or selecting candidates for chemotherapy, must certainly incorporate assessments of cost-effectiveness.[363] However, relying only on algorithms may overlook the important ability to adapt, which allows healthcare practitioners to balance standard guidelines with the unique details of each case.[364] The ethical imperative dictates that these measures prioritise the well-being of the patient, rather than favouring the mere advancement of technology or catering to the business rationale.[365]

---

361 Against the widely held belief that AI is morally neutral, having neither inherent goodness nor badness, Carter et al. argue that this is not accurate (Carter, Rogers, Win, et al., 2020: 27). They claim that even though AI certainly can yield both favourable and unfavourable outcomes, in the modern era of 'new AI', nearly every algorithm inevitably embeds values, whether they are made explicit or, more commonly, implicit.

362 Lupton, 2018: 3.

363 European Commission – Directorate-General for Justice and Consumers, 2019.

364 Imagine an AI system designed to predict the potential for extended hospital stays when assessing patient admissions. While such a system could be useful for managing hospital resources efficiently, it also carries the risk of bias, especially against the most vulnerable patients, such as the elderly or those with chronic conditions who are more likely to need longer care. For instance, if the system flags these patients as 'high-risk' for extended stays, hospitals may deprioritise their admissions or allocate fewer resources to them, unintentionally marginalising those who actually need the most attention. This could lead to inequitable care, where patients who are deemed more 'efficient' or 'cost-effective' receive priority, while the frailest patients face obstacles in accessing necessary care. See Blasimme and Vayena, 2020: 709.

365 LaRosa and Danks, 2018: 214.

CHAPTER 4

Medical AI and Standard of Care

1 The Standard of Care in Medical Practice

1.1 *The Concept of Standard of Care*

Throughout history, ethical principles have been embedded within medicine, as evidenced by ancient practices like the Hippocratic Oath.[1] Medical practice is shaped by enduring standards, which play a crucial role in laying the groundwork for fostering trust in the contemporary medical field. They shape what is currently known as the 'medical standard of care', basically a result of the conjoined assessment of lawfulness and culpability. This evaluation involves assessing the reasonableness of a doctor's actions and comparing them to those of a reasonably prudent doctor within the relevant medical community or practice area.[2] The medical standard of care can thus be defined as the care that a reasonably competent and skilled healthcare professional, with similar knowledge and under the same conditions,[3] would have provided.[4] In common law, this is known as the Bolam test:[5] if the doctor acted according to the practice generally accepted by a responsible body of medical doctors, he/she will not be deemed negligent.

Ethically, it refers to a wrongdoing that can constitute a violation of medical deontological rules. Legally, and in conjunction with the criterion of the *bonus pater familias*[6] – which is the traditional criterion to assess negligence in continental law (see Section 1.1.2, Chapter 4) – it identifies forms of conduct that are wrongful and culpable and therefore can lead to medical liability.[7]

1.1.1 Optimal Care, Ethical Care and Standard of Care

The standard of care is not synonymous with optimal care.[8] Optimal care refers to the highest standard of medical care achievable, given the current knowledge,

1 Askitopoulou and Vgontzas, 2018:1481–1490; Askitopoulou and Vgontzas, 2018: 1491–1500.

2 Vanderpool, 2021: 50–51.

3 The fundamental question of what actions a reasonably prudent doctor would have taken is generally treated as a factual inquiry, while determining the group of doctors deemed comparable is predominantly a legal consideration (Froomkin, Kerr and Pineau, 2019: 53).

4 Moffett and Moore, 2011:109–112; Raposo 2019: 401–416; Vanderpool, 2021: 50–51.

5 De Cruz, 2001.

6 Ataíde and Craveiro, 2020: 343–355.

7 Raposo, 2013: 45–47.

8 Vanderpool, 2021: 50. See also Schwartz, 1989: 241–280.

 HTTPS://DOI.ORG/10.1163/9789004543980-006

resources, and circumstances. It encompasses providing treatments and interventions that are most effective and appropriate for a patient's condition while considering individual patient needs and preferences. Optimal care aims to maximise positive health outcomes, ensure patient safety, and deliver compassionate, patient-centred care. Instead, the standard of care represents a spectrum ranging from barely acceptable care to the pinnacle of excellence.[9] From a malpractice liability perspective, doctors must position themselves somewhere along this continuum, even if they are closer to the lower end of acceptable care (still, doctors should strive towards providing care that aligns with the ideals of optimal care).[10]

In some situations, the conduct that protects doctors from liability does not coincide with the standard of care as the genuine ethical conduct. Doctors, fearing medical malpractice lawsuits, tend to engage in practices of defensive medicine,[11] which frequently deviate from the best standard of care.[12] For instance, nowadays it might be considered reckless to use AI in certain scenarios because courts, traditionally more conservative, might consider it as risky conduct that endangers patients' safety and thus a potential cause for a medical malpractice conviction. Nonetheless, from the perspective of the 'right thing

9 Vanderpool, 2021: 50. The standard of care for early-stage prostate cancer might involve options like active surveillance, surgery (prostatectomy), or radiation therapy. These options are generally accepted and widely practised by the medical community based on clinical guidelines and evidence. For instance, a doctor might recommend active surveillance for a patient with low-risk prostate cancer, which involves regular monitoring without immediate intervention. This approach is considered acceptable and falls within the standard of care, as it is evidence-based and aligns with common medical practices. Optimal care, however, would take a more personalised approach, considering the patient's specific circumstances, preferences, and overall health. For a patient who is highly anxious about the cancer progressing or who has a strong preference for definitive treatment, optimal care might involve a more proactive approach, such as robotic-assisted surgery, which has fewer side effects, and a quicker recovery time compared to traditional surgery. Optimal care would also involve a multidisciplinary team, including urologists, oncologists, radiologists, and perhaps a mental health professional to support the patient's emotional well-being. Additionally, the doctor would ensure that the patient fully understands all risks, benefits, and alternatives, thus supporting informed decision-making.

10 Vanderpool, 2021: 50.

11 This book uses the concept of defensive medicine provided by the Office of Technology Assessment (OTA): 'Defensive medicine occurs when doctors order tests, procedures or visits, or avoid high-risk patients or procedures, primarily (but not necessarily solely) to reduce their exposure to malpractice liability' (U.S. Congress, 1994). An analysis of this – and other definitions of defensive medicine – in Raposo, 2019: 402–404.

12 La Puma, Schiedermayer, Toulmin, et al., 1988: 121–124.

to do', in terms of patient well-being, using AI might be the most beneficial for the patient.

1.1.2 Standard of Care and Medical Leges Artis

In continental Europe, the concept of 'medical standard of care' is not widely used; instead, this same idea is referred to as medical *leges artis*, a concept with ethical and legal connotations.[13] *Leges artis* refers to the technical and ethical rules that doctors are obligated to follow in their practice. These rules are primarily derived from professional guidelines, scientific protocols, and codes of ethics, but they may also be informed by the common practices of the medical community.[14] A breach of *leges artis* indicates wrongful conduct, which is a prerequisite for establishing legal liability under civil law. It also suggests culpability, another key element for legal liability.

To assess whether the *leges artis* have been violated, continental law systems resort to another very specific concept, derived from Roman law: the '*bonus pater familias*'.[15] The latter is used to represent the standard of a reasonable person, someone with average intelligence who exercises ordinary diligence and care. This standard, similar to the 'reasonable person' criterion in common law,[16] serves as a benchmark in assessing a defendant's culpability in legal cases. If the court determines that the defendant acted in a manner consistent with how a reasonable person would have behaved in the same circumstances, no culpability – and thus no liability – is found. In medical malpractice cases, this standard shifts from a general 'reasonable person' to a 'reasonable doctor', reflecting the expectations of a doctor with average knowledge, skills, and care.[17] Instead of comparing the doctor's actions to those of a layperson, the court assesses whether the doctor acted as a typical medical professional would in the same situation.

1.1.3 Origins of the Standard of Care

The kind of behaviour that is expected from healthcare providers in any given situation results from several sources. Deontological codes play a major role in

13 About the similarities and differences between the 'medical standard of care' and the 'leges artis' see Raposo. 2019: 401–416.

14 Raposo, 2013: 45–46.

15 Rego, 2025.

16 Jorgensen, 1974–1975: 333.

17 Naik and Basu, 2022 (with a critique to the concept of 'reasonable doctor').

the definition of the standard of care.[18] Professional guidelines[19] (recommendations developed by professional bodies) and medical protocols[20] (rules and procedures resulting from scientific evidence) are also sources of the medical standard of care. Likewise, the common practices of the medical profession,[21] resulting from the consensus among healthcare professionals, define the medical standard of care. Although none of them are legal norms, they can be used to interpret legal norms. Moreover, the interpretation of those norms[22] and codes of conduct by courts and legal scholars[23] helps to densify their content. Even though neither case law nor legal literature is, *per se*, sources of the medical standard of care, they are key to understanding the possible meaning and outcomes of the norms that compose the medical standard of care.

1.1.4 Mutability of the Standard of Care

The standard of care is not an immutable concept. On the contrary, it is constantly evolving with new realities.[24] It often develops through the practices of healthcare providers, but it can also be influenced by practice guidelines established by professional medical organisations or legislative actions.[25]

It varies according to the geographical location. Two prevailing approaches have emerged to determine the standard of care in this regard: the national approach and the locality approach. These approaches aim to establish a benchmark against which doctors' actions are measured, taking into account the norms and practices of the surrounding professional community.[26]

In scenarios that involve the use of new technologies, a major obstacle to the definition of the standard of care is that there is no previously established practice.[27] In those cases, there is no single form of conduct within the standard of care, but multiple ones can fit into it. This is precisely the situation with the use of AI in medicine, where clinical practice is still evolving and no uniform benchmark has yet emerged.

---

18 Curca, 2010: 161–169.
19 Panteli Legido-Quigley H, Reichebner et al., 2019: 233–263.
20 Marcos, Balser, Ten Teije, et al., 2003: 132–141.
21 When defendants attempt to establish a standard of care, they typically present expert testimony describing the prevailing pattern of medical practice.
22 Moffett and Moore, 2011: 109–112; Naik and Basu, 2022. Note that the role of case law is especially important in common law jurisdictions.
23 Maestre, 2016: 36–39.
24 'The legal standard of care is key to liability for medical AI, but it is not forever fixed. Over time, the standard of care may shift' (Price II, Gerke and Cohen, 2019: 1765).
25 Aaron, Robertson, King et al, 2025: 1161.
26 Giuffrida and Treece, 2020: 114–118.
27 Harned, Lungren and Rajpurkar, 2019: 6.

1.2 *Standard of Care and Innovative Practices*

1.2.1 The Standard of Care as Predominantly Traditional

Historically, the legal system has shown great respect for customary practices among doctors when assessing malpractice cases. Essentially, any medical care considered normal or common (in a specific region) tends to be officially deemed as the 'reasonable conduct'.[28] As long as a doctor's treatment adhered to professional customs, it is generally considered satisfactory and not seen as a breach of the duty of care.[29] Thus, medical malpractice laws tend to absolve doctors of liability when they follow customary practices.

Consequently, for a significant period, doctors were granted unique protections from professional negligence liability through two rules: the ability to invoke custom as a defence ('it has always been done like this') and the locality rule ('here, we tend to do it like this'). These rules effectively shielded doctors from liability as long as their treatment fell within the common practices of their community.

This liability shield led to an obvious consequence: the discouragement of innovative practices. This reliance on tradition and local practice norms created a conservative medical culture where adherence to the status quo was prioritised over progress and adaptability, potentially stifling advances in medical science and limiting the evolution of care standards to the detriment of patient well-being.[30]

Over time, the medical profession and legal standards have evolved to place greater emphasis on evidence-based practices and patient-centred care, moving beyond the constraints of custom and locality to better reflect the dynamic nature of medical knowledge and technology.

Still, even now, current legal incentives encourage doctors to rely on AI merely for confirmation rather than embracing its full potential.[31] While AI can streamline processes by quickly suggesting the standard of care, its ability to propose superior treatments might remain untapped. This limited approach reduces the value and adoption of medical AI, as it discourages doctors from following nonstandard recommendations, which could otherwise lead to more significant benefits and advancements in healthcare.

---

28 Moffett and Moore, 2011: 112. About deviations in consolidated conducts, see Gvozdenovic, 2022.
See also Raposo 2019: 401–416.

29 Greenberg, 2009: 428.

30 Earl, 2019.

31 Mennella, Maniscalco, De Pietro, et al. 2024.

This approach has faced criticism for hindering medical innovation. Defining the standard of care based on custom means that any deviation from customary practices exposes doctors to malpractice liability, even if the innovation is ultimately beneficial on average.[32] Medical malpractice laws can act as a hindrance to medical innovation, creating barriers in the market for new technologies. This occurs despite the potential societal advantages such technologies may offer, such as improved clinical outcomes and decreased administrative expenses.

This book is founded on the belief that the day will come – and maybe very soon – when the standard of care in medicine generally coincides with the use of AI.[33] When that day arrives, doctors will be more protected from medical malpractice lawsuits by including AI in their practice.

1.2.2 The Suspicion of Courts

The court's approach to the use of new technologies in healthcare is often marked by a cautious yet open-minded stance. While courts recognise the potential benefits of innovative medical techniques and technologies, they also remain vigilant about ensuring patient safety and adherence to established standards of care. Judges typically require that any new technology or method must be supported by a responsible body of medical opinion that logically endorses its use.[34] This involves weighing the risks and benefits of the innovation and considering whether it meets the level of care expected from the medical community. Ultimately, the court seeks a balance between fostering medical advancements and protecting patients from unproven or inadequately supported practices, as some court cases illustrate.

The dynamics of the standard of care involving EHRs witnessed a shift in the Johnson v. Hillcrest Health Center case.[35] This case involved an incident where a patient's laboratory results were accessible to consulting doctors through the hospital's computer system, but were not included in the patient's physical medical chart; instead, they were mistakenly placed in the wrong chart. The consulting doctor, relying solely on the physical chart, found no indication to keep the patient hospitalised and decided to discharge them. Unfortunately,

32 Other paths are possible. Gary Chan, for instance, suggests that to achieve a better balance between innovation and patient safety, when a medical innovation is involved, the doctor should be required to demonstrate that the proposed treatment is beneficial for the patient (Chan, 2021: 180).

33 Stöger, 2021: 35.

34 Mehta, 2024: 125–142.

35 Johnson v. Hillcrest Health Ctr., Inc., 70 P.3d 811,814 (Okla. 2003). A comment to this case in Carter, 2020: 396–398.

shortly after being discharged, the patient suffered a heart attack and died. The Oklahoma Supreme Court introduced a significant modification to the standard of care. Before this case, doctors could largely rely on the information presented within traditional paper charts to fulfil their duty of care and mitigate potential liability. However, the Johnson v. Hillcrest Health Center case marked a departure from this precedent. While the court did not go as far as imposing an exclusive reliance on EHRS, it did introduce a crucial requirement: doctors became expected to undertake an additional step, notably consulting the EHRS, in order to meet the standard of care.

Another paradigmatic case that shows how the existence of a standard of care – or a consensual medical practice – cannot prevent the introduction of novel technologies and procedures can be found in the Singaporean case Gobinathan Devathasan v SMC.[36] The plaintiff filed a lawsuit against her vascular surgeon, claiming nerve damage to her legs following a surgical procedure intended to treat superficial reflux disease. She alleged medical negligence in several areas: the advice given, consent obtained for the surgery, the execution of the operation, and the post-operative care provided. The case involved the use of a novel surgical approach, where the defendant surgeon combined endovenous laser therapy with foam sclerotherapy in an unconventional manner. This novel medical treatment, not yet accepted by the medical community, was nonetheless considered beneficial for the patient. Expert testimony indicated that this innovative technique carried minimal risk and could even reduce the potential for injury. The court found that there was no proof that a responsible body of medical opinion, logically supported, would reject such innovation. This reasoning reflects an innovation-friendly interpretation of the standard of care. Paradigmatically, the court concluded that 'the fact that a novel technique is employed or that an accepted technique is used in a novel way does not, without more, give rise to a breach of duty' (par. 127).

1.2.3 The Suspicion of Healthcare Providers

Another main reason why healthcare innovation often progresses at a sluggish pace is the prevailing mentality within the sector, as well as among medical practitioners. Healthcare professionals often exhibit reluctance or outright resistance to adopting novel technologies, driven by various factors. These factors encompass real or perceived threats to their professional stature and independence and the intricate legal and ethical quandaries surrounding

---

36 Rathanamalah d/o Shunmugam v Chia Kok Hoong [2018] 4 SLR 159 [127] (Aedit Abdullah JC) (https://www.elitigation.sg/gd/s/2017_SGHC_153). A comment in Mehta, 2024: 134–145.

accountability.[37] One must bear in mind that doctors are instilled with the principles of adhering to protocols, implementing evidence-based medicine, streamlining procedures, and minimising deviations.[38] Any departure from the established norms that leads to unfavourable outcomes heightens the likelihood of facing malpractice lawsuits.

While many entrepreneurs exhibit a willingness to embrace substantial risks, doctors tend to lean toward risk aversion.[39] The inclination within the healthcare culture to shun risk is unsurprising, given the paramount significance of safeguarding patients' well-being. Doctors' inclination to mitigate risk is a natural outcome of their education and training. The cautious stance is grounded in the notion that venturing into unverified practices in healthcare delivery can potentially result in adverse outcomes for patients. While this approach is advantageous when tending to patients' immediate needs, it can inadvertently foster resistance to innovations in healthcare delivery due to the aversion to uncertainty.[40]

The same is true for healthcare facilities.[41] Apprehension due to litigation concerns could stifle the advancement and development of ground-breaking medical innovations. This cautious approach can create a barrier to innovation, as healthcare facilities might prefer to stick with established methods that are perceived as legally safer rather than exploring novel solutions that could expose them to legal risks.

1.2.4 Other Reasons Justifying the Exclusion of AI from the Medical Standard of Care

While AI offers promising advancements in healthcare delivery, there are several additional factors that justify the exclusion of AI from the medical standard of care, beyond the ones already analysed.

The recognition that a general practitioner may not possess the same level of expertise as a specialist, particularly within the specialist's domain, contributed to this framework. Doctors within traditional medical institutions and trained under more traditional methods would struggle with a standard of care that easily embraces novel practices.[42]

Similarly, the limitations of small rural practices, which lack the same resources as large urban teaching hospitals, may also be taken into consideration

37 Safi, Thiessen and Schmailzl, 2018.
38 Comer, 2025: E21–26.
39 Oussedik, Anderson and Feldman, 2017: 1–2.
40 Singhal and Patel, 2025.
41 Randolph, 2020.
42 Longoni, Bonezzi and Morewedge, 2019: 629–650.

by many courts. Traditionally, there was a prevailing sentiment that it would be unfair to expect rural practitioners to be as up to date as those affiliated with major medical institutions,[43] nowadays associated with AI.

A further issue to be considered is that of costs. In the early stages of AI integration in clinical care, AI systems are likely to be costly and accessible primarily to high-end academic medical centres and prestigious clinics. Consequently, it may be deemed inequitable to hold providers at these elite institutions to the same standards as those at facilities without access to cutting-edge AI systems. However, as the cost of AI systems decreases and their availability becomes more widespread, the rationale for establishing a national standard of care becomes increasingly compelling.[44]

The adoption of AI into the standard of care can carry both risks and benefits. On the one hand, it may lead to increased use of AI in medical practices. However, some experts warn that doctors might become overly reliant on AI recommendations, making them susceptible to losing their own skills and knowledge over time.[45] Moreover, if AI performance degrades due to changes in patient populations or patterns of care, it could leave the health system in a worse state than before, with doctors unable to rectify the situation.[46]

1.2.5 The (Still) Exclusion of AI from the Medical Standard of Care

As clearly stated in a WHO report, '[w]hether a doctor uses AI also depends on the prevailing standard of care. If AI technologies are viewed as deviating from or are not recognized as meeting the standard of care, doctors will be discouraged from using them, since, otherwise, meeting the standard of care defends (although not absolutely) medical error. If the standard of care requires use of AI technologies, doctors would essentially be mandated to integrate their use into clinical practice'.[47]

Price argues that the standard of care can evolve organically over time or be established by professional organisations as practice guidelines. These guidelines may recommend a standard of care that considers the level of risk involved. In cases where the risk and consequences are high, doctors might need to seek greater validation or inquiry before accepting an AI's determination. Greenberg supports this idea and suggests that when doctors encounter new medical devices, the standard of care might demand additional steps to

---

43 Mello, 2001: 655, citing a 1993 document from the Joint Commission.

44 Marchant and. Tournas, 2019: 40.

45 For instance, Froomkin, Kerr and Pineau, 2019: 33.

46 In cases where the protocol for treating any medical conditions undergoes modifications, like fine-tuning medication dosages to suit individual patients, if the AI system fails to maintain updated information, it can result in adverse effects on the patient's well-being.

47 World Health Organization, 2021: 77.

enhance understanding. This could include reading all clinical trial information, following label instructions, and seeking relevant training.[48]

Presently, as AI is not yet part of the medical standard of care, a possible defence for doctors is to look for additional grounds for any decision they take.[49] For instance, combining multiple algorithms to reach a so-called 'consensus opinion' (resulting from ensemble learning).[50] However, given the unparalleled velocity at which AI-driven technological advancements are progressing, healthcare stakeholders can no longer rely solely on the notion of 'accepted medical practice'. Instead, they must remain vigilant in continuously monitoring the latest breakthroughs in order to ascertain the optimal and safest course of treatment.

1.2.6 The Progressive Inclusion of AI into the Medical Standard of Care

With AI development, there is a potential for it to become indispensable in healthcare delivery, especially when the algorithm consistently outperforms human doctors. We might even envision a scenario where the conventional medical standard of care demands the use of AI, similar to how customary practices have evolved to incorporate other diagnostic technologies like MRIs or CT scans.[51]

Some scholars argue that this approach could lead to embracing AI systems (or, at least, some AI systems) as a means to achieve the gold standard of evidence-based practice,[52] because of the compelling appeal of evidence-based practice and also because of the higher success rate of AI systems over human experts. In this new paradigm, while a doctor's application of AI in medicine may deviate from conventional medical practices, such divergence does not automatically constitute negligence. The connection with evidence-based medicine is certainly appealing and convincing,[53] but if such strict criteria were imposed, the integration of any medical innovations into clinical settings would become impractical, as at the beginning of any new technological advancement we risk lacking enough medical evidence to ground its use.

---

48 Greenberg, 2009: 4444.

49 Banja, Hollstein and Bruno, 2022: 818.

50 'Ensemble learning' is an ML technique where multiple models are trained and combined to improve the accuracy and robustness of predictions. The consensus opinion resulting from ensemble learning refers to the collective prediction made by the ensemble of models (Ecord, 2023).

51 Harned, Lungren and Rajpurkar, 2019: 7.

52 Millar and Kerr, 2016, pp. 116–117.

53 Ferreira, 2023: 384–389.

I believe the adoption of AI will be the natural result of AI progression. As AI algorithms continue to improve and consistently offer statistically superior recommendations, it is challenging to argue that doctors are negligent for following them, even if their suggestions end up causing harm to the patient. Consequently, the conventional concepts of 'doctor negligence' and 'recklessness' associated with malpractice may become less applicable.[54] The advancement of medical AI surpassing human capabilities statistically will likely lead to the legal system treating machine diagnosis as the new 'standard of care'.[55] This designation will hold doctors and hospitals accountable for not using it (unless they can invoke a valid reason), exposing them to significant malpractice liability if any harm occurs. Moreover, insurance companies and hospitals will likely mandate the use of AI, at least as a tool for assisting doctors, as the failure to do so could result in legal consequences when unfavourable outcomes arise.

Even a fundamental rights perspective might ground the obligation to use AI. The protection of fundamental rights in Europe, particularly the rights to life (Article 2 Charter of Fundamental Rights of the European Union – CFR) and privacy (Article 8 CFR), places an obligation on the state to ensure that healthcare practices adhere to the established 'standard of care' or 'state of the art' in medicine. Stöger and colleagues claim that this obligation encompasses the prohibition of medical treatment methods that cannot meet the necessary quality standards without the involvement of AI.[56]

The timeline for this shift may depend on factors such as cost considerations and whether courts continue to differentiate between different types of medical practices and situations, such as teaching hospitals, rural hospitals, and solo practitioners. However, these factors primarily affect the speed and specifics of implementation rather than the overall trend.[57]

In any case, even though AI tools can provide valuable insights and guidance, they cannot supplant the critical role of doctors in making sound medical decisions based on their expertise and the unique circumstances of each patient. While doctors might be encouraged or even required to consult AI tools as part of their decision-making process, the final judgment rests with the doctor.[58] Their independent assessment and decision to follow or deviate

54 Thomas, 2017.

55 'Is not a question of "if" but "when" AI will become part of routine clinical care', argues Reddy, Allan, Coghlan, et at., 2020: 493.

56 Stöger, Schneeberger, and Holzinger, 2021: 35.

57 Froomkin, Kerr and Pineau, 2019: 50.

58 Kempt, Heilinger and Nagel, 2023: 1413; Kim, 2020: 3.

from AI-generated recommendations remain crucial in determining potential liability.

2 The Standard of Care within AI Medicine

2.1 *Ethical AI Medicine*

In the sphere of biomedical ethics, a seminal work is the 'Principles of Biomedical Ethics', authored by Beauchamp and Childress.[59] Their ground-breaking contribution introduced four core principles: autonomy, beneficence, non-maleficence, and justice.[60] Autonomy presupposes that rational agents are capable of making informed and voluntary choices.[61] It encompasses a negative obligation to refrain from interfering with an individual's choices and a positive obligation to provide sufficient information, enabling informed decision-making, under the principle that individuals are free to make their own decisions, provided their actions do not harm others. The primary objective of any treatment provided is to promote the patient's well-being, a principle known as beneficence.[62] The treating doctor is obligated to consistently prioritise the patient's best interests and ensure that the treatment provided avoids causing harm ('do not harm'), adhering to the principle of non-maleficence.[63] Justice is about fairness in distributing material and social goods, rights, and terms of cooperation.[64] These principles guide and shape how we approach ethics and medical care.

Central to patient-centred care is the concept of shared decision-making, which strives to identify the most appropriate treatment for an individual's unique circumstances.[65] This process involves a candid dialogue between the patient and the doctor, where the doctor imparts knowledge about potential outcomes, and the patient delves into their personal values and priorities, fostering a harmonious partnership in care. Incorporating AI into healthcare can enhance patient-centred care by supporting and enriching the principles of shared decision-making.[66] AI tools can provide valuable data and insights to help both patients and doctors understand potential outcomes more clearly.

59 Beauchamp and Childress, 2013.
60 Beauchamp and Childress, 2013. About these principles, see Gillon 2015: 111–116.
61 Varelius, 2006: 377–388.
62 Beauchamp, 2019.
63 Sokol, 2013.
64 Pu, 2021.
65 Council of Europe, 2023: 37–41.
66 Currie, Rohren and Hawk, 2024: 343–362.

This technology can facilitate more informed discussions, allowing patients to better weigh their options based on their individual values and needs.

Likewise, at the core of AI in medicine lies the concept of patient-centred care.[67] AI technologies should respect patients' rights to make informed decisions and remain actively involved in their care (autonomy), prioritising the promotion of patient well-being by tailoring interventions to their unique needs and goals (beneficence). These systems must also aim to minimise harm by incorporating patient preferences and ensuring safe and evidence-based recommendations (non-maleficence). Furthermore, AI should uphold fairness and equity by addressing the diverse needs of each patient and providing unbiased, equal consideration in its outputs and applications (justice).

2.2 *How AI Influences the Standard of Care*

Traditionally, medical malpractice claims have been centred on negligence, employing the reasonable doctor standard.[68] In essence, the court has assessed the actions of other doctors in similar specialities under comparable circumstances. This determination has often been facilitated by expert witnesses' testimonies. If the court has found that the defendant (the healthcare provider) has not follow that behaviour without full justification, it has generally concluded that they (or the medical institution) failed to meet the required level of care. Conversely, when medical devices or equipment have been implicated, product liability principles have come into play, encompassing strict liability standards for manufacturing defects, design flaws, and failure to provide adequate warnings.[69] However, the advent of AI has disrupted this well-established framework.

AI systems have the potential to supplant human doctors in certain capacities. A notable trend in the academic literature highlights the importance of evaluating the reasonableness of doctors' decisions, especially when it comes to their choices about using AI tools,[70] emphasising that the main goal nowadays is ensuring that healthcare providers deliver competent and skilled medical care, no matter what specific treatments are used.

This shift will probably lead to a different perception by courts: in situations involving the use of AI in medical practice, it is anticipated that courts

---

67 Epstein and Street, 2011: 100–103.

68 Raposo, 2013: 13 ff.

69 In the EU, it applies Directive (EU) 2024/2853 of the European Parliament and of the Council of 23 October 2024 on liability for defective products and repealing Council Directive 85/374/EEC.

70 Lee and Yoon, 2021: 271.

will examine various factors. This may include how the doctor integrates the AI tool into their practice, the rationality of employing AI under the given circumstances, and whether the medical guidance provided aligns with what a reasonably prudent doctor would have recommended.

In this new scenario, there arises a heightened risk of doctors facing malpractice liability when they persist in personally undertaking tasks that can be surpassed by AI capabilities. As a result, it is likely that more medical tasks will gradually be taken over by AI systems, which may eventually replace doctors in this role.

2.3 *The Difficulties in Defining the AI Standard of Care*

Typically, courts assess the medical standard of care by comparing the doctor's actions to what a reasonable peer with similar technical expertise would have done. In medical malpractice cases, the benchmark for assessing liability is often the conduct expected from a reasonable doctor possessing equivalent technical knowledge, training, and expertise as the defendant. Understanding the precise definition of this standard of care in a specific malpractice case typically requires an examination of evidence regarding customary clinical practices within the medical field. In essence, to determine what a reasonable doctor would do in a particular situation, we rely on typical practices and patterns observed within the medical community,[71] i.e., something established in the medical community and accepted by a relevant group of practitioners.[72]

However, when novel technologies – like AI – enter the medical practice, the standard of care can become ambiguous for doctors adopting these innovations. As a result, in this era of newly deployed AI, doctors face some uncertainty, especially in the early stages, regarding how and when to use AI appropriately to align with the standard of care.[73]

If a medical professional can demonstrate that his/her actions were in line with widely accepted practices within the medical community, they will not be found liable for negligence, even if the patient has suffered harm as a result of their care. This is a commonly accepted criterion, both in civil law and in common law (in the latter known as the Bolam test).[74] The standard of clinical conduct is typically deemed non-negligent under this criterion when it aligns with a responsible body of opinion, meeting the criteria set by other competent medical professionals. It has, however, been subject to criticism, as it has

71 Greenberg 2009: 428.

72 Not necessarily unanimously nor even by the majority of practitioners.

73 Schweikart, 2021: 13–14.

74 Bolam v Friern Hospital Management Committee, [1957] 2 All ER 118.

been argued that it can be used to justify dangerous or outdated practices, as long as a sufficient number of medical professionals continue to support them (it could serve, for instance, to exclude AI from the medical standard of care).[75]

One main difficulty arises from the impossibility of assessing the expected medical behaviour. The determination of the 'reasonable person' standard of care within a specific situation has a subjective nature.[76] It relies on the retrospective assessment conducted by judges, with the assistance of medical experts, who collectively represent reasonable individuals. Their evaluation encompasses personal experiences, judgment, a thorough examination of the defendant's actions, and the consideration of any pertinent evidence regarding how individuals in similar circumstances would react. Therefore, it is normally a complex procedure.

However, in AI scenarios it becomes even more complex because we do not know how a reasonable, prudent doctor (the *bonus pater familias*, applicable to medicine) will behave in extremely novel scenarios, as happens with medical AI. The same difficulty is present in other jurisdictions that follow slightly different standards, for instance, in the UK, where medical negligence is defined as the conduct where the doctor deviated from what is considered acceptable by 'a responsible body of medical opinion'.[77]

2.4 *A More Demanding Standard of Care for Doctors*

2.4.1 A Shift in the Standard of Care

The standard of care is often forgiving, not mandating doctors to provide optimal care or exhaustively research every technological advancement.[78] As mentioned by Hoffman and Podgurski, '[p]hysicians are not required to provide optimal care in order to avoid liability, but rather they are required to provide the level of care that could ordinarily be expected'.[79]

---

75 Chan, 2021: 179.

76 Jackson, 2013: 654–658.

77 The Bolam test is a legal test used in the United Kingdom to establish whether a medical professional is liable for negligence in their care of a patient. The test is named after the case Bolam v Friern Hospital Management Committee, [1957] 1 WLR 582. The Bolam test requires that the medical professional have acted in accordance with a practice that is accepted as proper by a responsible body of medical professionals. See Keown, 1998: 235–273.

However, the Bolitho case (Bolitho v City and Hackney Health Authority, [1998] AC 232) established that the relied-upon standard must have a logical foundation. See Brazier and Miola, 2000: 85–114.

78 Wiley and Terry, 2016: 79–80. See also Giuffrida and Treece, 2020: 115.

79 Hoffman and Podgurski, 2009: 1534.

However, this might change with AI. With the growing use of AI by doctors, the benchmarks for ascertaining their legal accountability will undergo adaptation in response to their altered practice dynamics. It is anticipated that doctors employing AI might face elevated expectations compared to their counterparts not using AI.[80] This anticipated shift stems from the fact that doctors using AI have access to an extensive reservoir of information, setting them apart.[81] AI possesses the capacity to harness an extensive breadth of data, surpassing the limits of human doctors. Consequently, doctors using AI may find themselves subject to heightened expectations and scrutiny due to their privileged access to supplementary knowledge and data, potentially setting a higher standard for their professional conduct in comparison to their counterparts who do not use such advanced technological resources.[82]

Similar to how specialist doctors are held to higher standards of care due to their advanced training and expertise,[83] one can expect elevated expectations from AI-using doctors as well.[84] This expectation stems from the advanced training and higher levels of competence that specialists possess. Specialist doctors undergo extensive education and training in their particular area of expertise, often dedicating several additional years beyond general medical training to hone their skills and knowledge. This specialised training equips them with a deeper understanding of complex medical conditions and the latest advancements in their field, enabling them to provide a higher level of care. Because of this advanced expertise, specialists are expected to stay current with emerging medical research, technologies, and treatment protocols within their speciality. They are often seen as the ultimate authority in their field, providing guidance and oversight not only to patients but also to general practitioners and other healthcare professionals. The common theme is that with greater knowledge and more advanced tools comes greater responsibility. Doctors using AI must meet heightened care standards. They are expected to stay informed about emerging AI technologies and integrate these advancements into their practice to ensure the best possible outcomes for their patients.

This emerging scenario comes with some drawbacks. Imposing a higher standard on AI machines may inadvertently deprive patients of superior

80 Braum, Hummel, Beck, et al., 2021. Such expectations are especially high in AI medical imaging. See Strohm, Hehakaya, Ranschaert, et al., 2020: 5526; Coppola, Faggioni, Gabelloni, et al., 2021.

81 Komorowski, Celi, Badawi, et al., 2018: 1719.

82 Giuffrida and Treece, 2020: 116.

83 Harrold, Field and Gurwitz, 1999: 499–511.

84 Allain, 2013: 1064; Wegman, Stannard and Bal, 2012: 1382.

care by discouraging their use, fearing malpractice lawsuits, and thus prevent the healthcare system from benefiting from potential cost savings offered by AI systems. Conversely, applying a lower standard of care to AI machines could inadvertently incentivise the delivery of substandard care. A careful balance is thus required.

2.4.2 Validation of AI Results

As referred to in the previous section, AI will most likely make the AI standard of care more demanding. Still, in my view, what will be expected from doctors will depend on the level of risk of the specific medical act.[85]

Take the case of the validation of AI recommendations.[86] Low-risk interventions – like some medical exams (blood collection, e.g.) or taking widely used low-side-effect drugs (take the case of aspirin, which is a widely prescribed drug worldwide) – will not require any validation of AI recommendations. When it comes to medium-risk interventions – such as off-label prescription[87] – doctors will be asked to double check the AI recommendation. This might imply consultations with a senior colleague or a confirmation collected from common established practices. In contrast, for riskier interventions, such as high-risk drug dosages or especially risky off-label prescriptions (the off-label prescription of a drug to a pregnant woman), additional validation will be necessary before relying on AI recommendations, such as detailed validation in evidence-based medicine.

Note, however, that this additional request for validation will only persist until AI becomes an integral part of the standard of care. When – not 'if' – AI is fully accepted and even demanded by the medical standard of care, such additional validation will no longer be necessary.

2.4.3 A Constant Requirement of 'Being Up-to-Date'

The malpractice standard today significantly differs from that of twenty years ago, even ten years ago, primarily due to developments in medical knowledge and the introduction of new technologies, rather than changes in the legal framework alone. As medical knowledge and technology progress, the standard of care continuously adapts to incorporate emerging practices.[88]

---

85 Cf. Greenberg, 2009: 444.

86 See Price II, 2018: 301–302.

87 Off-label prescription is not necessarily considered a risky intervention, much less a form of medical malpractice. Still, it carries a higher degree of risk than on-label prescription. More on this issue in Raposo, 2014: 35–75; Raposo, 2017b: 1–21.

88 For instance, during the 1980s, anaesthesia-related errors resulted in numerous fatalities and injuries globally. The causes of these mishaps ranged from inadequate preoperative

In any case, healthcare providers bear the responsibility of staying updated on emerging techniques and advancements.[89] This responsibility encompasses continuous education, understanding AI mechanisms, interdisciplinary collaboration, and prioritising patient safety and ethical considerations. In this regard, Greenberg claims that doctors are bound by a 'duty to stay abreast'[90] of advancements in medical care, that encompasses the responsibility to stay informed about emerging practices and incorporate new scientific knowledge into their medical practice. This obligation highlights the importance of doctors keeping up to date with evolving developments and utilising the latest scientific insights to provide appropriate and effective medical care.

2.4.4 Assessment of AI's Performance

Another potential AI-specific obligation pertains to the oversight of AI systems as a mechanism to assess their performance.[91] Unlike traditional medical tools, AI systems are dynamic and can evolve over time through ML algorithms, potentially leading to changes in their decision-making processes. Therefore, healthcare providers and institutions must implement robust oversight mechanisms to ensure that these systems consistently deliver accurate, reliable, and safe outcomes.[92] This oversight includes regularly updating AI algorithms, validating their outputs against clinical standards, and auditing their performance to detect any deviations or biases that may arise.

Moreover, it has been debated whether healthcare providers and hospitals should assess the quality of black-box algorithms and ML technologies before integrating them into patient care. This requirement would indeed face several challenges. The necessary information to conduct thorough evaluations is often not readily available. Black-box algorithms, in particular, are notoriously difficult to scrutinise because their internal processes are not transparent. This lack of information makes it hard to understand how decisions are made or to predict the outcomes reliably.[93]

---

preparations to improper techniques and insufficient postoperative care. Responding to these concerns, the American Society of Anaesthesiologists introduced novel approaches for monitoring oxygen levels during surgery, establishing a fresh benchmark of care within the field. Cf. See, 2021: 427.

89 Mennella, Maniscalco, De Pietro, et al., 2024.

90 Greenberg, 2009: 430.

91 Laux, 2023.

92 Williamson and Prybutok, 2024.

93 Green, 2022.

However, these challenges are not insurmountable. Healthcare providers and hospitals should develop their own methods for evaluating AI technology. This might involve establishing rigorous validation protocols, collecting extensive performance data, and collaborating with AI developers to gain deeper insights into the functioning of these systems. By creating robust frameworks for AI evaluation, the healthcare industry can ensure that these advanced technologies are both safe and effective for patient care.[94]

2.4.5 Digital Skills

The digital capacities of human players are also at stake. Much like the mandatory certifications required for aeroplane pilots or healthcare practitioners, it is essential to ensure that individuals responsible for operating and prescribing AI systems are genuinely deserving of the trust that comes with those roles.[95]

The concept of being a 'bearer of trust' could serve as a foundational framework for establishing initial confidence in healthcare AI systems.[96] Just as patients trust their doctors due to their extensive education, certifications, and experience, they should also trust that the AI systems used in their treatment are managed by professionals with the requisite skills and knowledge. This would involve implementing certification programs for AI operators, ensuring they understand the intricacies of the AI systems they use, including their limitations and potential risks.[97] In the context of healthcare, this means doctors and other healthcare providers would need to be proficient not only in medical skills but also in understanding and managing AI technologies. They would need to undergo specialised training to assess the performance, reliability, and safety of AI systems, much like how they currently evaluate new medical treatments or devices.

2.4.6 Less Medical Autonomy

The growing role of AI in healthcare raises concerns about its potential to undermine the independence of medical professionals.[98] Many doctors are apprehensive that AI could reduce their ability to make autonomous decisions, posing a challenge to their professional authority.

---

94 Alowais, Alghamdi, Alsuhebany, et al., 2023.
95 European Health Parliament, 2016.
96 Feldman, Aldana and Stein, 2019: 414.
97 Yampolskiy, 2024.
98 Lorenzini, Ossa, Shaw, et al., 2023: 426; Funer and Wiesing, 2024.

At the heart of clinical practice is the doctor's autonomy,[99] which is essential for exercising sound judgment in patient care. Doctors rely on this freedom to make informed decisions tailored to the unique needs of their patients. However, this autonomy is not without responsibility; it comes with the obligation to ensure the highest standards of care.

The link between a doctor's autonomy and their responsibility stems from the societal value placed on this independence.[100] Granting doctors the ability to make independent decisions is not just for their benefit but also because it ultimately serves the greater good by promoting quality healthcare outcomes. In this sense, a doctor's professional freedom is intertwined with their duty to provide the best possible care for their patients.[101]

2.5 *AI and Defensive Medicine*

AI has the potential to significantly improve the accuracy and precision of medicine, including medical diagnoses and treatment decisions. Thus, if doctors become less likely to make mistakes, the logical conclusion is that the possibility of litigation will decrease and so the chances of defensive medicine.[102]

This, in turn, can lead to reductions in the cost of healthcare, as doctors would no longer need to order as many tests, or administer unnecessary treatments.[103] For example, if an AI algorithm is able to accurately diagnose a patient's condition with a high degree of certainty, then the doctor may be less likely to order unnecessary tests, reducing the overall cost of healthcare.

That being said, some potential challenges need to be addressed before AI can be used to reduce defensive medicine. For example, it is important to ensure that the AI algorithms are accurate and reliable. Additionally, it is important to ensure that doctors are comfortable using AI in their practice; otherwise, AI risks having an opposite effect and fuelling defensive medicine instead.[104]

A major concern in this regard relates to over-confidence in AI technology, as a shield to defend practitioners from malpractice lawsuits. A paper from Bernstein et al. investigates the influence of incorrect AI results on radiologists'

99 McAndrew, 2019: 198–203.

100 Timmermans, 2005: 490–501. See also Funer and Wiesing, 2024.

101 McAndrew, 2019: 198–203.

102 Defensive medicine is a practice in which doctors order more tests and treatments than necessary to protect themselves from being sued for malpractice. This can lead to increased costs for patients and the healthcare system as a whole. Cf. Raposo, 2019: 401–416. See also Thomas, 2017.

103 Khanna, Maindarkar, Viswanathan, et al., 2022.

104 Banja, Hollstein and Bruno, 2023.

performance and explores whether optimising human factors can minimise errors in a study conducted among radiologists.[105] The data indicate that participants were more hesitant to disagree with the AI output, possibly because they felt threatened by the prospect of their decision to override the AI algorithm being documented and discoverable.[106] If that is indeed the case, AI may turn out to be just another tool at the service of defensive medicine.

2.6 *A Specific Standard of Care Applicable to AI*

2.6.1 The Justification of an AI Specific Standard of Care

In the event that AI is formally recognised as a healthcare provider[107] – whether equivalent to human providers or with a distinct legal status – it could require the creation of a unique standard of care tailored specifically to AI systems, rather than merely applying the existing standards to the human professionals who operate or supervise them. [108] For instance, a doctor might be considered unreasonable for failing to read a patient's medical record, especially when the record was concise, no other patients were waiting, and the patient's condition was stable enough to allow the time needed to review the report. However, these human-centric considerations do not apply in the same way to AI, as AI systems are designed to process information systematically without the contextual nuances that influence human judgment. This approach acknowledges the fundamentally different nature of AI, which functions based on algorithms and data processing, rather than human intuition, experience, and ethical considerations.

Vladeck argues that if AI were granted a form of personhood status, the standard of care for AI would likely diverge from that applied to human healthcare providers. Unlike human practitioners, AI systems could be evaluated based on their design, coding, data integrity, and algorithmic decision-making

105 Bernstein, Atalay, Dibble, et al. 2023: 8263–8269.

106 About this study, Banja, Hollstein and Bruno, 2023.

107 This hypothesis rests on a fundamental prerequisite: that AI is granted some form of independent legal status, potentially similar to that afforded to humans. Although this book will not delve deeply into this issue (for an in-depth discussion, see Kostenko, Bieliakov, Tykhomyrov, et al., 2024: 1683–1693; Raposo, 2024a: 1049–1072), some relevant considerations will be explored in subsequent sections.

108 'The Oviedo Convention acts as a minimum standard for the protection of human rights which requires translation into domestic law. On this basis, there is an opportunity to make specific, positive recommendations concerning the standard of care to be met in AI-mediated healthcare' (Mittelstadt, Allo, Taddeo, et al., 2016: 56).

processes.[109] This would establish a separate set of expectations that account for the technological capabilities and limitations of AI.

Chung, on the other hand, draws a parallel between AI and medical students, noting that both operate under a form of delegated authority, perform specific tasks within a defined scope, and are subject to supervision.[110] This comparison suggests that the operational dynamics and oversight requirements of AI in healthcare are sufficiently analogous to those of medical students to consider granting AI a quasi-personhood status. Such a status would involve distinct responsibilities and accountabilities, similar to how medical students are trained, evaluated, and gradually entrusted with greater autonomy as they demonstrate competence.

2.6.2 How to Define the AI-Specific Standard of Care?

2.6.2.1 *The Asimov Laws and the AI Standard of Care*

Assuming there is a specific medical standard of care for medical AI the question is, therefore, which rules and principles will guide that standard. In the middle of the last century, Isaac Asimov formulated the three laws of robotics: 1) A robot may not injure a human being or, through inaction, allow a human being to come to harm; 2) A robot must obey the orders given it by human beings except where such orders would conflict with the First Law; 3) A robot must protect its own existence as long as such protection does not conflict with the First or Second Laws.[111]

The first principle (the 'first law') emphasises the paramount importance of AI tools delivering the most accurate diagnoses possible. This pursuit of excellence in diagnosis aims to enhance patient healthcare outcomes. However, it also recognises the potential risks associated with AI failures or inaction, which could potentially harm the patient. The second principle (the 'second law') stresses the necessity of thorough and appropriate training for AI systems. Moreover, it emphasises the critical role of human doctors in supervising the learning process of AI software. This supervision ensures that the outputs generated by the AI system are not only technically accurate but also clinically applicable and relevant. The third principle (the 'third law') draws attention to the issue of obsolescence that AI software may face over time. The rapid evolution of medical AI introduces new modalities and advancements. Consequently, the specific training of AI software can become inadequate

109 Vladeck, 2014: 150.

110 Chung, 2017: 39.

111 Asimov, 1950: 40. About these laws and their relationship with ethical principles, Anderson, 2008: 477–493.

when new developments or modalities are introduced into clinical practice. These principles collectively underscore the need for continuous evaluation, adaptability, and a commitment to patient safety when using AI in healthcare. By adhering to these principles, the potential benefits of AI technology can be maximised, while risks are mitigated in the dynamic landscape of medical imaging.[112]

2.6.2.2 *The 'Reasonable AI Algorithm'*

What we can consider the 'reasonable algorithm'[113] – the AI version of the 'reasonable doctor' – must meet certain requirements.[114]

For instance, AI systems in healthcare are expected to comply with existing medical protocols and regulatory frameworks.[115] The standard of care could be tied to AI's ability to follow well-established medical guidelines, such as those set by health organisations[116] (e.g., the World Health Organization or national medical boards). Any deviations from these protocols would need to be justified and documented. This requirement is upheld by the AIA, which demands that AI systems demonstrate compliance with specific regulatory standards for safety and efficacy (Articles 11 and 12 of the AIA).[117]

Even if AI meets the technical standards of care, ethical considerations play a key role in healthcare. AI's standard of care should include principles such as respect for patient autonomy (ensuring that AI systems facilitate informed consent by clearly explaining diagnoses and treatment options);[118] equity in care (avoiding biases in AI algorithms that could lead to discrimination based on race, gender, socioeconomic status, or other factors);[119] and patient-centred care (striving to provide recommendations that are not just technically correct but also in line with the patient's values, preferences, and broader well-being).[120] In this sense, AI's standard of care would need to integrate ethical frameworks that prioritise the patient's best interests, just as human doctors are expected to do.

---

112 European Society of Radiology, 2019: 5.
113 Chagal-Feferkorn, 2020: 213–261.
114 Note that the level of accuracy required for AI systems' entry in the market is on the rise (Verdicchio and Perin, 2022).
115 Bouderhem, 2024.
116 Kitsios, Kamariotou, Syngelakis, et al., 2023.
117 Lazcoz and De Hert, 2023.
118 Lorenzini, Ossa, Shaw, et al., 2023: 424–429.
119 Chinta, Wang, Palikhe et al., 2025; Sasseville, Ouellet, Rhéaume et al., 2025.
120 Rothlisberger, 2024: 255–266.

Last but not least, like human doctors, AI systems should be expected to avoid harm and minimise risks.[121] This means that in cases where AI makes a recommendation, the system should clearly communicate the level of uncertainty or confidence in its predictions and be transparent about limitations, so healthcare providers and patients can make informed decisions based on both AI and human input. In parallel, it should be able to flag cases where human intervention or a second opinion might be necessary, especially in ambiguous or complex situations. The standard of care might also require AI systems to have fail-safes in place, meaning protocols to prevent or mitigate harm if the system makes an erroneous or harmful recommendation. The demand for risk management systems is embraced by Article 9 of the AIA.[122]

Ultimately, defining the standard of care for AI will require a balance between ensuring technical excellence and preserving the ethical, emotional, and interpersonal dimensions of healthcare that AI cannot fully replicate.

2.6.2.3 *Examples of 'Reasonable' AI Algorithms in Various Medical Scenarios*

What is considered 'reasonable' for an AI algorithm depends on the specific tasks being performed and expectations around what is scientifically viable and realistic (considering, of course, that the assessment concerns AI systems, not humans).

In CDSS, for instance, a reasonable AI would offer suggestions that align with accepted medical practices and improve diagnostic accuracy without replacing the doctor's judgment. By contrast, in radiology AI for image analysis, a reasonable AI would highlight potential areas of concern, such as tumours or fractures, with high accuracy, reducing the likelihood of missed diagnoses but still allowing the radiologist to make the final call. In predictive analytics for patient outcomes, a reasonable AI would use patient data to make predictions that help guide treatment decisions, ensuring that its suggestions are evidence-based and within the scope of current medical knowledge. In the case of NLP in EHRS, that is, AI systems that extract and organise relevant information from patient records to streamline documentation and improve care coordination, a reasonable AI would accurately process and present information without misinterpretation or significant errors that could impact patient care. In personalised medicine, a reasonable AI would provide recommendations that reflect current understanding of genetic factors and treatment efficacy, offering options that are scientifically sound and clinically validated. Likewise, in robotic surgery assistance, a reasonable AI would enhance precision and

121 Díaz-Rodríguez, Del Ser, Coeckelbergh, et al., 2023.

122 Novelli, Casolari, Rotolo et al., 2024.

control, offering support that complements the surgeon's skills while adhering to safety standards. In all these examples, a 'reasonable AI' would operate in ways that are consistent with established professional standards, enhance human decision-making, and remain within the boundaries of its intended purpose without introducing significant risks or errors.

2.6.3 The Benchmark to Assess AI

2.6.3.1 *Humans as the Benchmark*

2.6.3.1.1 How Humans Operate as Benchmarks

To determine if an algorithm is 'reasonable',[123] a comparison between its outcomes and reasonable human behaviour is an option.[124] If we equate AI with a human doctor, in assessing the standard of care for AI, we will likely compare its performance to that of a 'reasonably skilled and knowledgeable human doctor'. For that, we need to analyse AI's performance over time, comparing it to the human counterpart. AI systems could be evaluated based on criteria such as the accuracy of diagnoses compared to standard diagnostic tests and expert medical opinion and the error rates and how these compare to those of human practitioners in similar fields.[125] The AI's standard of care could involve constant benchmarking against both the state of the art in AI and the average performance of human healthcare providers.[126] Under this reasoning, algorithms should aim to outperform and enhance safety, avoiding algorithms that pose more dangers than the humans they replace.[127]

Given that AI can handle enormous datasets, recognise patterns, and analyse information more quickly than any human doctor, it might be reasonable to hold AI to a 'higher standard' in areas where data processing and predictive analysis are critical (e.g., diagnosing complex diseases).[128] However, in areas involving judgment, empathy, or patient communication, AI might not be able to match the expectations placed on human doctors.[129]

---

123 Chagal-Feferkorn, 2020: 213–241.

124 Borghetti, 2019: 98.

125 For instance, in the specific field of triage and diagnosis, Baker, Perov, Middleton, et al., 2020.

126 Chung, drawing an analogy between AI and medical students, suggests that there are enough similarities in terms of authority, tasks, and oversight to justify granting AI personhood and treating the 'AI person as a medical student'. See Chung, 2017: 39.

127 For a comparison between algorithms and humans, see Chagal-Feferkorn, 2020: 235–248. For the domains in which AI outperforms humans, see Whittlestone, Nyrup, Alexandrova, et al., 2019: 28–35.

128 Chen, Beam, Saria, et al., 2019: 91.

129 Montemayor, Halpern and Fairweather, 2022: 1353–1359.

It might also be the case that the mere presence of a human is considered preferable simply because they are human, meaning that AI would need to outperform human capabilities in order to be accepted. All things being equal – that is, if AI performs at the same level as a human – there would be no compelling reason to replace the human with AI, unless the human perceives the task as too insignificant to warrant human involvement (e.g., packing boxes).

There are two caveats, though, to the use of humans as benchmarks.

First, comparing AI to human decision-makers is problematic since AI operates in a different logic, devoid of consciousness and mental awareness.[130] When assessing human fault a comparison with a model human standard is valid because humans are assumed to possess the same kind of rationality to determine reasonable actions. However, this approach cannot be directly applied to AI since its functioning differs from that of humans.[131]

Second, while it is conceivable to design studies that confirm whether such software consistently reaches the same conclusions as human doctors, it is crucial to recognise that AI systems will not always be benchmarked against human doctors.[132] Some products will be engineered to process types and volumes of information that exceed human capacity and push the boundaries of established medical practices.[133] So, at some point, eventually in the very near future, humans could not serve as benchmarks anymore.[134]

2.6.3.1.2 The Arnold v. Reuther case

In the middle of the last century, the case of Arnold v. Reuther,[135] heard by an intermediate appellate court in Louisiana, raised the question of whether autonomous cars should be held to higher standard of care than human-driven cars. The court then claimed: 'A human being, no matter how efficient, is not a mechanical robot and does not possess the ability of a radar machine to discover danger before it becomes manifest. Some allowances, however slight, must be made for human frailties and for reaction, and if any allowance

130 Piletsky, 2019: 66–71; Rawas, 2024.
131 Nishant, Schneckenberg and Ravishankar, 2024: 19–40.
132 'When diagnostic AI ignores problems with inclusivity and bias yet still manages to deliver better results than unaided human observation for many or most patients, the patients who do suffer an injury may not have a tort remedy under a negligence standard' (Evans and Pasquale, 2022: 29)
133 As an example, see Najjar, 2023.
134 Some studies have already concluded that algorithms outperform humans in decision-making. Cf. Mahmud, Islam, Ahmed, et al., 2022.
135 92 So. 2d 593, 596 (La. Ct. App. 1957).

whatever is made for the fact that a human being must require a fraction of a second for reaction and cannot respond with the mechanical speed and accuracy such as is found in modern mechanical devices, it must be realized that there was nothing that Reuther, a human being, could have done to have avoided the unfortunate result which the negligence of Mrs Arnold brought upon herself'.

The statement acknowledges that humans, unlike machines, cannot anticipate danger with the same precision and speed. It highlights the need for some leeway when evaluating human actions, considering the natural limitations and reaction times of humans and the importance of understanding and accommodating human limitations in situations where quick responses are required.[136] All things considered, the Court considered that it is not be suitable to use the standard of care established for humans to assess AI and vice versa.[137]

2.6.3.2 *Another AI Algorithm as the Benchmark*

Another viable approach to evaluating an algorithm is through comparison with other algorithms that perform similar tasks.[138] The idea is that if both systems, which are designed to function at the same level of reliability, make the same error in a given situation, it implies that the error might not be due to negligence but rather an inherent limitation or challenge in the task itself. However, if the other AI system has not made the same mistake, it may indicate that the system being used is faulty or inadequately designed.[139] This method involves assessing the performance, reliability, safety, and outcomes of the algorithm in question against those of other established algorithms in the same field.

For instance, consider the use of AI algorithms in radiology for detecting lung nodules on chest CT scans. Multiple algorithms exist that are designed to identify these nodules, each with varying levels of accuracy, false-positive rates, and processing times. To evaluate a new algorithm, one could compare

136 'These kinds of differences in basic structure, speed, connectivity, updatability, scalability, and energy consumption will necessarily also lead to different qualities and limitations between human and artificial intelligence. Our response speed to simple stimuli is, for example, many thousands of times slower than that of artificial systems. Computer systems can very easily be connected directly to each other and as such can be part of one integrated system' – Korteling, Van de Boer-Visschedijk, Blankendaal, et al., 2021.

137 Couture, Roy, Ganache, et al., 2021; Vayena, Blasimme and Cohen, 2018.

138 There are indeed several papers comparing algorithms that perform similar task. For instance, Ahmad, Sheikh and Aved, 2020.

139 See Holm, Stanton and Bartlett, 2021: 171–188.

its performance against established algorithms that are already in use and considered reliable. Suppose the new algorithm detects lung nodules with a sensitivity of 95% but has a higher false-positive rate compared to other leading algorithms, which have sensitivities around 93% but with significantly lower false positives. A comparative evaluation would consider whether the increased sensitivity justifies the trade-off with the false-positive rate, or whether adjustments are needed for the new algorithm to be deemed reasonable and safe for clinical use.

This approach not only helps in benchmarking the new algorithm against the industry standard but also encourages continuous improvement and optimisation across all algorithms, fostering a competitive environment that drives advancements in technology. It ensures that the benchmark is not static but evolves with technological progress, maintaining high standards of care and effectiveness in healthcare applications.

Another problem is the scarcity of providers within a specific clinical domain and the dominance of tech giants in the AI sector. Both factors pose practical obstacles to this approach.[140]

Moreover, even AI systems with similar accuracies may differ in the patients they misclassify due to their distinct mathematical models.[141] As a result, the existence of differing classifications among effective AI systems does not necessarily indicate an error.[142] This process requires careful consideration, as merely comparing outcomes in similar situations will not suffice. Instead, a comprehensive analysis of their overall results is necessary, considering the different rationalities by which they operate.[143]

2.6.3.3 *Humans Plus AI Algorithms as the Benchmark*

Chagal-Feferkorn proposed a model that integrates both human and algorithmic reasonableness to assess whether an algorithm's actions were appropriate.[144] This approach involves two key components.

First, the court would evaluate the reasonableness of a human in a similar decision-making scenario.[145] However, this comparison does not account for the fundamental differences between humans and algorithms, and it falls

140 Khanal, Zhang and Taeihagh, 2025: 52–69.

141 Holm, Stanton and Bartlett, 2021: 182–183.

142 Holm, Stanton and Bartlett, 2021: 183.

143 LaCroix and Luccioni, 2025: 4029–4047.

144 Chagal-Feferkorn, 2020: 256–258.

145 For instance, if an algorithm had administered penicillin to a patient with an allergy, this action would be compared to the decision-making process of a human doctor under the same circumstances.

short of distinguishing situations where an algorithm might be held to a higher standard than a human or where an algorithm might be disadvantaged and unable to perform as well as a human.

To address this gap, the second component of the model shifts the focus to the reasonableness of the programmers or manufacturers behind the algorithm by examining the precautions taken by manufacturers to minimise harm and assessing whether these measures were reasonable.[146] The adequacy of these safety measures will vary depending on the specific context and circumstances. The less an algorithm's behaviour aligns with that of a reasonable human, the more stringent the expectations placed on the manufacturer to ensure the decision is considered reasonable. Thus, a 'reasonable algorithm' evaluation includes both the algorithm's reasonableness (focused on the specific decision and harm caused) and the manufacturer's reasonableness (centred on general safety measures to mitigate risks). Importantly, a finding of reasonableness does not necessarily require both assessments to be positive. The outcome of the first component will influence the level of reasonableness required in the second. If the algorithm's actions are deemed reasonable compared to a human (meaning a human would likely have made the same error), then the manufacturer's standards of reasonableness may be less demanding. Conversely, if the algorithm causes harm in a situation where a human would not, the manufacturer will be held to a higher standard of care regarding the precautions they should have taken to avoid such risks, in order for the decision to be deemed reasonable and to avoid liability.

2.6.3.4 *Analogy between AI and a Human Doctor*

One possibility to address AI-caused harm is to equate AI with a human health practitioner. Assuming that the AI device is accurate and explainable, it appears that consulting with it could be interpreted as a usual medical behaviour, similar to a doctor's advice from a clinical expert.

For example, consider the elements that constitute a consultation with a radiologist to diagnose pneumonia: (1) consultation with a radiologist who has received adequate training; (2) the radiologist can describe how the patient's medical history allowed him to make the diagnosis of pneumonia; (3) can identify similar cases; (4) can point to the exam to show what he finds relevant. Similarly, if a doctor consults an AI system to assist with this examination, they are consulting with an expert who: (1) has received adequate training, albeit

146 For example, if a medical algorithm were responsible for administering medications, the scrutiny would involve evaluating what safety protocols were implemented to prevent harm, such as safeguarding against administering drugs to allergic patients.

somewhat different from what a human radiologist receives; (2) can list clinical history and findings that led to the diagnosis of pneumonia (using counterfactual explanation); (3) can identify similar cases (using case-based learning methods), and (4) can 'point' to parts of the scan that are relevant (using heat maps). Given these similarities, a doctor consulting an AI device could be considered to be obeying medical custom. This is because the four legitimate elements of an expert consultation are present both in the case of consulting an expert doctor and consulting an accurate and interpretable AI machine.[147]

There is a significant caveat, though: AI is not like a human. AI's reasoning is different.[148] AI's reasoning is based on processing vast amounts of data, using patterns and statistical analysis to make decisions.[149] While this can make AI incredibly efficient at diagnosing conditions based on symptoms, scans, or other medical data, it lacks the intuitive, holistic, and ethical judgment that human doctors bring to the table.[150] Human reasoning involves understanding the patient's context, their emotions, and even the unspoken aspects of communication, which AI is ill-equipped to handle. Doctors rely not only on medical data but also on personal experience, empathy, and ethical considerations when making decisions.[151]

2.6.4. *Technical Features of the Standard of Care Applicable to AI*

A technical aspect of the standard of care applicable to AI involves risk management. AI systems (just like human doctors) are expected to avoid harm and minimise risks. In cases where AI makes a recommendation, the system must clearly communicate the level of uncertainty or confidence in its predictions;[152] be transparent about its limitations, so healthcare providers and patients can make informed decisions based on both AI and human input;[153] and be able to flag cases where human intervention or a second opinion might be necessary.[154] The standard of care might also require AI systems to have fail-safes in place, that is, protocols to prevent or mitigate harm if the system makes an erroneous or harmful recommendation.

---

147 Harned, Lungren and Rajpurkar, 2019: 7.

148 Witkowski, Okhai and Neely, 2024.

149 Xu, Liu, Cao, et al., 2021.

150 McStay, 2024: 133–163.

151 For instance, a doctor may choose a different treatment approach for a patient with a terminal illness based on their psychological state, family dynamics, or personal preferences – factors that AI, governed strictly by data, would likely overlook or mishandle.

152 Prabhudesai, Yang, Asthana, et al., 2023: 379–396.

153 Prabhudesai, Yang, Asthana, et al., 2023: 379–396.

154 Methnani, Tubella, Dignum et al., 2021.

Compliance with industry regulations and medical protocols should also serve as critical criteria for AI systems in healthcare. AI technologies must meet specific regulatory standards for safety and efficacy, similar to other medical devices or pharmaceuticals that are approved by regulatory bodies. Under the AIA, high-risk AI systems, including most AI applications used in healthcare, are required to adhere to stringent technical standards (see Section 2.4, Chapter 3). These standards are essential for ensuring that the AI systems are eligible to receive the CE marking of conformity (Article 48 AIA), which is a prerequisite for launching the product on the market. The CE marking demonstrates that the AI system complies with the relevant safety, performance, and regulatory requirements.[155] The standard of care applicable to AI must be tied to AI's ability to follow well-established medical guidelines, and any deviations from these protocols would need to be justified and documented.[156]

2.6.5. *Ethical Features of the Standard of Care Applicable to AI*

Even if AI meets the technical standards of care, ethical considerations play a key role in healthcare. AI's standard of care should include respect for patient autonomy, ensuring that AI systems facilitate informed consent by clearly explaining diagnoses and treatment options; equity in care, to avoid biases in AI algorithms that could lead to discrimination based on race, gender, socioeconomic status, or other factors; and patient-centred care, which strives to provide recommendations that are not just technically correct but also in line with the patient's values, preferences, and broader well-being. In this sense, the AI's standard of care would need to integrate ethical frameworks that prioritise the patient's best interests, just as human doctors are expected to do.[157]

The applicable standard of care for AI in healthcare would likely be a hybrid model that combines elements of both technical performance and human ethical principles. On the one hand, AI should be held to a high technical standard, given its ability to analyse vast amounts of data and make evidence-based recommendations. On the other hand, it would also need to respect human-centric values, including patient autonomy, fairness, and the ethical principles that guide human doctors.[158]

155 Thelisson and Verma, 2024: 113–121.

156 For instance, in radiology, medical protocols include specific recommendations for interpreting lung nodules, assessing their size, and determining the likelihood of malignancy, which AI must comply with. See Potočnik, Foley and Thomas, 2023: 376–385.

157 Lorenzini, Ossa, Shaw, et al., 2023: 426.

158 Farhud and Zokaei, 2021.

2.6.6 A Duty of Care for AI

Traditionally, in cases of medical malpractice, where a treating doctor consults with another doctor, the plaintiff can only establish liability against the consulting doctor if there is a doctor-patient relationship between them.[159] Considering the AI system as a 'consulting doctor', the issue of its duty to patients becomes significant in terms of liability. While AI plays a vital role in providing information to the treating doctor, it will not engage in direct patient interaction. Existing legal precedents indicate that a consulting doctor who does not directly interact with the patient has no duty towards the patient and it may be challenging to establish a breach of duty required for medical malpractice claims.[160] Thus, even if courts draw an analogy between AI and a consulting doctor – which, in itself, is a far-fetched perspective – the former will not owe a legal duty to patients that would lead to a medical malpractice claim.[161]

This assessment may become more complex when AI systems gain the capability to directly interface with medical treatment, control life support equipment, and administer medicines. If AI evolves and gains the ability to directly treat patients by interfacing with medical equipment, its involvement in the diagnosis and treatment process would be much more intimate than that of a consulting doctor who had no direct interaction with the patient. Until that point, it is expected that courts will struggle to establish a legal duty on the part of AI towards patients.[162]

159 Roberts, Fowler and Appelbaum, 2022.

160 Lupton, 2018: 5.

161 Eldakak, Alremeithi, Dahiyat, et al., 2024: 3–4.

162 This same conclusion in Allain, 2013: 1073.

CHAPTER 5

Allocation of Liability for Damages Caused to Patients

1 Professional Liability versus Product Liability

1.1 *Different Perspectives and Different Scenarios*

Introducing an AI system into the doctor-patient relationship complicates identifying the primary cause of harm, namely, to know if the case was an error made by the doctor (a 'mistake') or an error in the AI system (a 'malfunction').[1] To solve this conundrum, it is also important to establish whether the AI system is acting as a kind of healthcare provider on its own merit, or merely assisting the doctor in patient care, as a tool.[2]

Envision a situation where a doctor employs an AI diagnostic technology that fails to provide the correct diagnosis. Subsequently, the doctor relies on the AI's recommendation and prescribes a harmful treatment to the patient. In such cases, the injured patient typically sues the doctor and/or the hospital under medical malpractice laws. Moreover, the patient – and eventually also the doctor or hospital – may then pursue legal action against the AI system producer, seeking accountability for the AI's role in the process.[3]

The argument underlying the manufacturer's liability is grounded in the representation of AI as a tool capable of misleading humans.[4] The focus of product liability regimes typically centres on safeguarding consumers who experience harm due to defective products. However, in the case of advisory AI systems, the patient rarely acts as the direct consumer, but as an indirect one. This distinction raises unique challenges in assigning liability when AI systems provide guidance in healthcare settings.[5]

---

1 I am using the distinction coined by Kempt et al. between mistakes and malfunctions (Kempt, Heilinger and Nagel, 2023: 1409).

2 For a discussion on whether product liability or medical liability should be applied in these cases, see Skounakis v. Sotillo, A-2403-15T2 (N.J. Super. Ct. App. Div. Mar. 19, 2018).

3 Bertolini, 2020: 111–116.

4 In the US, for instance, computer code has been viewed as a service rather than a product, which raises ambiguity regarding the application of design claims to programmers (Vladeck, 2014: 132).

5 Holm, Stanton, Bartlett, et al., 2021: 181.

 | HTTPS://DOI.ORG/10.1163/9789004543980-007

When AI systems serve as tools, a possible focus lies with the individual employing them. However, responsibility would generally be lifted from a person lacking control over the actions of an autonomous machine, mirroring how one would not be held accountable for the wrongful deeds of another person beyond their control. The concept of attributing responsibility to an individual for another's actions is overloaded with notions of unfairness and lacks justification under commonly held perspectives. Not only does it clash with the principles of equity, but it also corrodes the motivational force wielded by the law.

Moreover, when individuals find themselves in a position to prevent harm caused by another, their motivation to intervene lessens, given the certainty of being implicated by mere association.[6] The same reasoning can be expanded to other players who risk being held responsible for their mere association with AI: who will worry about adopting a duty of care if the mere fact of being involved in AI can serve as a basis for their responsibility?

If we consider AI as a healthcare provider, a possible categorisation of the kind of liability involving harm caused by AI lies in the cause of such harm. Under this framework, when an AI machine makes errors in its judgment similar to an error from a human doctor, it should be held accountable under medical malpractice liability standards, potentially falling under the purview of state Medical Boards for regulation.[7] This would constitute a case of medical liability; however, the liable party would be the AI itself – treated here as equivalent to a human – rather than the human healthcare provider or the healthcare institution (such as the hospital or clinic). However, if the fault lies with other aspects such as manufacturing defects, design flaws, or inadequate warnings associated with the AI machine, the principles of traditional product liability should apply.[8]

The determination regarding whether the system is classified as a medical device, medical procedure, or service could impact the claims and the standards applicable to assess liability. Depending on the classification, fault and damages assessments may differ, influencing the plaintiff's legal options and arguments in the case.

Let's consider an example involving an AI system designed to formulate therapeutic plans. Suppose the system diagnoses for a specific type of cancer and formulates a treatment plan that ultimately proves to be incorrect, leading to harm to the patient. In such a case, the decisions made by the system should be assessed based on the negligence framework. It is important to note that the

---

6 Chinen, 2016: 357–358.

7 TerKonda and Fish, 2023.

8 A similar distinction in Marchant and Tournas, 2019: 38.

machine itself is not held accountable for every instance of harm, but rather for situations where the standard of care falls below what is expected (leaving open the question on how such a standard shall be defined). This approach enables patients to receive appropriate compensation when the expected level of care is not met, regardless of whether the professional judgment was made by a human or a machine. Conversely, if the AI system is found to have a poorly constructed component that directly causes harm to the patient, the evaluation of liability would be conducted under the principles of product liability. In this context, the system manufacturer would be subject to strict liability, wherein accountability is established irrespective of fault, based on factors such as manufacturing defects or substandard design. Therefore, depending on the nature of the harm and the factors involved, the legal analysis would differentiate between negligence, focusing on the standard of care, and product liability, emphasising strict liability principles.[9]

In many cases, the most suitable option will be to adopt a hybrid approach, where AI developers are responsible for the safety and efficacy of their products, while healthcare providers are accountable for their competent application and integration into patient care. This dual responsibility ensures that both the creators and users of AI systems uphold high standards, ultimately safeguarding patient well-being.

1.2 *A Case Study: the Taylor v. Intuitive Surgical Case*

In the case of Taylor v. Intuitive Surgical[10] – which involved a patient who suffered injuries during procedures utilising Intuitive Surgical's da Vinci robotic surgical system – the plaintiff filed a product liability claim against the manufacturer of a robotic surgical system that incorporated AI. The central issue in this case, as determined by both the trial and appellate courts, revolved around assigning responsibility for the injury: whether it lay with the operating surgeon or the machine itself.

Despite potential errors on the part of the surgeon, plaintiff attorneys sought to structure the lawsuit as a product liability claim against the manufacturer, leveraging the advantageous aspects of such suits. This approach persisted even when it was apparent that the surgeon had made a mistake. The Da Vinci System provided a manual to the doctor, clearly stating that the recommended maximum body mass index (BMI) for robotic surgery was 30. However, the patient in question had a recorded BMI of 39. The surgeon acknowledged that

9 Marchant and Tournas, 2019: 39.

10 Taylor v. Intuitive Surgical, Inc., 355 P.3d 309 (Wash. Ct. App. 2015). Case commented in Alshanteer, 2020: 45–46; Marchant and Tournas, 2019: 36–37.

the patient was not an ideal candidate for robotic surgery but proceeded with the procedure regardless. Complications arose during the surgery, resulting in a diminished quality of life for the patient, including respiratory issues requiring ventilator support, neuromuscular damage leading to impaired mobility, and incontinence. The patient passed away four years after the surgery due to complications.

The plaintiff argued that the manufacturer was ultimately responsible, primarily for allegedly failing to adequately train or warn the doctor. Nevertheless, the jury ruled in favour of Intuitive Surgical, determining that the company had fulfilled its duty to warn the doctor about the nature of the robotics system and had provided sufficient warnings. The court of appeals confirmed this decision, concluding that Intuitive Surgical did not have a separate duty to warn the hospital. However, on further appeal, the Washington State Supreme Court expanded the company's duty to warn beyond the operating surgeon, to include the hospital as well, stating that the manufacturer had provided sufficient warnings to the surgeon, fulfilling its duty to disclose information about the system's nature. The court found that, even though the company had trained and warned the doctor, manufacturers also had a duty to inform hospitals, as purchasing agents, about the potential risks associated with their products. Consequently, the lower courts' decisions in favour of the manufacturer were vacated, and the case was remanded for further proceedings.

One intriguing question is how the outcome of the case might have differed if the plaintiff had framed their argument from the perspective of medical malpractice rather than product liability. Had the plaintiff pursued a medical malpractice claim against the surgeon, the jury might have sided with the plaintiff, as in a medical malpractice claim, the central issue would be whether the surgeon's decision met the standard of care expected in the medical community. The plaintiff could argue that performing robotic surgery on a patient with a BMI of 39 deviated from these standards, thus constituting negligence. The surgeon chose to proceed with the surgery despite the patient's BMI exceeding the recommended limit outlined in the da Vinci System manual. This decision might be scrutinised more intensely in a medical malpractice case, emphasising the surgeon's responsibility to adhere to established guidelines and standards of care.

1.3 *Selecting the Best Option*

The allure of pursuing product liability claims in cases involving AI in healthcare is particularly strong for plaintiff attorneys, largely because product manufacturers typically have deeper financial pockets compared to individual doctors. This financial disparity makes manufacturers a more attractive target, as they are better positioned to pay substantial settlements or judgments.[11] Moreover, product liability lawsuits offer plaintiffs more favourable legal doctrines that can significantly enhance the likelihood of success and the potential financial recovery. One of the key advantages in product liability claims is the doctrine of strict liability, which allows plaintiffs to hold manufacturers accountable for defects in their products without the need to prove negligence. In strict liability cases, the focus is on whether the product was defective and caused harm, rather than on the behaviour or intentions of the manufacturer. This contrasts sharply with medical malpractice lawsuits, where plaintiffs must establish that the doctor failed to meet a standard of care, which is often a more complex and nuanced argument to make.[12] Additionally, product liability claims can include, in some jurisdictions, the potential for punitive damages, designed to punish particularly egregious conduct by the manufacturer and deter future misconduct.[13] Punitive damages can substantially increase the amount of compensation awarded, providing a strong incentive for plaintiff attorneys to pursue these claims. In contrast, medical malpractice cases typically do not include punitive damages, and in many jurisdictions, there are caps on the amount of compensatory damages that can be awarded, further limiting the potential financial recovery. The combination of these factors shifts the focus from proving a doctor's negligence to challenging the safety and performance of the technology itself, thereby opening up new opportunities for holding manufacturers accountable in a rapidly evolving healthcare landscape.

However, there are several counterarguments to consider.[14] Firstly, the perception that manufacturers have deeper pockets overlooks the fact that large settlements and verdicts in product liability cases can be appealed – and frequently they are, as big harm can count on big legal teams at their service and money is not an issue – leading to prolonged legal battles and uncertainties about eventual payouts. Secondly, product liability cases are often more complex and costly to litigate than medical malpractice cases. They require extensive expert testimony, technical evidence, and rigorous scientific analysis to establish that a product is defective and caused harm. This complexity can result in longer litigation times and higher legal expenses, which might offset the potential financial gains.

11 Ledley, McCoy, Vaughan et al., 2020: 834–843

12 Usemahu and Panjaitan, 2024: 130–140.

13 Fahim, 2024: 41–78.

14 Garber, 2013; Del Riego, 2021: 387–447.

1.4 *The Liability Gap*

Given the intricate and inseparable interactions between humans and algorithms, a significant concern emerges regarding the allocation of liability, which cannot be effectively addressed by current liability law.[15] Due to the novelty of medical AI and the lack of legal precedence, there is currently a severe lack of clarity in defining responsibilities in this domain.[16] Thus, there is a need for new mechanisms and structures to ensure adequate accountability in medical AI and manage complaints, compensation and sanctions, when necessary, as well as to ensure that the acts are not repeated.

In light of classic civil liability standards, we trace liability to the producers of the system or to those who use it (doctors, hospitals). However, if the AI entity has the capacity to learn, sometimes in ways that even those who developed it cannot explain, and if it manages to act autonomously and often unpredictably, will this legal framework still be adequate?

Chances are that the requirements of culpability and causation will fail. This became known as the 'AI responsibility gap'.[17] The gap stems from the difficulties in adapting existing legal frameworks and accountability concepts to the emerging landscape of AI applications in the medical field. The challenges revolve around:[18]

a) The complex network of stakeholders: The presence of multiple participants in medical AI makes it intricate to assign responsibilities among the various contributors engaged in the development, implementation, and utilisation of medical AI. This includes AI developers, data managers, healthcare professionals, patients, healthcare administrators, and more.[19]

b) Unclear attribution of the cause of errors: Determining the precise source of a medical error associated with AI can be elusive. It may result from the AI algorithm itself, the data used in its training, or its misapplication and misinterpretation in clinical practice.[20]

---

15 Mehta, 2023. Viljanen (2017: 1278) refer to it as 'cyborg turn in law', a phenomenon denoting the fusion of humans, non-humans, individuals, and objects, leading to a new and unique legal realm beyond the conventional distinctions of human and non-human entities.

16 Asaro, 2012: 183.

17 Johnson, 2015: 708.

18 Raji, Smart, White, et al., 2020: 33–44.

19 European Parliament, 2022: 26–27.

20 Bleher and Braun, 2022: 747–761.

c) The lack of foreseeability:[21] The actions undertaken by AI after deployment are highly unpredictable for its creators. Consequently, not only is establishing developer liability complex within the framework of current tort regulations, but it could also be deemed illogical.

d) AI autonomy: As AI gains increasing autonomy, the feasibility of enforcing vicarious liability doctrines diminishes, owing to hospitals' limited influence over the actions of AI systems.[22]

e) Diverse governance frameworks and ethical norms: The AI industry is characterised by a multitude of governance structures and a lack of standardised ethical and legal principles. Many private entities rely on ethical codes and responsibility standards that are often criticised for their ambiguity and the difficulties in translating them into practical applications.[23]

1.5 *Mechanisms to Address the Liability Gap*

1.5.1 Mixed Models

Smith and Fotheringham put forth the idea of shared liability for damages related to medical AI devices, involving both technical and medical teams as joint contributors to the overall risk.[24] However, enlarging the scope of accountable individuals might not be the best solution. Given the expansiveness of the concept and the potential for various entities to be categorised as AI operators, along with the persistent manufacturer's liability whenever accidents transpire, introducing an additional liability rule would not simplify the current framework by establishing a clear singular entry point for primary legal action. On the contrary, it would introduce one or more entities, among which liability could theoretically be distributed, requiring an even more intricate evaluation of individual accountabilities and contributions.[25]

Similarly, Allain proposes a solution where restitution to the claimant is equally divided between the stakeholder supplying the AI system and the doctor or hospital adopting it. This equitable burden-sharing approach ensures that the risk of loss is spread across all actors, leading to reduced economic

---

21 So much so that Schütte and Majewski claim that '[t]he unpredictability of AI can itself be deemed foreseeable' (Schütte and Majewski, 2022: 133). Based on this, the authors argue that harm caused by AI should be considered recoverable even without the necessity of foreseeing a specific type of harm. This is particularly relevant when the nature and type of harm are not deemed highly extraordinary within the given context.

22 Allain, 2013: 1078.

23 Walter, 2024.

24 Smith and Fotheringham, 2020: 147.

25 This critique in Bertolini and Episcopo, 2021: 655.

disincentives. Consequently, AI manufacturers would be encouraged to improve system accuracy, and hospital management teams would have the reassurance that they will not bear the entire cost without appropriate contributions or reimbursement from their technical partners.[26] However, the solution to these cases will depend on legislative action, judicial practice and guidelines established by professional organisations, which risks making this system overly complex, attracting the same criticism as in the previous case.

1.5.2 Strict Liability

One approach to managing the uncertainties surrounding medical AI is the implementation of strict liability, a common path when it comes to product liability. This approach shifts the focus from proving fault to simply establishing that the AI system caused harm (causation), thereby simplifying the legal process for plaintiffs. Applying this legal framework to medical practitioners would mean that they would be held accountable for any harm caused by their actions, regardless of whether negligence or fault can be proven.

It has been suggested by the Expert Group on Liability and New Technologies that the operator's strict liability could be a favourable option in specific, more precisely defined instances compared to that of the manufacturer.[27] The operator's responsibility would not nullify the applicability of the PLD; however, the report does not elaborate on how these two systems should coexist. A possible way to overcome this issue is to establish processes to identify the roles of AI developers and clinical users when doctors who are assisted by AI decisions harm individuals.[28]

However, the effectiveness of strict liability in enhancing patient outcomes remains a contentious issue.[29] Stringent liability laws – of which strict liability is among the most rigorous – may lead to defensive practices among healthcare providers and AI developers, such as overtreatment or reluctance to adopt innovative technologies that could ultimately benefit patients. These defensive practices can increase healthcare costs without necessarily improving patient care or outcomes.[30] Moreover, it places an excessively heavy burden on healthcare practitioners, as under strict liability, the focus is solely on the occurrence of harm, not on whether the healthcare provider followed appropriate standards of care or took reasonable precautions, thus

26 Allain, 2013: 1077.

27 European Commission, 2019: 42–43.

28 Panel for the Future of Science and Technology, 2022: 27.

29 Raposo, 2019: 401–416.

30 Bertoli and Grembi, 2017; Maliha, Gerke, Cohen, et al., 2021: 629–647, Martin-Lapoirie, 2022: 251–282.

imposing an excessive burden on them.[31] Ultimately, while strict liability offers a straightforward mechanism for addressing harms caused by medical AI, its impact on patient outcomes and the entire healthcare delivery system is complex, to say the least.

1.5.3 No-Fault and the Socialisation of Risk

Another option, potentially more interesting,[32] is the socialisation of risk (See Chapter 5, Section 7).[33] This refers to the process of distributing the financial or social burdens of risks across a broader community, rather than having individuals or specific groups bear the entire impact alone.[34] The concept is often applied in contexts such as healthcare,[35] insurance, and public policy, where the costs of risks – illness, accidents, or economic downturns – are shared collectively. This approach spreads the financial risk of medical costs across a larger group, making healthcare more affordable and accessible for individuals.

For medical AI, it has been suggested that a compensation fund be established, which is a financial mechanism designed to spread the burden of damages or losses across a larger group, rather than leaving it to be borne by individual parties directly involved.[36] This type of fund is typically financed through contributions from various stakeholders, such as industries, companies, or even individuals, and is used to compensate victims for damages or losses without the need for lengthy legal battles. By pooling resources, the fund can cover large claims or damages that would be financially devastating for any single member. This approach mitigates the risk for individual contributors and provides a reliable source of compensation for claimants. The fund operates under predefined rules that outline the conditions under which compensation will be paid. Typically, a third-party organisation or a governing board manages the fund, ensuring transparency, accountability, and fairness in its

31 Maliha, Gerke, Cohen, et al., 2021: 634.

32 Marchisio (Marchisio, 2021) suggests the no-fault model for AI manufacturers and for medical practice in general (Marchisio, 2020: 87–118).

33 Strict liability and no-fault models have some similarities, but they are not the same (Nguyen, 2019). Under strict liability, a party (usually a manufacturer or provider) is held responsible for damages or harm caused by their product or actions, regardless of whether they were negligent or intended to cause harm, but liability is established. By contrast, in a no-fault system, compensation for damages or harm is provided without the need to establish fault or blame on any party because there is no liability involved.

34 Raposo, 2016b: 57–69.

35 In healthcare, the socialisation of risk is already frequent through mechanisms like health insurance, where premiums paid by many are pooled together to cover the medical expenses of those who need care.

36 Holm, Stanton and Bartlett, 2021: 171–188.

distribution. Thus, contributors may be incentivised to adopt safer practices to minimise their contributions to the fund or avoid penalties. A very clear practical advantage is that it can provide quicker compensation to victims compared to traditional legal processes, which can be lengthy and costly.[37]

1.6 *The Difficulties in Assessing Liability Grounded on AI Specificities*

1.6.1 Black-Box AI and Lack of Explainability

Under traditional tort paradigms, establishing liability becomes extremely challenging if the rationale behind an AI's decision-making remains opaque (see Chapter 3, Section 3.4).[38] For instance, when the AI's reasoning is obscure, it becomes arduous to prove a breach of duty.

Liability-wise, the lack of explainability can lead to two possible scenarios. On the one hand, courts might exclude liability for medical practitioners in cases involving black-box AI, considering that a person cannot be held responsible for something they do not understand.[39] On the other hand, the lack of explainability can create ambiguity and uncertainty, potentially fuelling medical liability and holding doctors liable for using obscure technology.[40] For instance, introducing an explainability function in AI devices may increase a doctor's responsibility and liability. Doctors might be expected to scrutinise the explanations provided by the AI system to avoid potential liability. Failing to understand or analyse the explanations to the expected level of scrutiny could be seen as a breach of the doctor's duty of care.[41] Whether courts will recognise this extended duty of care, including the need to meticulously examine AI-generated explanations, remains uncertain.

1.6.2 Lack of Adequate Information and Lack of Comprehension of That Information

A major challenge in proving the elements of legal liability in court is the scarcity and complexity of the information involved, even in white-box AI. This is not a matter of lack of explainability but a lack of transparency.

Transparency issues can arise from several factors. First, proprietary concerns may lead companies to withhold details about their AI systems to protect

37 Raposo, 2016b: 59–61; Wallis, 2017: 38–39.
38 Schweikart, 2021: 4–5.
39 Durán and Jongsma, 2021: 333.
40 Price II, 2018: 295–306.
41 Price II, 2018: 301.

intellectual property or trade secrets.[42] Second, even when some level of transparency is offered, the technical nature of AI models can be difficult for non-experts to comprehend,[43] limiting effective oversight and accountability.

That is, even if a plaintiff is granted access to the information required to file a lawsuit, it is improbable that they could adequately prove their entitlement to compensation solely by accessing data.[44] The data must be filtered to identify the relevant part related to the incident and then understood. This process of interpretation is undeniably intricate and expensive, demanding the involvement of a skilled engineer or technician. This not only adds to the cost but also underscores the importance of having specialised knowledge to handle such tasks.[45] It also proves that providing transparency or explanations (see the previous section) from a system does not automatically enable us to effectively challenge its decisions.[46]

1.6.3 AI Autonomy

1.6.3.1 *Autonomous and Unpredictable AI Systems*

Traditionally, developers and software engineers had control of the machine's behaviour in every detail, to the extent that they could explain its general design and function to a third party. This traditional conception of responsibility in software design assumed that the developer could reflect on the likely effects of the technology and the potential for malfunction and make design choices to achieve the most desirable outcomes according to the functional specification.[47]

However, existing complex AI systems often make decisions that do not align with their creators' original intentions.[48] AI systems can sometimes go beyond their creators' expectations, leading to unforeseen decisions that even the creators could not predict.[49] Unpredictability creates a challenge: some

42 Foss-Solbrekk, 2023.

43 Yu, 2023: 140.

44 Methnani, Tubella, Dignum et al., 2021.

45 Bertolini and Episcopo, 2021: 653.

46 Lyons, Sycara, Lewis, et al., 2021.

47 Mittelstadt, 2021: 19–20.

48 This becomes particularly evident when examining AI that learns on its own, like a student learning from a teacher. Think of the AI as a student who receives some instructions from a teacher. However, the student figures out their own ways to follow these instructions. This is like the student exploring various paths to solve a problem, even if those paths were not directly taught by the teacher.

49 Bathaee, 2018: 907. A well-known example is AlphaGo, an AI that played the game Go against human experts and won. The way AlphaGo played seemed unusual to people at

actions are hard to predict, making it difficult to prove the manufacturer's intent, whether it be recklessness or gross negligence. This arises due to the prerequisite of some level of risk awareness or predictability, a critical foundation for proving such states of mind and, therefore, culpability. When this happens, a problem arises because AI ends up making decisions that its creators did not anticipate or plan for.[50] If AI designers cannot predict how AI will act once it is released into the world, holding them liable becomes a challenge. In turn, if the legal system absolves AI developers of responsibility because AI's actions are unpredictable, injured patients may have less opportunity for redress.[51]

1.6.3.2 *Autonomy and Causation*

In cases involving complex AI systems, it becomes arduous and costly for an injured party to pinpoint a specific cause that led to the damage. Consequently, the claimant faces an unfair burden of proof to establish the defendant's fault.

The challenge of establishing a clear causal link between the harm and the defendant's actions is further complicated when dealing with self-adapting, autonomous AI systems. These AI systems can modify themselves based on new data, shifting from deterministic to probabilistic computing. Additionally, emerging technologies often enter the market incomplete and may undergo updates or interact with other systems and data sources, which further complicates the causation aspect. For claimants, proving causation becomes intricate due to a multitude of potential factors, such as defects in the initial product, flaws in subsequent updates, interactions between the AI and other systems, failures by operators to install required updates by operators, and various other possibilities.[52]

1.6.3.3 *Liability within Autonomous AI Systems*

Traditionally, vicarious liability holds employers accountable for the actions of their employees within the scope of their employment. However, as AI systems become more independent and capable of making complex decisions without direct human oversight, hospitals find it challenging to exert the same level of influence or control over these systems as they do over human staff. Consequently, when an AI system makes a decision that leads to harm, it

times because it made moves that human players would not typically consider. This demonstrated that the AI was making its 'own' choices, rather than strictly following the given instructions (Ribeiro, 2016).

50 Developing this problem, Bathaee, 2020: 145–149.

51 Scherer, 2016: 353–400.

52 Whittam, 2022: 252.

becomes difficult to attribute that decision to the hospital in the same way one would to a human employee. The hospital might not have had any meaningful way to predict or prevent the AI's specific actions, especially if those actions are based on real-time data processing and autonomous decision-making capabilities. This growing autonomy of AI means that its actions are often the result of intricate algorithms and ML processes rather than direct human input.[53]

This highlights the complex legal landscape that surrounds the accountability of AI technology. Blame can only be justified when the actor has some degree of control and intentionality in carrying out the action. Delegating decision-making to AI can shift responsibility away from human decision-makers. The traditional model of accountability fails because nobody has enough control over the machine's actions to be able to take responsibility for them.

This acknowledgement does not necessarily dismiss liability as we know it. Manufacturers should be held accountable precisely because they were aware they were introducing an inherently unpredictable tool or device into society for widespread use, making it particularly difficult for claimants to demonstrate the requirements of liability.[54] This could involve ensuring that AI developers and manufacturers bear some responsibility for the 'behaviour' of their creations, or establishing rigorous standards and protocols for AI use in healthcare settings to minimise the risk of harm.[55] For professional users, new models of accountability and liability might be necessary, potentially focusing on the deployment and oversight of AI systems rather than the actions of the AI itself.[56] In this context, the legal framework needs to evolve to address the unique challenges posed by AI autonomy.[57] This conundrum can be addressed by applying strict liability or similar faultless liability schemes.[58]

2 Medical Liability within AI Medicine

2.1 *Basic Concepts in Civil Medical Liability*

Medical liability, like tort liability in general, serves two purposes: first and primary, to compensate injured parties for their damages;[59] and second, to deter potentially harmful behaviour (even though this aim is more closely connected

53 Mills and Sætra, 2024: 590–592.
54 Lior, 2020: 1135.
55 Mills and Sætra, 2024: 592–593.
56 About the different approaches to liability see Novelli, Taddeo and Floridi, 2023.
57 Novelli, Taddeo and Floridi, 2023.
58 Mittelstadt, 2021: 20.
59 Goldberg, Sebok, Zipursky, et al., 2021.

to criminal liability).[60] In order to achieve these objectives, the patients who have suffered the damage, or someone on their behalf, can claim in court compensation for property damage and/or moral damage, based on the fact that the provision of the medical service has not met the expected standard of care.

Medical liability requires the existence of damage,[61] caused by illicit and culpable conduct by the doctor, embodied in a deviation in the healthcare provider's conduct from what would be the expected behaviour. These are the three requirements of responsibility:[62] illegality, culpability, and causation.[63]

Therefore, not every harm (damage) is compensable. To determine liability, three key factors must be considered. Firstly, the violation of the standard of care (see Chapter 4). Secondly, it must be evaluated whether the defendant can be blamed for the conduct in question, assessing the alignment between the individual's behaviour and the standards of care, intent or risk awareness required by the legal system (culpability)[64]. Depending on the context, culpability may rest on intention, knowledge, recklessness or negligence, reflecting different levels of fault. Thirdly, one must examine whether a reasonable person in the defendant's position could have foreseen the damage at the time of the breach. It entails assessing whether the type of harm suffered by the claimant was reasonably foreseeable, irrespective of the actual extent of the injury. This is called the remoteness test in common law[65] and the foreseeability test in civil law.[66] Legal causation hinges on establishing an unbroken chain of causation between the defendant's culpability and the claimant's injury. In other

60 Popper, 2011: 181–204.

61 Besides physical harm, the uses of AI have the potential to impact privacy, reputation, and dignity in various ways. Consider the scenario of an AI system erroneously diagnosing cancer or dementia, which could significantly affect an individual's sense of dignity and overall well-being, irrespective of any physical harm caused. Similarly, the integration of AI within the human body, such as in the case of prosthetic limbs or software regulating vital organ functions, carries the potential for profound consequences on personal privacy, reputation, and dignity (Soyer and Tettenborn, 2022: 394).

62 The terminology varies. In common law, these elements are often framed differently (duty, breach, causation), while in civil law they are more explicitly articulated as unlawfulness, fault, and causal link.

63 Raposo, 2013: 145–162.

64 Selbst suggests that to better align with psychology research, the law should reconsider the definition of a 'reasonable person'. If psychological studies indicate that the average individual is incapable of performing certain actions, the law should not expect the 'reasonable person' to accomplish them (Selbst, 2020: 1348–1349).

65 Mulheron, 2020: 487–499.

66 Melchiori, 2020.

words, there must be a clear link connecting the defendant's actions and the harm suffered by the claimant.[67]

When demonstrated in court that the doctor followed the applicable *leges artis* and fulfilled the duty of care that he/she is charged with, any damages that the patient may suffer cannot be imputed to him/her for lack of culpable conduct and therefore, there is no liability (most jurisdictions refuse strict liability for medical practitioners). The occurrence of an unfavourable outcome following a medical procedure does not automatically indicate the presence of malpractice.[68]

The challenge lies in the fact that within AI medical care, some of the requirements of tort liability might be difficult to assess. Take, for example, what happens to intentions or, more generally, culpability.[69] Humans leave traces of evidence, such as emails, letters, and behaviour. The AI, however, makes predictions and decisions as humans do but is often unable to communicate its reasons for doing so. A similar difficulty happens with causation, that is, proof of the causal link.[70] When AI is a black box, this exacerbates the difficulty due to its opacity, which makes it difficult to establish cause-effect relationships between a decision and an outcome. Thus, the liability framework must be fine-tuned to the AI scenario.

2.2 *Medical Liability in Civil Law and in Tort Law*

The principles governing the liability of doctors for negligence require the satisfaction of several essential elements discussed in the previous heading. Despite a basic common ground, they slightly differ in civil law and common law.[71]

In civil law, medical liability is established upon several key components.[72] First and foremost, there must be evidence of unlawful behaviour on the part of the healthcare provider. Secondly, that behaviour must also be culpable, meaning that the doctor did not act according to what the *bonus pater familias* (the criterion to assess the conduct of the reasonable person in civil law)[73] would have done. These two criteria, taken together, reflect the violation of

67 Clarke, 2014: 217–243.

68 Paterick, 2019: 15–22. It is important to recognise that the standard of care in medicine is not uniform and can vary depending on the specific circumstances. In many medical scenarios, multiple treatment options can align with the principles of reasonable care.

69 Bartlett, 2023: 241–270.

70 Holzinger, Langs, Denk, et al., 2019.

71 For a general overview of the differences see Nys, 2005; Pereira, 2013: 591–404.

72 Raposo, 2013: 45–145.

73 Hinteregger, 2023: 65–66.

the medical *leges artis*.[74] Moreover, there must be harm. This requires showing that the provider's misconduct directly resulted in the injury or damage experienced by the patient. Lastly, there must be a clear and demonstrable causal link between the breach of duty and the harm sustained by the patient and that the injury was a foreseeable outcome of the provider's negligent actions, rather than an improbable or indirect consequence.

By contrast, under common law, the first two components encompass the presence of a duty of care held by the doctor toward the patient who suffered harm and the consequent violation of that duty by the doctor.[75] These elements have been firmly established in common law and serve as fundamental criteria for establishing liability in cases of medical negligence.[76] In the landmark case of Donoghue v Stevenson,[77] the concept of the 'duty of care' was established, stating that individuals have a universal responsibility to exercise reasonable care to avoid causing harm to others, referred to as their 'neighbours' in legal terms. This means that the provider's actions must deviate from the accepted standard of care (the *medical leges artis* of civil law)[78] in a manner that is both wrongful and blameworthy.

2.3 *The Allocation of Liability between Stakeholders*

Practical challenges arise when a patient seeks to bring a claim involving AI. Determining whom to sue becomes a significant hurdle in such cases. The options include the healthcare facility (publicly or privately owned), the members of the staff directly responsible for ensuring safety and quality in healthcare, the healthcare providers and the developers and/or manufacturers of the AI system.[79] Who will be held liable? The intricate nature of AI can make it difficult for the claimant to ascertain who might be at fault and, consequently, whom to sue.[80]

---

74 Raposo, 2013: 87–96.

75 Newdick, 2005: 132–138.

76 Kloss, 1984: 66.

77 *Donoghue v Stevenson* [1932] AC 562.

78 On the connection between standard of care and medical leges artis see Raposo, 2019: 404–405.

79 Finding a responsible party is puzzling, as so many different entities – software developers, hardware engineers, and product designers – collaborate on creating AI systems.

80 Sullivan and Schweikart, 2019: 163; Yeung, 2018: 10. It has also been suggested that healthcare organisations and AI providers could establish contractual agreements (Lupton, 2018: 2–7) to outline how liability will be apportioned in the event of malpractice verdicts or settlements.

This difficulty is fuelled by the obscurity of many AI models. Determining who is responsible depends on whether the failure lies in the design of the AI system or the misuse of the technology by the doctor. Without explainability, it becomes nearly impossible to ascertain whether a system's decision was based on sound medical reasoning or influenced by errors, biases, or flaws in the algorithm.[81] The more opaque and mysterious the AI algorithm, the harder it becomes to identify the cause of errors and attribute responsibility.[82]

An additional problem in current accountability schemes related to AI is its increased autonomy. According to Chinen, as machines achieve more autonomy, it will become more difficult to attribute and distribute legal responsibility for their behaviour to human beings. As the AI system becomes more autonomous, humans (i.e., doctors, healthcare organisations and AI designers) have less control over it.[83] The impossibility of holding a person (human or legal) accountable will have a direct impact on compensation for damages, eventually leading to models of strict liability, in which responsibility is reallocated to third parties who control, benefit from, or are institutionally connected to the harmful activity (such as in the principal/trustee relationship).

Overall, the introduction of AI in the loop raises a crucial dilemma: if a decision made by an algorithm appears to be clearly erroneous, is it justifiable to dismiss one claimant's complaint while allowing another claimant, in an identical position, to proceed with their suit simply because their determination was made by a human? This discrepancy challenges the principles of equality and fairness in the pursuit of fair and consistent compensation.

In cases of AI-related harm, it is probable that the injured party will pursue claims against various entities, each based on different legal grounds. Li and Schütte distinguish five distinct legal bases for a patient to seek compensation:[84] (a) filing a strict product liability claim against the developer of an AI system, (b) holding the developer accountable under fault-based liability, (c) alleging negligence against the hospital, such as failure to provide adequate supervision, (d) targeting the doctor who made an incorrect judgment based on AI-generated recommendations, and (e) accusing the doctor of misusing the AI device, in a way that leads to inaccurate guidance. In such

81 Xu and Shuttleworth, 2024: 52–57.

82 Schweikart, 2021: 4–5.

83 Chinen, 2016: 342.

84 Li and Schütte, 2024: 146.

cases, the provider, the hospital, and the doctor are likely to be held jointly and severally liable for the harm caused.[85]

Gerke et al. emphasise that the determination of the optimal liability framework is heavily contingent upon on one's understanding of the underlying 'problem' at hand.[86] If the primary concern revolves around the potential risks to patients associated with the deployment of AI-based technology in the clinical domain, maintaining the current medical malpractice regime, which seeks to fulfil the dual objectives of deterring misconduct and compensating victims, may be preferred. Conversely, if there is a belief that the use of AI ultimately enhances patient well-being in the majority of cases, it could pose a challenge if doctors exhibit hesitancy in relying on these algorithms, particularly the more opaque ones, due to the lingering liability they bear for any resulting adverse outcomes. Such circumstances might prompt policymakers to consider alternative models, such as the no-fault model.[87]

2.4 *Holding the Human Doctor Accountable*

2.4.1 Trust in AI and Medical Liability

In general terms, two main situations can lead to liability. One is when an AI system provides a recommendation aligned with the standard of care, yet the doctor disregards it, and the patient experiences harm. The other is when the AI system suggests a course of action that deviates from the standard of care, but the doctor follows it, and the patient is harmed as a result.

In the eyes of the courts, the evaluation of the doctor's use of the AI tool will encompass several factors. These factors include the reasonableness of employing AI given the specific context, as well as an assessment of whether the medical recommendation given by AI matches what a careful doctor would typically provide. External structural and organisational factors, such as the doctor's workload and the establishment of reliance on automation, will influence the decision to put the recommendations into practice and, as such, must be considered by the court.

Doctors hold the responsibility to familiarise themselves with the tools they use, including AI systems' potential risks. However, if the AI system's algorithm is a 'black box' with limited transparency, doctors may be unable to adequately

---

85 However, since the AILD exclusively addressed fault-based liability and requires that the fault directly impact the AI system's output, only claims under categories (b), (c), and (e) would fall within the scope of the AILD.

86 Gerke, Minssen and Cohen, 2020: 313–314.

87 Holm, Stanton and Bartlett, 2021: 171–188; Marchisio, 2021: 54.

assess and mitigate these risks.[88] When evidence surfaces indicating that a human doctor could have spotted the anomaly that escaped the notice of AI, a strong argument for causation can be constructed.[89] Note, however, that the likelihood of healthcare providers detecting all or most errors made by AI systems during routine clinical practice is notably low. This is particularly true when healthcare providers believe that the AI system outperforms their own judgment or when they are instructed to rely on the system with the premise that 'it performs better than you do'.[90] Consequently, while the frequency of errors leading to patient harm may decrease if doctors adhere to the guidance of highly accurate AI systems, the patient's ability to seek compensation when demonstrable errors occur may diminish.

As algorithms continue to surpass the average doctor in terms of accuracy, it becomes questionable to solely hold doctors accountable for medical outcomes. If the algorithm consistently offers statistically superior suggestions, it becomes challenging to argue negligence on the part of the doctor when they follow the algorithm's recommendations. Even if the algorithm's suggestion proves to be incorrect and leads to harm, the statistical superiority of the algorithm makes it difficult to assign blame to the doctor.[91]

Doctors might find themselves trapped in a 'moral breakdown zone', where they become liable for AI-based decisions only because they chose to use AI.[92] While on the one hand, it makes a strong case for assigning full legal responsibility to doctors, on the other hand, doctors become responsible for an element over which they may have no control.[93]

In sum, the doctor cannot be held *a priori* accountable just because of the use of an unconventional approach, such as the current use of AI. Conversely, the mere fact that the doctor relied on AI that is fully certified (as a medical device) is not enough to exclude medical liability.

2.4.2 The Relevance of Human Oversight

Under Article 26 of the AIA, it is the deployer's (usually, the hospital) responsibility to make sure that individuals tasked with human oversight possess the required skills, training, authority, and adequate support. If any of these individuals believe the AI system presents a risk, they are obligated to notify

88 Smith and Fotheringham, 2020: 135–136.
89 Harvey and Gowda, 2022: 236.
90 Rahman, Victoros, Ernest, et al., 2024.
91 Also in this sense, see Thomas, 2017.
92 Elish, 2019: 40.
93 Smith and Fotheringham, 2020: 134.

the provider or distributor, as well as the relevant market surveillance authorities, and halt the system's use. This same procedure must be followed if the deployer encounters a serious incident.[94]

One potential scenario where oversight could impact the output is if previous oversight had identified the AI system's tendency to underperform in a given aspect (biases, for instance), leading the developers or other entities to correct the model's design before the relevant application of the system. However, this scenario appears unlikely. Even if argued that proper oversight would have detected an erroneous output, prompting an immediate redesign of the system and the generation of a correct output, the fact is that the instantaneous redesign of the AI model is highly implausible.

2.5 *Causation*

2.5.1 Causation in AI Systems

Contemporary AI models, while impressive, are limited in their ability to perform a thorough analysis or deep investigation. They mainly focus on data associations, forming correlations through passive observations and post-hoc data analysis.[95] Inevitably, this strategy imposes limits on the extent of achievement. The core reason lies in the foundational nature of these AI algorithms. Within this scenario, the potency to uncover genuine cause-and-effect relationships remains elusive, as observational data, unless meticulously gathered through randomised controlled experiments, fail to illuminate the intricate causal pathways.[96]

This weakness also jeopardises the claimant's role. While the claimant is expected to establish causation according to the general legal principles, the Expert Group on Liability and New Technologies proposed reducing this requirement by considering factors such as the technology's potential role in the harm, the presence of a recognised defect risk, the ease of retracing and understanding the technology's operations afterwards, the accessibility and understandability of data produced by it, and the type and extent of the harm

94 However, there remains a lack of clarity regarding what constitutes a 'risk' or a 'serious incident' in the medical field, making it unclear when reporting is mandatory (Solaiman and Malik, 2024).

95 A trailblazer in this arena is Judea Pearl (2018), who coined the term 'New Science of Cause and Effect' to express the emergence of a novel AI capability: the competence to deduce causal relationships from observational data, all within the constraints of the conditions under which real-world systems operate. It represents a significant stride toward unlocking the mysteries of cause and effect within AI.

96 Baldwin and Arata III, 2021: 97–98.

involved.[97] The Expert Group on Liability and New Technologies assumes that simply shifting the burden of proof would satisfactorily enhance consistency and foreseeability in legal applications, yet diverse national level interpretations could paradoxically result in the same legal and economic fragmentation that the Report aims to prevent.

2.5.2 *Res ipsa loquitur*

In situations where establishing causation becomes challenging, the legal doctrine of *res ipsa loquitur* – 'the facts speak for themselves'[98] – can sometimes be employed to address malpractice claims.[99] This doctrine allows culpability to be inferred against a specific defendant when the accident leading to physical harm aligns with the kind of accidents typically caused by negligence within a particular group of actors.

According to Laurie and colleagues, this theory proved to be particularly useful in cases involving machines or complicated processes, of which the patient has little understanding, and no explanation is offered by the defendant.[100] Thus, the doctrine appears almost perfectly suited to the application of AI; in a dispute, it would place the onus of providing a meaningful explanation on the decision-maker to explain the use of AI.[101]

However, the use of the *res ipsa loquitur* doctrine faces two major obstacles in this domain. First, 'if the harm in question is unexplained, untraceable and rare, then the elements of res ipsa loquitur probably cannot be satisfied'.[102] That is, for *res ipsa loquitur* to apply, it must be clear that the injury or damage would not typically occur without some form of negligence, and that the instrumentality causing the harm was under the defendant's exclusive control. This leads us to the second requirement: that the defendant has exclusive control over the cause of the damage. It happens, however, that AI systems involve a large number of actors, without any of them having exclusive control over the system.[103]

---

97 European Commission, 2019: 49–50. As argued by Bertolini and Episcopo (Bertolini and Episcopo, 2021: 653), the means of achieving the reduction in the claimant's proof burden is left unspecified by the Report, whether it should be achieved through standardised European regulations or individual Member States should decide.

98 Prosser, 1936: 241–271.

99 Sustaining the use of this doctrine, Schönberger, 2021: 197; Schweikart, 2021: 12–13.

100 Laurie, Harmon and Porter, 2016: 878–879.

101 Bartlett, 2024: 295–326.

102 Jackson, 2019: 59.

103 Pointing out this issue, Schweikart, 2021: 13.

2.6 *The Liability of the Human Healthcare Provider*

2.6.1 Complementary AI and Substitutive AI in Medicine

2.6.1.1 *Complementary AI*

2.6.1.1.1 The Role of AI as a Complement

The dominant position among scholars is that in the healthcare domain, AI will serve primarily as a complement to the doctor's activity, functioning mainly to confirm and support their decisions.[104] In fact, most AI solutions currently under development are designed to assist rather than replace human professionals. This approach ensures that the critical aspects of patient care remain under the control of trained medical experts. Complementary or assistive AI can thus be defined as one that does not replace the human but simply works as an aid, providing additional resources for them to do their job.[105]

In practice, this means that AI systems will analyse data, provide diagnostic suggestions, and recommend treatment options, but the final decision-making authority will rest with the doctor. This collaborative model leverages the strengths of both AI and human expertise: AI's ability to process and analyse vast amounts of data quickly and the doctor's experience, judgment, and ability to understand the nuances of patient care.[106]

The reason is that the majority of AI solutions are not designed for complete autonomous deployment due to current technological limitations and ethical considerations.[107] Fully autonomous AI systems in healthcare would need to achieve a level of reliability and accountability that is currently beyond reach. Additionally, the ethical implications of allowing AI to make autonomous medical decisions without human oversight are profound,[108] raising concerns about accountability, patient safety, and the doctor-patient relationship.

Therefore, AI's role in healthcare settings is typically of the 'augmented' type, meaning it is fully integrated with doctors in the practice of medicine and serves as a tool to aid patient treatment.[109] This role is also mirrored in the designations currently given to this technology. AI is widely understood to be the abbreviation of 'Artificial Intelligence', but in its 2018 policy recommendations, the American Medical Association (AMA) preferred the expression

104 Alowais, Alghamdi, Alsuhebany, et al. 2023.

105 Distinguishing between the complementary and the substitutive options when referring to the intervention of AI in healthcare, Chen, 2020: 245–267.

106 Alowais, Alghamdi, Alsuhebany, et al., 2023.

107 Alowais, Alghamdi, Alsuhebany, et al., 2023.

108 Inglada Galiana, Corral Gudino and Miramontes González, 2024: 178–186.

109 Crigger, Reinbold, Hanson, et al., 2022.

'augmented intelligence'.[110] In a supporting document,[111] the AMA noted that, in the health area, the latter is a more appropriate term because it reflects the enhanced capabilities of human clinical decision-making when coupled with these methods and computational systems. This labelling highlights the idea that the doctor, not Al, will have primacy.[112] While AI is expected to significantly enhance healthcare delivery, it is envisioned as a tool that supports and augments the doctor's role rather than replacing it. This perspective ensures that AI serves as a valuable ally in improving patient outcomes while maintaining the essential human touch in healthcare.

Therefore, if an AI solution were to fail and cause harm to a patient, such as a missed diagnosis or unnecessary treatment due to a false positive, the doctor responsible for the patient's care would still be exposed to potential negligence claims. Ultimately, in assistive AI, the responsibility for any potential medical errors falls upon doctors, as they retain the role of final decision-makers, given their expertise and experience as human professionals. The AI software is perceived as an aid to doctors, who decide independently, as the system only aims to provide the doctor with more information. In light of this logic, the doctor should be subject to the exact same pattern of malpractice regardless of whether or not an AI software is used, because any treatment decisions or actions are ultimately the responsibility of the healthcare provider.[113]

2.6.1.1.2 The Liability Framework for Complementary AI

Complementary AI can be framed in two different ways.

AI can be equated to mere information; in which case the manufacturer's liability is most likely not implicated. Drawing a parallel, if a doctor proceeds with a course of action based on incorrect information from a book, the patient cannot hold the author of the text accountable through a successful lawsuit. Likewise, if a doctor acts upon inaccurate information provided by AI, the patient should not have grounds to sue the programmer, placing the burden of responsibility solely on the doctor's choices and actions.[114]

AI can also be equated to a 'consulting doctor'[115] who, according to current legal principles, is not obligated to have a duty towards the patient, as they do not directly interact with them. In simpler terms, doctors who only provide advice

110 AMA, 2018.
111 American Medical Association, 2018.
112 Terry, 2019: 137.
113 Babic, Gerke, Evgeniou, et al., 2019: 1202–1204.
114 Hodge, 2022: 468.
115 Wood and Castro, 2023.

without examining or speaking to patients do not establish a doctor-patient relationship (what common law scholars call 'the duty of care'),[116] which could lead to a recognised medical malpractice claim.[117] Advisory models do not independently make diagnoses or treatment decisions. Instead, they offer guidance to healthcare professionals, who ultimately retain decision-making authority, equating such AI systems to consulting doctors.

Based on this understanding, some AI manufacturers are shifting responsibility for system safety entirely onto doctors.[118] This situation creates what Smith and Fotheringham[119] – quoting Ellish[120] – call a 'moral crumple zone' for doctors, where they bear the moral burden of AI's errors due to their choice to use the technology.

Indeed, a healthcare provider who relies upon an AI system to devise a treatment recommendation will still bear accountability if the actions undertaken fall short of the accepted standard of care. Doctors carry an independent obligation to employ the appropriate standard of care, regardless of the output generated by the AI algorithm.[121] As an illustration, legal authorities have permitted malpractice claims to proceed when patients assert that doctors relied on literature containing inaccuracies or relied on an intake form lacking a comprehensive medical history. Likewise, most courts have exhibited reluctance to accept medical guidelines as unequivocal evidence of the duty of care, instead demanding a tailored assessment in such circumstances.[122]

Still, it would be unfair to hold doctors fully and solely liable for errors, and a growing trend is emerging against this notion.[123] This is especially true if AI becomes fully integrated into the medical standard of care. This integration will require doctors to not only incorporate AI into their practice but also adhere to its recommendations. To ensure fairness for both the claimant and the doctor, a more equitable solution could involve considering joint liability and sharing the risk between the doctor and the AI's manufacturer in the court's assessment (see Chapter 5).

116 Blake, 2012: 403–406.

117 Chung and Zink 2018: 70.

118 IBM's Watson for Oncology, for instance, emphasises that it provides recommendations based on evidence, leaving the ultimate decisions to doctors (Hagen and Narozniak, 2019).

119 Smith and Fotheringham, 2020: 134.

120 Elish, 2019), 40–60.

121 Maliha, Gerke, Cohen, et al., 2021: 632.

122 Hodge, 2022: 456.

123 Lawton, Morgan, Porter, et al., 2024.

2.6.1.2 *Substitutive AI*

2.6.1.2.1 AI as the Sole Decider

Unprecedentedly, AI advancements have changed the way we think. The abilities of AI systems allow them to autonomously reach intuitive conclusions, all without human intervention.

Assuming that AI is functioning autonomously, questions of fairness might prevent courts from holding the professional user solely liable for errors or malfunctions.[124] This is because the provider may not have complete control over the technology's actions, and the technology may have been programmed with errors that the deployer may not have been aware of.

If and when AI replaces doctors, substituting them entirely, the sole responsibility of the manufacturer is more appropriate.[125] Manufacturer's liability is a strict liability, as follows from the statutory regime for defective products, which in the EU (and in many jurisdictions) is a strict liability regime.

To illustrate this scenario, picture a Computerised Physician Order Entry (CPOE) that decides everything regarding the prescription of drugs, preventing any intervention from the human.[126] Suppose, for example, that the AI system prescribed drug A, while the best solution for that particular patient and that particular medical condition would be to prescribe drug B.[127] If the CPOE autonomously makes all decisions regarding the prescription of the pills, this would hinder the doctor from co-prescribing them, relinquishing responsibility, as they are unable to override the system. Under the principles of strict liability, if a preventable adverse event occurs, the manufacturer, distributor, and retailer of the product could be held accountable, even without direct fault.[128]

Substitutive AI provides a liability shield for doctors based on the legitimate expectation principle, but only insofar as several conditions apply: i) the system is certified under the full quality assurance procedure; ii) the human expert has good reasons to rely on the system's correct performance; iii) the applicable legal framework recognises the intervention of AI in healthcare delivery. This procedure involves rigorous testing, validation, and continuous

124 Thomas, 2017.

125 In 2018, the United States Food and Drug Administration (FDA) granted approval for the first algorithm capable of autonomously analysing diagnostic images without doctor intervention (Reardon, 2019). As no medical professional is involved in the process, the company behind the algorithm assumes legal accountability for any potential medical errors.

126 The AI policy of the AMA encompassed the assertion that creators of independent AI systems bear the responsibility and liability for any potential harm arising from the diagnostic system they generate. Cf. AMA Board of Trustees Policy Summary, 2019.

127 Pasquale, 2020.

128 Pasquale, 2022: 5.

monitoring to ensure that the AI system consistently performs correctly and reliably.[129] It not only addresses technical aspects but also factors in ethical considerations and potential risks associated with the system's operation.

2.6.1.2.2 Liability of the Healthcare Provider within Substitutive AI

At first glance, it may seem that human responsibility is absent in substitute AI. However, even in those cases, there are exceptions.

Professional malpractice liability might still arise because while the AI developer and advisory system lack direct contact with the patient, the doctor acts as the gateway to the AI system's influence on treatment decisions. Therefore, even if harm occurs due to inappropriate AI recommendations, the doctor's conduct becomes a crucial factor, and patients frequently bring claims against doctors. The reason why this happens is because the medical intervention is necessary to establish the link between AI and the patient's harm, that is, the doctor becomes the necessary and essential cause of the damage.[130] It is likely – and reasonable – that in those cases the healthcare provider will not be held liable. Still, there are exceptions.

Liability may still arise for the healthcare provider if the latter – either a human doctor or a healthcare facility – selects an unreliable AI system. This could occur if the choice of AI system goes against the recommendations of professional societies or IT experts.[131] In such cases, the responsibility for any harm caused by the AI system's decisions could revert to the healthcare provider due to the initial poor choice of the AI system (for instance, due to ignorance or seduced by a lower price). In a sense, this case is not fundamentally different from a situation in which a doctor, while leading a medical team, selects an inappropriate team member for a specific task (amounting to a case of *culpa in eligendo*). This highlights the importance of due diligence in the selection and implementation of AI technologies in medical practice, ensuring that they meet established standards and guidelines to safeguard patient safety.[132]

2.6.1.3 *Cooperation between Human Doctors and AI*

The 'shared decision-making model'[133] in the context of AI and human professionals refers to a collaborative approach where both AI systems and human experts contribute to the decision-making process. The model combines

129 Umoh, 2024: 74–85.
130 Smith and Fotheringham, 2020: 135.
131 Jobin, Ienca and Vayena, 2019: 389–399.
132 Dhillon, Grewal, Buddhavarapu, et al., 2024.
133 Discussing the application of this model, Lagioia and Contissa, 2020: 269–279.

the strengths of AI, such as data analysis and pattern recognition, with the expertise and judgment of human professionals to arrive at the best possible outcomes.[134]

This model is based on the idea of a joint cognitive system. Research indicates that when humans and AI systems collaborate towards a common goal, it is more accurate to describe them not as separate interacting components, but as part of a unified cognitive system.[135] In this system, control is shared between the human cognitive system and the AI system. This joint cognitive approach means that the standard of care is a combination of the traditional medical practice standard, and the standards derived from AI-generated medical diagnoses,[136] in which 'new modes of clinical care would need to be derived that utilise the best aspects of human doctors and artificial systems, implement appropriate safety and resilience checks, and minimise the weaknesses and implicit biases of both agents'.[137]

2.6.2 Human Trust in AI

2.6.2.1 *Automation Bias*

Doctors, being human, are prone to various cognitive biases that impact their ability to render precise judgments. Of notable relevance is 'confirmation bias', wherein doctors assign undue importance to evidence that aligns with their presumed diagnosis while disregarding contradictory evidence.[138]

AI is strongly connected with this phenomenon. There is a well-known cognitive bias associated with decision support systems, where doctors rely on automated systems' guidance and halt their search for confirming evidence, possibly relinquishing decision-making responsibility to the machine. The notion of 'automation-induced complacency' comes into play,[139] wherein individuals using imperfect AI systems are least likely to detect errors, which can result in a lack of due diligence in evaluating the operations of technological systems.[140] This phenomenon occurs most strongly when they are using a generally dependable system, dealing with multiple concurrent tasks, and nearing the end of their shift.[141]

134 Bignami, Montomoli, Bellini, et al., 2023.
135 Lagioia and Contissa, 2020: 278.
136 Lagioia and Contissa, 2020: 278.
137 Mittelstadt, 2021: 64.
138 Mendel, Traut-Mattausch, Jonas, et al., 2011: 2651–2659.
139 Merritt, Bryant, Staley, et al., 2019.
140 Arnold, 2021: 125.
141 Challen, Denny, Pitt, et al., 2019: 234.

To illustrate, envisage the case of a junior doctor in a busy accident and emergency department evaluating a patient with chest pain. Relying on a newly acquired AI system for chest pain diagnosis and treatment advice, the doctor usually receives recommendations with impressive diagnostic accuracy. The AI system suggests a 76% likelihood of condition A, prompting treatment X, and varying probabilities for other conditions. Considering the patient's history and symptoms, the doctor decides on treatment X. Unfortunately, the AI system's advice proves to be erroneous, as the patient actually has condition Y, making X inappropriate and risky. While it is true that the junior doctor technically makes the diagnosis and treatment decision, the circumstances surrounding her decision-making contribute to the potential error.

2.6.2.2 *Excessive Trust and Liability*

Excessive trust in AI does not necessarily mean that the healthcare providers acted in a reprehensible manner, i.e., that there was a violation of the standard of care, especially in the case of young doctors, as in the example provided. Overruling the AI system's advice may feel akin to disregarding guidance from a senior colleague.[142] In a high-pressure environment, such as an accident and emergency department, junior doctors often rely heavily on the expertise and recommendations of more experienced colleagues.[143] Similarly, AI systems, which are designed to assist in clinical decision-making with high accuracy, can be perceived as authoritative sources of guidance. For a junior doctor, the prospect of overruling the AI system can be daunting.[144] The AI's recommendations, backed by extensive data and sophisticated algorithms, might seem more reliable than the doctor's own judgment, especially when under time constraints and dealing with complex cases. This psychological pressure can lead to a heavy reliance on the AI system, making the doctor less likely to challenge its recommendations, even if there are signs that an alternative diagnosis should be considered.

The main question is whether it was reasonable to trust the AI's judgment given the situation. When a professional user – in our case, a healthcare provider – leans on an AI's evaluation without human intervention or review, it shows confidence in the AI. If substantial deference is not justified – like when

142 Holm, Stanton and Bartlett, 2021: 176–177. The authors emphasise that healthcare providers must exercise critical thinking when considering advice from AI systems, but they do not propose disregarding the validity and reliability of the AI-generated recommendations. Nonetheless, negligence can still occur if a healthcare provider unquestioningly accepts and acts upon clearly flawed AI advice.

143 Goddard, Roudsari and Wyatt, 2012: 121–127.

144 Bartlett, 2023: 252.

the AI lacks robust testing or unbiased training data – this might serve as a basis for assigning liability. The key question is whether the oversight of the AI was reasonable, similar to the standard applied in cases of negligent supervision. This becomes a concern when there is uncertainty about how the AI makes decisions or how it performs in real situations. Relying only on the AI without human review might show that the AI user did not meet the expected standard of care.[145]

Bathaee claims that even if an AI's medical diagnosis is accurate over half the time, unwavering deference might not be justified. Some patients could face too high a risk if the AI misdiagnoses. However, deferring to a model with less than 50% accuracy in diagnosing a disease might suffice if the condition being diagnosed is relatively minor, such as a common cold.[146]

Froomkin et al. argue that if a patient suffers harm because a healthcare professional relied on a predictive AI, the key factor in determining liability is whether that reliance was justified based on the AI's performance history or its adherence to medical standards. Essentially, if the AI's track record supports its use and it aligns with accepted care practices, the healthcare provider is more protected from liability.[147] Thus, the reasonableness of the healthcare provider's reliance on the AI is crucial in deciding liability.

2.7 *The Liability of the Healthcare Facility*

2.7.1 Arguments in Favour of and Against Hospital Liability

Under the corporate negligence theory, hospitals have an independent duty to protect the health of their patients. Prince and Cohen put forward three main arguments in favour of hospital liability instead of the manufacturer's liability.

First, hospitals have a much closer connection to patients than developers do.[148] While developers may know how the AI was built, hospitals possess detailed knowledge of their patients, workflows, and institutional capabilities. Hospitals are also best positioned to monitor the AI's performance during regular use, adapt the system to changes in patient demographics or illness patterns, and ensure that it aligns with their operational capacities. Under this view, hospitals are in the best position to recognise and address that gap.

Second, hospitals are already central to doctor malpractice cases.[149] They are best equipped to conduct systemic evaluations of how AI performs within

145 TerKonda and Fish, 2023.
146 Bathaee, 2020: 155.
147 Froomkin, Kerr and Pinea, 2019: 45.
148 Price II and Cohen, 2024: 360.
149 Price II and Cohen, 2024: 360–361.

their specific environment. While individual doctors may be involved in cases of misuse or injury, determining whether these incidents stem from improper AI use or broader systemic issues can be complicated. Assigning liability to hospitals simplifies this by focusing responsibility and incentives in one place. Moreover, hospitals are increasingly recognised for their role in coordinating patient care and managing AI systems to guide treatment is a natural extension of this responsibility.

Third, from an administrative perspective, concentrating liability in hospitals simplifies matters.[150] It can be challenging to separate different factors contributing to AI failures, such as inadequate training, poor data, or improper integration into clinical workflows. Additionally, patients are unlikely to be aware of who, between hospitals and developers, is responsible for AI errors behind the scenes. If they expect anyone to be accountable for their care, it is likely to be the hospital where they receive treatment.

However, some counterarguments must also be considered.

For a start, placing liability for AI-related injuries primarily on hospitals might deter the adoption of innovative AI technologies. Hospitals, concerned about the potential for lawsuits, may be hesitant to integrate new AI systems into patient care, fearing that any failure – whether due to system errors or human oversight – could lead to significant legal repercussions. This hesitancy could slow down the implementation of potentially life-saving technologies.

Moreover, hospitals often lack the in-depth technical understanding of how AI systems operate.[151] Unlike developers, who are intimately familiar with the design, algorithms, and limitations of the AI, hospitals may struggle to maintain, monitor, or troubleshoot these systems effectively. This knowledge gap could make it unfair to hold hospitals solely responsible for AI-related failures, as they may not be equipped to recognise or prevent technical issues that lead to harm.

Other counterarguments include the complexity of determining whether an error originated from the AI system itself, from human oversight, or from a combination of factors. In some cases, it may be unclear whether the hospital's failure to monitor the AI system or a flaw in the AI's design caused the damage. Placing liability solely on hospitals might oversimplify these complex interactions, potentially leaving developers less accountable for issues related to the AI's design or performance.

150 Price II and Cohen, 2024: 361.

151 Even Prince and Cohen seem to recognise this factor (Price II and Cohen, 2024: 363).

Lastly, there is the concern that centralising liability within hospitals could create an undue financial burden on healthcare institutions, further straining resources that could be better used for patient care.

2.7.2 Hospital Duties regarding AI Use

In addition to the direct interaction between the patient and human healthcare provider, and the respective responsibility of the latter, the responsibility of the healthcare institution (public or private) where the medical act is carried out may also arise.[152] Institutional duties encompass several dimensions, including the provision of adequate resources, ensuring a safe and sterile environment, and maintaining up-to-date equipment and technology.[153] Hospitals play a central role in promoting the appropriate advancement, integration, and oversight of AI systems. It is their responsibility to establish and uphold protocols and best practices for the use of AI technology in healthcare settings.[154] This entails providing comprehensive training programs, implementing standardised protocols, and disseminating best practices related to the use of AI. Hospitals should also prioritise patient education and ensure individuals are well-informed about the implementation and implications of AI in their care.[155] Additionally, hospitals should actively participate in the development of measures that enhance the reliability and robustness of AI systems, such as implementing procedures for simultaneous diagnosis and crosschecking by both doctors and AI algorithms.[156] By taking on these responsibilities, hospitals contribute to the responsible and effective integration of AI in healthcare.[157]

Court judgments and legal scholars have consistently upheld hospitals' obligation to uphold facilities and equipment that are secure and sufficient in their operations.[158] However, courts have yet to address a critical matter: whether healthcare facilities are obligated to conduct autonomous evaluations of the quality of AI processes and algorithms before doctors use them in inpatient

152 On institutional liability, Babic, Gerke, Evgeniou, et al., 2019: 1202–1204. See also Maliha, Gerke, Cohen, et al., 2021: 632–633.

153 Pallikarakis and Bliznakov, 2016: 47–61. On the liability of the hospital for providing inadequate equipment, see Ambrose v. Saint Joseph's Hosp. of Atlanta, Inc., 325 Ga. App. 557, 559 (1), 754 S.E.2d 135 (2014).

154 Rajpurkar, Chen, Banerjee, et al., 2022: 36.

155 Bhagat and Kanyal, 2024.

156 Varnosfaderani and Forouzanfar, 2024.

157 Schiff and Borenstein, 2019: 142–143.

158 For instance, Josephine A. Casco, Petitioner, vs. National Labor Relations Commission, Sixth Division, Capitol Medical Center and/or Thelma N. Clemente, Respondents, Supreme Court of Philippines, [G.R. No. 200571, February 19, 2018]. In legal doctrine see Parker, 1999: 1053–1066.

treatment.[159] As we project forward, it stands as a logical expectation that hospitals will also bear the responsibility of thoroughly evaluating the quality of AI tools in terms of both safety and efficacy.[160] According to the AIA, AI developers must comply with a very strict set of requirements, both technical and legal, in order to receive the CE marking of conformity and thus commercialise their products (see Chapter 3, section 2.4.2).[161] In theory, the law could require healthcare institutions to conduct a separate assessment before integrating AI systems into their healthcare delivery processes.[162] However, this option faces two caveats: i) it is not imposed by the AIA nor any other existing law;[163] ii) at present, the information required to make these evaluations and provide answers to these questions is largely unavailable.[164]

2.7.3 Direct Liability of the Healthcare Institutions

2.7.3.1 *Liability for Adopting Substandard AI Systems*

2.7.3.1.1 Scenarios Leading to Liability

Individual doctors, especially when employed by the hospital (it is slightly different when they work as independent contractors), rarely have the authority to purchase and use their own AI systems. Instead, it is the institution that selects, installs, trains, and operates the AI system, and thus bears direct or vicarious liability for any faults.[165]

One approach to establishing hospital liability concerning AI adoption can involve the concept of 'negligent credentialing'.[166] Just as hospitals are typically held liable if they fail to thoroughly review the credentials and past practices of healthcare workers and doctors, they might be expected to uphold a similar

159 Suppose, for example, that an AI system for diagnosing heart problems was trained based on data from a predominantly young and healthy population, whereas a different AI system, trained on another type of data, should be used for an elderly population. The hospital, however, only acquired the first one, using it for all sorts of patients. If the system fails to detect cardiac problems in older patients, the institution will be liable for negligence.

160 Greenberg, 2009: 438–439.

161 Gornet, 2024.

162 As suggested in Farhud and Zokaei, 2021.

163 Note that the AIA's obligations are addressed mostly to AI developers rather than to AI deployers (i.e., professional users).

164 Price II, 2015: 449.

165 Kamensky has expressed the view that liability may arise from inadequate scrutiny of AI before its implementation in clinical settings (Kamensky, 2020: 11). Hodge claims that such an outcome would not be unexpected, considering that hospitals already bear the responsibility of providing safe equipment and facilities (Hodge, 2022: 458).

166 Maliha, Gerke, Cohen, et al., 2021: 631.

duty when incorporating AI systems into their practices.[167] That is, a similar responsibility will likely apply to cases of AI systems that perform below what would be required.[168] In this context, hospitals will probably be burdened with the responsibility of examining past errors related to adverse events caused by the use of AI.[169]

This understanding is based on the recognition that it is up to the institutions to acquire material (technology, medication, instruments, furniture) that preserves patient safety. If such material has been purchased below standard or has not been well preserved and maintained, the responsibility may lie with the institution. Outside of direct doctor-patient interaction, healthcare systems, hospitals and practices are also responsible for the well-being of patients. This way, hospitals can be encouraged to exercise due diligence in selecting AI technologies and ensuring their effective use, thus minimising the potential for legal repercussions, assessing the certification process, and ensuring a seamless integration of AI with the existing workforce, much like the process of hiring a human employee.[170] To establish liability, a plaintiff needs to demonstrate that the defendant hospital did not exercise reasonable care, which is the level of care typically exercised by an average hospital, in determining the doctor's competence.

Likewise, the hospital's authority and oversight in acquiring and evaluating medical AI systems, managing its workforce, and ensuring the effective collaboration between AI technology and human staff make it an ideal case study for demonstrating the advantages of enterprise liability. The proper integration of AI into healthcare transcends the limitations of individual doctor practice, medical specialities, departments, and even the professional divisions between doctors, nurses, and other healthcare workers. Hospital coordination is essential, especially considering the long-term perspective, unlike individual doctors, who may not have the same ability to factor in these long-term investments.[171]

---

167 This theory places a responsibility on hospital systems to assess the competence and performance history of doctors both during admission to the medical staff and on a regular basis, usually every two years thereafter.

168 Price II, 2018: 295–306.

169 Licensing agreements between AI vendors and healthcare organisations will play a pivotal role in assigning liability and providing indemnification as the landscape of AI-assisted healthcare continues to evolve.

170 Lee and Yoon, 2021.

171 Price II and Cohen 2024: 359.

2.7.3.1.2 Challenges to this Theory

There are, however, some problems with this theory.[172]

First, there is the requirement of systematic control. In the case of AI, examination cannot be a one-time affair but needs to be continuous or periodically conducted.[173] Requiring hospitals to supervise each AI recommendation made by doctors in real-time seems impractical and might overly deter the adoption of medical AI.[174]

Secondly, determining the extent of the duty imposed on hospitals to investigate the safety of medical AI is a complex matter.[175] It remains uncertain whether such an obligation promotes or hinders public policy, because if hospitals were to duplicate the efforts of the authorities in charge of approving AI systems, it could lead to inaccurate inferences and ineffective safety enhancement.

Thirdly, another challenge lies in determining negligence based on comparisons with the degree of care exercised by other hospital systems, especially in the early stages of AI integration in healthcare. As AI technology is still relatively new in the medical field, there is a lack of standardised practices and benchmarks for its use.[176] This variability complicates the process of establishing what constitutes reasonable care and due diligence when implementing AI systems. Moreover, the novelty of AI in healthcare means that legal precedents are still evolving. Courts and regulatory bodies are in the process of developing frameworks to address issues of liability and negligence concerning AI,[177] making it difficult to define the hospitals' obligations in this regard.

2.7.3.1.3 The Use of Sub-Optimal AI Systems

Tricky situations are those in which a hospital uses a sub-optimal or obsolete AI system to provide advice and where there is a higher performing, more accurate (and usually more expensive) system on the market. The challenge here lies in the fact that the system is not itself defective, but it is a low-performing one when compared with other products.

In instances where healthcare institutions persist with sub-optimal AI systems while better alternatives are available, culpability lies with the organisation rather than the staff or the manufacturer. The failure to update technology

172 Pointed out by Price II, Gerke and Cohen, 2024: 150–166.
173 Procter, Tolmie and Rouncefield, 2023: 3.
174 Shachar and Gerke (2023: 283) make a similar remark with regard to the identification of AI biases in healthcare provision.
175 Hall, Shuck and Canlas, 2020: 436–445.
176 Krishnan, Singh, Pathania, et al., 2023.
177 Knoedler, Vogt, Alfertshofer, et al., 2024.

in such cases leads to errors attributable to the organisation rather than any defect of the product, based on a well-defined duty to provide safe equipment and facilities and to train their staff to use equipment provided by the health system.[178]

2.7.3.1.4 AI Validation

Even disregarding the potential obligation of the hospital to test AI systems, the healthcare facilities must still consider the sub-obligations of AI validation: the hospital must verify that the AI tool has been rigorously tested and meets regulatory standards before it is deployed.[179] This differs from the obligation previously described; here, I am asserting that the hospital must verify that the AI systems have been tested, whereas the previously mentioned obligation required the hospital itself to conduct the testing of the AI system.

2.7.3.2 *Liability for Lack of AI Maintenance*

AI systems in healthcare, much like any other technology, require regular maintenance to function correctly.[180] Failure to perform these tasks can result in malfunctions, outdated recommendations, or inaccurate diagnoses, all of which can compromise patient safety.[181]

This duty includes, for instance, software updates, as AI systems often receive updates to improve functionality, fix bugs, or enhance security, and to reflect the latest medical guidelines and research findings.[182] Neglecting these updates can leave the system vulnerable to errors or cyber-attacks, which can directly affect patient care. Algorithm recalibration is another dimension of this duty. Medical knowledge and guidelines are constantly evolving and, as such, AI algorithms must be recalibrated to reflect these changes. If a hospital fails to recalibrate, the AI may provide outdated or incorrect medical advice. Hardware maintenance is also an issue.[183] The physical components of AI systems, such as servers and diagnostic machines, need regular checks and maintenance because hardware failures can disrupt the system's functionality, leading to potential delays or errors in patient care.

---

178 In the US case law see, for instance, Berg v. United States, 806 F2d, 978 (10th Cir 1986).
179 High-Level Expert Group on AI 2019: 22.
180 Mahmood, Chan, Drukker, et al., 2024.
181 Li, Mao and Zhang, 2022.
182 Nagwani and Suri, 2023.
183 Zamzam, Hasikin and Wahab, 2023.

2.7.3.3 *Liability for Lack of AI Oversight*

The hospital must provide adequate oversight of the AI system's functioning.[184] The hospital should implement ongoing monitoring mechanisms to regularly assess the performance and accuracy of AI systems. This includes tracking outcomes, error rates, and any anomalies that could impact patient care. It is essential to define who within the hospital is responsible for overseeing the AI systems. It could also involve creating dedicated roles or teams tasked with managing AI operations, ensuring accountability, and providing a clear point of contact for related issues. It is difficult to envision a single doctor being responsible for that level of monitoring. Large hospital systems, however, are much more likely to take on such responsibilities.[185]

Moreover, hospitals need to identify and mitigate risks associated with AI, such as biases in decision-making, data breaches, or technical failures. A robust risk management framework should be in place to address potential issues swiftly.[186]

Likewise, establishing feedback mechanisms through which healthcare professionals can report on the AI's performance is crucial.[187] In practical terms, this entails an organisational obligation for hospitals to implement structured reporting channels, ensure that such feedback is systematically reviewed by designated clinical and technical governance bodies, and require that identified issues trigger documented corrective measures, updates, or, where necessary, the suspension of the AI system's use.

2.7.3.4 *Liability for Lack of Staff Training*

Hospital duties extend beyond merely providing facilities and equipment; they include ensuring that the healthcare providers who work within their walls are competent and diligent.[188] Hospitals must rigorously vet and monitor the qualifications and ongoing training of their staff,[189] especially when new and potentially complex technologies are involved, and develop robust protocols for the introduction and use of AI systems, ensuring that all medical staff are thoroughly trained and competent in these new tools.[190]

For instance, if a hospital allows a doctor who has not received proper training on a sophisticated AI diagnostic tool to use it on patients, and this results

184 World Health Organization, 2021: 95.
185 Price II and Cohen, 2024: 360.
186 Božić, 2023: 63–80.
187 Tanaka, 2023.
188 Terry and Wiley, 2016: 77.
189 Esmaeilzadeh, 2024; Lambert, Madi, Sopka, 2023: 1–14.
190 Li, Mao and Zhang, 2022.

in harm, the hospital could be found liable. This liability arises because the hospital failed to uphold its duty of care by entrusting critical healthcare tasks to someone who was not adequately prepared to perform them safely.

In the case of Aldoroty v. HCA Health Services of Kansas, Inc.,[191] the court stressed the significant responsibility that hospitals bear in ensuring patient safety, particularly concerning the use of advanced technologies like AI systems and emphasised that hospitals could be held liable if they permit inadequately trained doctors to utilise AI systems, pointing to the broader principle of corporate negligence. The Aldoroty case reinforces the need for stringent oversight and comprehensive training programs. This includes continuous education and assessments to keep up with advancements and updates in AI technology.[192] The court highlighted that a hospital, under the corporate negligence theory, possesses an independent duty to safeguard the health of its patients by refraining from entrusting healthcare tasks to an independent contractor/doctor lacking competence and diligence.

This responsibility extends to providing clear guidelines to doctors regarding the proper use of these technologies to avoid errors that may result in legal consequences for both the hospital and the doctors involved.[193] The primary goal of these guidelines is to help doctors understand the appropriate use of technologies in their day-to-day practice, including specific instructions on operation, monitoring, and troubleshooting.

2.7.4 Vicarious Liability

2.7.4.1 *The* Respondeat Superior *Doctrine*

2.7.4.1.1 The *Respondeat Superior* Doctrine in Healthcare Delivery

It may happen that the institution is held responsible, not for something it has directly done or failed to do, but for the simple fact that healthcare is provided for its benefit and subject to its (general) guidance and to the principles of agency.[194] According to this principle, an employer can be held civilly liable for injuries caused by an employee's negligence within the scope of employment. This concept is known as vicarious liability,[195] but it is often referred

191 Aldoroty v. HCA Health Services of Kansas, Inc., 265 Kan. 666, 962 P.2d 501(1998).

192 'Under the corporate negligence theory, a hospital has an independent duty to its patients to ensure their health by not entrusting the work of health care to an independent contractor/physician who is not competent and careful' (Aldoroty v. HCA Health Servs. Kan., Inc., 962 P.2d 501, 511 (Kan. 1998)).

193 Alowais, Alghamdi, Alsuhebany, al., 2023; Esmaeilzadeh, 2024.

194 Smith, Stepan, Valdmanis, et al., 1997: 37–60; Shenoy, Shenoy, Shenoy, 2021.

195 Shenoy, Shenoy, Shenoy, 2021. Thompson, 1994: 1–42.

to as 'respondeat superior' or 'let the master answer'.[196] The rationale behind this imposition of liability is the idea of 'enterprise liability', where the entity that benefits from the actions of its employees assumes the risks and losses. Typically, such entities have more financial resources to compensate injured parties, which is also a compelling reason to ground it.[197]

For an agency relationship to give rise to vicarious liability, three elements must be present: (1) the parties must agree to an arrangement where the agent voluntarily acts for the principal's benefit and under their direction and control; (2) the agent must be acting on behalf of the principal within the scope of employment; and (3) the principal must have the power to control and direct the agent's work.[198] Courts have consistently interpreted vicarious liability broadly, extending liability to employers even in cases where the worker is not a direct employee. For instance, courts have used the notions of apparent or ostensible authority to attribute liability to hospitals, even when the doctor involved is an independent contractor.[199] Apparent authority arises when a principal represents, through explicit or implicit means – such as verbal communication or conduct – that another individual is acting as their agent. This representation can be addressed to the third party directly or communicated to the public through various means, such as signs or advertising.[200]

2.7.4.1.2 The *Respondeat Superior* Doctrine Applied to Hospitals and AI

In the medical field, hospitals can be held vicariously liable for the malpractice of their employees, including doctors. However, when it comes to AI, the question arises as to whether an AI system can be considered an 'agent' or 'employee' of a hospital for the purpose of assigning liability to the hospital as the 'principal'.[201] This question is challenging to answer because vicarious liability is based on agency theory, which assumes that the principal (in this case, the hospital) has control or power over the agent (such as an employee or potentially an AI system).[202] Determining whether an AI system can be considered an agent or employee of a medical facility, thus enabling the application of the *respondeat superior* theory, poses a challenging question. First,

196 Thompson, 1994: 1–42.
A current appraisal of this theory in Silver, 2024: 89–108.
197 Keating, 2022.
198 Thompson, 1994: 1–42.
199 Williams, 2015: 216, 226.
200 Hodge, 2022: 459–460.
201 Schweikart, 2021: 16.
202 Schweikart, 2021: 16.

agency theory assumes that an agent is a legal person, which does not apply to AI under the current legal framework. Second, considering the autonomous nature of AI systems,[203] many are designed to operate independently. Consequently, if it is judicially established that AI is completely autonomous, holding a medical facility vicariously liable for the harm caused by the use of AI becomes arduous, as the autonomous AI system operates outside the realm of control by the principal.[204]

Chinen suggests a distinction in this regard.[205] He analysed the level of sophistication of a machine, observing that individual human users dealing with less advanced machines (as opposed to fully autonomous ones) have much more control over it, which allows for a clear allocation of responsibility for any harm caused by such a tool under this theory. On the contrary, if a court determines an AI to be fully autonomous or, if not autonomous, under the dominion of its designers rather than the hospital that acquired and used it, holding the hospital vicariously accountable for any injuries caused by the AI becomes impracticable, as an autonomous AI would effectively operate beyond the control of the principal.

As seductive as this theory might seem, it has its flaws.

The theory hinges on a sharp distinction between 'less advanced' machines under human control and 'fully autonomous' machines beyond control. This binary approach oversimplifies the spectrum of AI capabilities and fails to account for the complex interactions between human operators and AI systems. Many AI systems operate in a semi-autonomous state where human oversight, decision-making, and control vary in degrees. This nuanced reality challenges the clear-cut responsibility allocation proposed by Chinen.

Moreover, it lacks a clear and practical definition of what constitutes 'full autonomy' in AI. Autonomy in AI is not an absolute state but rather a continuum that involves various levels of independence in decision-making, learning, and adaptation. Without precise criteria to determine when an AI is autonomous enough to absolve users or institutions of responsibility, the theory may lead to inconsistent or arbitrary legal judgments.

Last but not least, the theory suggests that responsibility shifts entirely away from the hospital to the AI's designers when an AI is deemed autonomous. However, hospitals are responsible for integrating, maintaining, and overseeing the use of AI systems in their operations. A blanket removal of responsibility

203 Schweikart, 2021: 16.
204 Chinen, 2016: 360–361.
205 Chinen, 2016: 360–361.

from the hospital could ignore crucial aspects such as the oversight of AI deployment, updates, and usage policies that are essential in preventing harm.

2.7.4.2 *The Chilling Effect of* Respondeat Superior *on AI Adoption*

Another form of liability can take place when hospitals refuse to adopt AI, fearing liability under the principle of *respondeat superior*. Indeed, the court's interpretation of the *respondeat superior* doctrine could significantly impact hospitals' decisions to invest in AI equipment. A hospital's aversion to risk might deter it from adopting new technologies if it fears bearing the consequences of any errors. The application of this doctrine could create an environment where hospitals and clinics are hesitant to invest in new technologies due to the potential for expensive litigation.[206] Given that the hospital bears the responsibility to ensure the technology's efficacy and the proper training of its staff in its use, the *respondeat superior* doctrine could potentially foster an atmosphere of reluctance among healthcare institutions to invest in innovative technologies, fearing expensive legal actions.

In this regard, Eldakak et al. make a distinction. They claim that the application of this doctrine appears highly compelling when dealing with dangerous products capable of causing severe bodily harm or fatalities. However, its relevance becomes less clear when considering medical devices, which are predominantly designed to improve treatment quality and minimise deaths and injuries. While it makes sense to apply stringent regulations to products that pose significant risks, the primary purpose of most medical devices is to enhance patient outcomes and ensure safety. Consequently, the rationale for imposing the same level of strict liability on medical devices may not be as strong, given their intended role in healthcare.[207]

In conclusion, while the principle of *respondeat superior* aims to protect patients by holding institutions accountable, it may also inadvertently slow down the adoption of potentially life-saving AI technologies in healthcare. To strike a balance, there may be a need for legal reforms that encourage the safe and responsible integration of AI in healthcare without exposing hospitals to prohibitive risks.

---

206 See, 2021: 429.

207 Eldakak, Alremeithi, Dahiyat, et al. 2024: 2.

3 Product Liability

3.1 *AI as a Product*

3.1.1 The Dichotomy Product versus Service

To address compensation issues related to patient harm, it is essential to determine whether the AI system functions as an independent healthcare provider on its own merit (akin to a human healthcare provider), or if it is simply a tool assisting the doctor in delivering patient care. The argument underlying the manufacturer's liability is grounded in the second perspective.[208]

Another possible categorisation of the kind of liability involving harm caused by AI lies in the cause of such harm. Under this framework, when an AI machine makes errors in its learned judgment based on ML principles, similar to an error from a human doctor, it should be held accountable under medical malpractice liability standards, potentially falling under the purview of state Medical Boards for regulation.[209] However, if the fault lies with other aspects such as manufacturing defects, design flaws, or inadequate warnings associated with the AI machine, the principles of traditional product liability should apply.[210]

Marchant and Tournas provide the following example involving an AI system designed to formulate therapeutic plans. Suppose the system makes a diagnosis for a specific type of cancer and formulates a treatment plan that ultimately proves to be incorrect, leading to harm to the patient. In such a case, the decisions made by the system should be assessed based on the negligence framework. For these authors, the machine itself is not held accountable for every instance of harm but rather for situations where the standard of care falls below that of a human doctor. This approach would enable patients to receive appropriate compensation when the expected level of care is not met, regardless of whether the professional judgment was made by a human or a

208 In the US, for instance, computer code has been viewed as a service rather than a product, which creates ambiguity regarding the application of design claims to programmers (Vladeck, 2014: 132).

The focus of product liability regimes typically centres on safeguarding consumers who experience harm due to defective products. However, in the case of AI systems used in healthcare settings to advise and assist doctors, the patient rarely acts as the direct consumer. Instead, it is the hospital or clinic that assumes the role of the consumer. This distinction raises unique challenges in assigning liability when AI systems provide guidance in healthcare settings (Holm, Stanton and Bartlett, 2021: 181).

209 TerKonda and Fish, 2023.

210 Marchant and Tournas, 2019: 38.

machine. Conversely, if the AI system is found to have a poorly constructed component that directly causes harm to the patient, the evaluation of liability would be conducted under the principles of product liability. In this context, the system manufacturer would be subject to strict liability, wherein accountability is established irrespective of fault, based on factors such as manufacturing defects or substandard design. Therefore, depending on the nature of the harm and the factors involved, the legal analysis would differentiate between negligence, focusing on the standard of care, and product liability, emphasising strict liability principles.[211]

In the current state of the art, I find it difficult to equate AI with a medical professional, regardless of the task it performs. Perhaps in the future, with the development of general AI – where AI can truly mimic human intelligence – we might consider applying the criteria of personal negligence to assess its performance. Until then, however, the appropriate framework to apply should be product liability.

Still, classifying software as a product can be an intricate task, given its diverse types and forms. Software integral to the functionality of a physical product is relatively straightforward to categorise as part of that product. However, standalone software can pose challenges when it comes to classification.[212]

3.1.2 The US and the European Perspectives

In the US, while the Supreme Court has yet to address this concern in the domain of medical AI, lower courts have already deemed AI in other domains as a service rather than a tangible product. The prevalent perspective characterises algorithms primarily as services, thus prompting some courts to bypass the application of product liability principles to algorithms.[213] This approach has been particularly pronounced in healthcare software cases, possibly driven by a desire to foster innovation and progress in the healthcare sector.[214]

In turn, the FDA's approach to approving AI platforms is inconsistent, with some platforms being granted medical device status while others are denied such recognition. It appears that certain AI platforms, like the one developed by Google researchers for detecting diabetic retinopathy, have received FDA approval as medical devices. On the other hand, a similar AI platform developed

211 Marchant and Tournas, 2019: 39.

212 Wade, Abraham, Coder et al., 2023.

213 In the very comprehensive study of Cusenza (2024) on how US courts address the use of AI by the government, the issue of manufacturer liability never came up.

214 See, 2021: 432.

by IBM researchers was treated as a medical procedure or service and is being used by healthcare practitioners without undergoing FDA approval or oversight. This disparity in regulatory treatment raises questions about the appropriate classification of AI in the medical domain and highlights the need for clearer guidelines to ensure consistent and appropriate regulation.[215]

In the EU, the issue has been controversial up until now. While the 1985 Product Liability Directive[216] (1985 PLD) boasted a broad product definition of 'product',[217] there were doubts on whether AI (and software in general) could be considered as such.[218] The revised Product Liability Directive (PLD),[219] in turn, very clearly classifies AI as a product, putting an end to the discussion. According to Article 4(1) of the revised PLD, a 'product' means 'all movables, even if integrated into, or inter-connected with, another movable or an immovable; it includes electricity, digital manufacturing files, raw materials and software'.

3.2 *The Liability of the Manufacturer*

3.2.1 The Concept of Manufacturer

The concept of manufacturer, in this domain, may encompass different players.[220] In cases of defective AI products, several parties could potentially be held liable under the manufacturer's liability depending on the circumstances. Key candidates include AI developer/software providers,[221] the manufacturer

---

215 Alshanteer, 2020: 44.

216 Council Directive 85/374/EEC of 25 July 1985 on the approximation of the laws, regulations and administrative provisions of the Member States concerning liability for defective products.

217 Article 2 of Directive 85/374/EEC – 'For the purpose of this Directive "product" means all movables, with the exception of primary agricultural products and game, even though incorporated into another movable or into an immovable. "Primary agricultural products" means the products of the soil, of stock-farming and of fisheries, excluding products which have undergone initial processing. "Product" includes electricity'.

218 Health Action International, 2023: 1–2.

219 Directive (EU) 2024/2853 of the European Parliament and of the Council of 23 October 2024 on liability for defective products and repealing Council Directive 85/374/EEC (revised PLD).

220 Ulfbeck, 2020: 196–199.

221 If the defect lies in the algorithm, data processing, or programming, the developer responsible for the AI software could be liable. This might include flawed algorithms, failure to foresee potential risks, or inadequate testing before deployment.

of the hardware,[222] system integrators,[223] data suppliers,[224] end-product manufacturers,[225] maintenance providers,[226] distributors/retailers,[227] and designers.[228]

Although the producer is one of the main targets of liability under European law, they may still be able to pass on their responsibility to other players. Chopard and Musy, for instance, claim that the manufacturer of an AI system could potentially hold the doctor liable for civil liability in the event of a defective AI system,[229] by asserting that the medical outcome depended on how the system was being used.[230] Indeed, unlike other products, AI systems have the distinctive feature of learning along with their use, in such a way that their user –

---

222 If the defect is due to a hardware issue (e.g., sensors, processors, or mechanical parts), the manufacturer of those physical components could be held liable. For AI systems integrated into devices, any malfunction caused by the hardware could point to the hardware producer.

223 Companies or entities responsible for integrating AI into a product (e.g., an autonomous vehicle or robotic system) may also bear liability. The integrator is responsible for ensuring that all components, including AI, hardware, and other systems, work together correctly.

224 If the AI system is dependent on data (for example, AI models trained on third-party data) and the data qualify as part of the product itself, liability might extend to the supplier of defective or biased data that lead to faulty AI decisions.

225 In cases where the AI product is part of a larger system (e.g., a self-driving car), the manufacturer of the final product could be held responsible if they failed to ensure the proper functioning of the integrated AI technology.

226 Where updates amount to a functional reconfiguration, the maintainer may effectively assume responsibility for the modified system, because the updated system can be considered a 'new' product for liability purposes.

227 Depending on consumer protection laws, distributors and retailers that sell defective AI products may also face liability if they knowingly sold defective products or failed to disclose known risks.

228 If the defect stems from the original design of the AI system, whether in the AI software or the hardware, designers might be held liable for failing to properly foresee risks or vulnerabilities in the product.

229 Chopard and Musy, 2023.

230 Picture the case of a dermatologist who has a tendency to overdiagnose certain types of skin lesions as malignant due to personal biases and a history of erring on the side of caution. This results in a higher number of false positives in his diagnostic records. When the dermatologist uses an AI system to assist in his diagnosis, he systematically 'approves' AI outcomes that indicate a positive diagnosis of cancer. Consequently, the AI begins to overdiagnose benign lesions as malignant, leading to an increased number of false positives. This not only causes unnecessary anxiety and treatment for patients but also undermines trust in AI technology within the medical community. Therefore, the doctor's biased actions have a cascading effect, compromising the reliability and effectiveness of the AI system.

in this case, the healthcare professional – contributes to shaping the product.[231] This is in line with Article 11(1)(c) of the PLD, which excludes the manufacturer's liability when the defect only appeared later.[232] Whether the role of the human agent using AI should be equated to that of the manufacturer and lead to the same type of liability is, however, a huge legal stretch.

3.2.2 The Rationale behind the Manufacturer's Liability

In the EU (the same applies in the US and some other jurisdictions), producers are liable for damage caused by their products, even if they acted without fault. This is called strict liability, and it is especially important for products considered particularly risky. AI-based medical devices have been labelled by some as 'unavoidably unsafe products' under the argument that they are inherently designed in a manner that prevents complete safety.[233] This assertion has been used to justify the thesis that the manufacturer of these products will bear the responsibility for absorbing any economic losses resulting from harm caused by their usage.

However, this is a very narrow perspective on medical AI. AI often acts as a valuable tool with the potential to significantly enhance patient safety in healthcare, especially in areas where current strategies are ineffective. This recognition highlights the importance of integrating AI into healthcare.[234] While AI does come with its own risks, ignoring the potential of AI in medicine also carries significant risks. Framing manufacturer liability solely on the basis that AI-based medical devices are inherently risky overlooks this broader risk–benefit balance and misconstrues the rationale of product liability.

Another rationale behind holding the manufacturer liable lies in the belief that the latter is ideally positioned to design the AI technology in a manner that ensures safety and minimises harm to users.[235] However, considering the evolutionary nature of AI (especially ML), this argument does not hold. If AI

---

231 Javaid, Haleem, Singh, et al., 2022: 58–73.

232 'An economic operator referred to in Article 7 shall not be liable for damage caused by a defective product if that economic operator proves any of the following (...) that it is probable that the defectiveness that caused the damage did not exist when the product was placed on the market, put into service or, in respect of a distributor, made available on the market, or that this defectiveness came into being after that moment'.

233 See the decision Taylor v. Intuitive Surgical, Inc., 187 Wash.2d 743, 769 (2017), involving the Da Vinci robot, in accordance with the Restatement (Second) of Torts: Special Liability of Seller of Product for Physical Harm to User or Consumer § 402A (AM. LAW INST. 1965), Comment K.

234 Bates, Levine, Syrowatka, et al., 2021: 1–8.

235 Buiten, De Streel and Peitz, 2023.

changes its performance over time, eventually 'behaving' in unpredictable ways (depending on the inputs it receives), how can the manufacturer be held accountable, inasmuch as they cannot control such inputs?

3.2.3 The Learned Intermediary Doctrine

The learned intermediary doctrine posits that manufacturers of prescription drugs or medical devices are responsible for informing the prescribing healthcare provider about the correct usage and potential risks of their products, rather than directly advising the patient or the general public.[236] This principle is based on two assumptions: i) that the prescribing doctor is best positioned to assess the appropriateness of a treatment and to communicate relevant warnings; ii) that manufacturers often lack practical ways to engage directly with patients, as doing so could disrupt the doctor-patient relationship.

Under this doctrine, if medical AI is considered a product, the manufacturer might deploy the learned intermediary doctrine as a protective measure against liability,[237] thus, exempting the manufacturer from bearing any direct obligation to the patient.[238] Manufacturers can fulfil their obligation to caution against potential product risks by directing their warnings to doctors, who are regarded as the ultimate consumers. Should a doctor then neglect to appropriately inform a patient and comprehensively disclose the product's risks and advantages, it is the doctor who will bear the burden of liability. Consequently, doctors are tasked with informing patients about all potential risks and side effects, which also underscores the significance of transparent communication between medical professionals and patients.[239]

This legal strategy allows manufacturers to avoid potential liability stemming from information flaws, as any failure by a doctor to adequately communicate risks and benefits to the patient would make them liable rather than the manufacturer.[240] On the doctor's side, in turn, there is an increased burden, as the learned intermediary doctrine exerts significant pressure on doctors to carefully assess the potential risks and benefits associated with specific AI applications.[241]

Moreover, when doctors conclude that the anticipated risks outweigh the expected benefits, they might choose to refrain from employing the product to

236 Hill, 2005: 362–379.
237 Plaza, 2020: 393.
238 Husgen, 2014: 479–480.
239 Contaldo, Pasceri, Vignati, et al., 2024.
240 Husgen, 2014: 480–481.
241 Thornton 2003: 359–361.

mitigate potential liability concerns. Conversely, if doctors decide to proceed despite being aware of the risks, they will bear responsibility for any resulting adverse outcomes.

This legal doctrine operates on the premise that healthcare providers are better positioned than manufacturers to communicate product risks to patients. However, the advent of digital healthcare and telemedicine is reshaping this landscape.[242] With doctor-patient interactions increasingly occurring through online channels after prescriptions are issued, the traditional concept of an 'intermediary' is undergoing transformation. Doctor warnings, in the context of digital interactions, begin to resemble the information available on product labels.[243] This shift raises essential questions about the efficacy and applicability of the learned intermediary doctrine.[244] In this context, striking a balance between online medical interactions, patient awareness, and the responsibility to communicate potential risks is becoming increasingly challenging.

In contrast, the learned intermediary doctrine never gained significant traction in Europe.[245] European legal systems place more direct responsibility on manufacturers to ensure that consumers are properly informed about the risks associated with their products. Even though doctors may also be held liable for failure to comply with informed consent impositions,[246] their liability is not a replacement for the manufacturer's liability.

3.3 *Defective AI Products*

Defective AI systems can be thought of as faulty software artefacts that do not work as intended. Useful distinctions exist between design errors (types) and operational errors (tokens) and, more importantly, between failure to operate as intended (dysfunction) and unintended side effects (malfunctions).[247]

Malfunctions are distinguished from mere negative side effects via the concept of 'avoidability', or by analysing the extent to which comparable systems are able to fulfil their intended functions without these negative effects.[248] Flawed programming introduces one of the most interesting aspects, as it

242 Stoumpos, Kitsios and Talias, 2023.

243 Karwaki, 2020: 538.

244 Schweikart, 2021: 15.

245 Duffourc and Gerke, 2023: 44.

246 The connection between the learned intermediary doctrine and the medical duty to comply with informed consent requirements in Simon and Kesselheim, 2024: 114.

247 Mittelstadt, 2021: 20.

248 Mittelstadt, 2021: 20.

raises the possibility that a loss of control could stem directly from a conscious design choice that produces unintended consequences.[249]

This phenomenon can occur when developers, while attempting to create a system that is robust and autonomous, inadvertently embed errors or biases within the AI's algorithms. These flaws can manifest in various ways, from incorrect data inputs and improper handling of edge cases to the reinforcement of pre-existing biases within the training data. Moreover, flawed programming can lead to a scenario where the AI system behaves unpredictably or in a manner that deviates from its intended purpose.[250] For instance, an AI system used in medical diagnostics might be programmed to prioritise speed over accuracy in analysing patient data. While the intention might be to provide quicker diagnoses, the result could be a higher rate of false positives or negatives, potentially leading to misdiagnoses and inappropriate treatments. These examples highlight the critical importance of rigorous testing, validation, and ethical considerations in the design and deployment of AI systems. Developers must ensure that their conscious design choices are thoroughly examined for potential unintended consequences and that robust mechanisms are in place to detect and rectify flaws before they can cause harm.[251]

3.4 *The PLD and the AILD*

3.4.1 A Comparison between the Two Models

The EU is developing legal frameworks centred on the harmed individual.[252] That is, when a particular technology makes it harder to prove liability beyond what could reasonably be expected, particularly concerning the link of causation and the presence of flaws or deficiencies in the product, there should be a provision for aiding victims with the burden of proof.[253] However, the consequences that these legal regimes will have on the development of innovative AI products is something that also deserves careful consideration.[254]

---

249 Schwartz, Schwartz, Vassilev, et al., 2022: 25.

250 Williams and Yampolskiy, 2021.

251 Buiten, 2024 :239–273.

252 'For situations exposing third parties to an increased risk of harm, compulsory liability insurance could give victims better access to compensation and protect potential tortfeasors against the risk of liability' – European Parliament – Expert Group on Liability and New Technologies, 2019: 4.

253 European Parliament – Expert Group on Liability and New Technologies, 2019: 48–57.

254 Buiten, De Streel and Peitz, 2023.

The EU decided to address the issue in two ways: by drafting a brand-new directive specifically targeting AI (known as the AILD Proposal)[255] and by revising the PLD, which now includes AI products. Meanwhile, the proposal for the AILD was withdrawn, although it remains unclear whether this is a definitive decision or merely a temporary pause. Anticipating the possibility that the proposal may return to the legislative agenda – or that a similar instrument may be introduced in its place – this book offers an analysis of the proposed legal framework. The considerations regarding the legal solutions embedded in the AILD remain not only significant in light of the rise of a similar regime, but also help to clarify several issues involved in liability for AI-related harm.

A crucial issue arises concerning the delineation of the scopes of these directives, as they are intended to be distinct and non-overlapping.[256] The AILD proposal applied solely to AI systems, whereas the PLD extends its reach to encompass software and products in general. Interestingly, some companies claim to use AI when they are actually using standard software or slightly more complex algorithms, falling short of true AI.[257] This discrepancy could potentially mislead victims when seeking compensation. They may choose the wrong compensation regime due to the confusion surrounding the classification of the technology involved. The broad scope of the PLD aims to address this issue.

Besides their scope, other differences arise between the two. The PLD is based on a strict liability regime that disregards the existence of culpability. In contrast, the AILD proposal adopted a liability approach based on fault. This implied that individuals who had suffered harm due to an AI system had to demonstrate that the operator of the AI system acted negligently or with intent while fulfilling their duty of care to claim compensation for the damages incurred.

While the PLD introduces amendments to both material and procedural aspects of product liability law, the AILD proposal focused solely on procedural matters. It aimed to harmonise rules pertaining to disclosure and the

---

255 Proposal for a Directive of the European Parliament and of the Council on adapting non-contractual civil liability rules to artificial intelligence (AI Liability Directive), Brussels, 28.9.2022. COM(2022) 496 final, 2022/0303(COD), https://eur-lex.europa.eu/legal-content/EN/TXT/?uri=CELEX%3A52022PC0496.

See Whittam (2022: 249–265), arguing in favour of a specific targeted legislation for liability within AI, due to the difficulties of the claimant to demonstrate the classic liability requirements.

256 On the distinction between these two legal regime and their different material scopes, see Hacker, 2022; Madiega, 2023.

257 Woollacott, 2024.

burden of proof in cases involving fault-based liability related to AI systems, as dictated by the laws of Member States.

The target defendants also differ. The PLD does not provide consumers with the option to seek compensation from the business user for harm resulting from an AI system. While the proposal does outline a roster of accountable economic actors typically involved in the supply chain, such as manufacturers, importers, authorised representatives, and distributors, it does not establish the liability of professional users employing the product. The AILD proposal, in turn, was targeted at providers, operators, or users of AI systems, as defined in the AIA.

The types of harm being addressed are another relevant difference. Under the PLD, the coverage extends to a closed list of damage categories caused by AI systems, encompassing harm to property, loss of life or personal injury, and data loss (Article 6(1) PLD). Non-material losses are covered only insofar as they result from these forms of damage and are compensable under national law (Article 6(2) PLD). Infringements on privacy, harm to one's reputation, or enduring a state of uncertainty as autonomous heads of damage do not fall within the ambit of the PLD. This limitation is especially noteworthy, given the propensity of AI systems to give rise to non-material damages. In contrast, because such restrictions were absent from the AILD proposal, one can assume that it covered all types of harm.

Both proposals share the common goal of ensuring effective compensation for individuals harmed by AI systems and preventing liability gaps that may arise due to the unique characteristics of AI technology.[258] The aim is to provide AI system victims with the same level of protection and ease of compensation as those injured in scenarios not involving AI.

3.4.2 A Plaintiff-Friendlier Regime

The strict liability regime, allied with the reversion of the burden of proof, could have significant positive effects on consumer confidence and the growth of the AI market.[259] Presently, patients may find it daunting and cumbersome to access and comprehend technical data required to establish a defect and causal link in AI-related damage claims.[260] This complexity undermines

258 It is not clear whether a plaintiff could use both directives at the same time. Hacker (2022: 8) says this could be a possibility and indeed there is no provision in any of the directives excluding it.

259 Whittam, 2022: 259.

260 Whittam, 2022: 251–253.

patient-consumer trust in AI products, leading to a decline in consumer willingness to invest in them. If the burden of proof is placed on producers and/or professional users, patients will feel more secure and protected, knowing that producers must demonstrate the safety and quality of their AI systems.[261] This increased confidence can translate into higher demand for AI products,[262] resulting in a boost to producers' profit margins due to consumers' willingness to pay more for products with enhanced legal protection, and addressing the existing informational asymmetry between AI manufacturers and those affected by it.[263]

It must be acknowledged that the traditional paradigm of liability – mostly focused on intent and causation – does not work well when applied to AI.[264] Consider the outcome generated by an AI model assigned to conduct patient diagnoses. These outputs constitute recommendations akin to decisions stemming from human judgment. Yet, there is a notable contrast, as no human presence exists to elucidate the intricate decision-making process that underpins these outputs. Unlike human decisions with a transparent intent, AI recommendations can demonstrate incongruity between their creators' design intentions and operational outcomes. It is necessary to create a different model of liability to help plaintiffs deal with this issue. It remains to be seen, however, whether the models put forward by the PLD and the (meanwhile withdrawn) AILD proposal are the most suitable ones.

3.5 *The PLD*

3.5.1 The Adequacy of the Strict Liability Regime

The PLD establishes a framework of strict liability, eliminating the need to prove fault.[265] It is commonly claimed that a liability system for AI that does not depend on establishing fault serves as a powerful deterrent, motivating

261 Group of the European Parliament, 2024.

262 Oyekunle, Matthew, Preston, et al., 2024: 94.

263 Whittam, 2022: 258–259.

264 Bathae, 2018: 890–891.

265 This is clearly declared in Recitals 2 and 3 of the PLD.
In reality, fault considerations resurface in various norms of the PLD, and similarly of the repealed 1985 PLD, especially when determining defectiveness (Raposo, 2018: 121–122). Nonetheless, this notion of fault does not pertain to intent or negligence but rather to an objective breach of duty. In this regard see, amongst others, Zech, 2021: 154; Raposo, 2018: 72.

companies to focus on creating inherently safe technology.[266] Proponents of strict liability argue that it ensures a higher standard of care by holding manufacturers directly responsible for the safety of their products.[267] They contend that this legal framework incentivises developers to rigorously test and validate their AI systems, prioritise patient safety, and swiftly address any identified risks or defects. Such an approach would play a decisive role in fostering social trust and acceptance of this technology.[268]

While there are advantages to implementing a strict liability regime in the context of AI, it is crucial to acknowledge its drawbacks. Yavar Bathaee highlights numerous flaws inherent in applying strict liability to AI-related damages. Bathaee's argument gains weight when we consider the AI black-box problem.[269] The author contends that strict liability assumes that the creators of AI systems can foresee and adjust for potential harm caused by their software and that they possess a certain level of control or predictive capacity over the decision-making process, which directly impacts the extent of physical damage inflicted on the world.[270] Without this control or predictive ability – which is difficult to guarantee in AI products – obtaining adequate insurance becomes challenging, rendering strict liability an inappropriate framework. Moreover, according to Bathaee, instituting a strict liability regime in the AI domain will create substantial entry barriers for companies venturing into the AI market. In practical terms, companies will carefully consider the potential repercussions of assuming responsibility for any harm caused by their AI entities.[271] Consequently, this can lead to a scenario where large corporations with the financial capacity to manage such liabilities can dominate the AI market. In contrast, smaller, newer companies might be hesitant to undertake the financial risks associated with developing unpredictable AI entities that could trigger strict liability. If this perspective holds true, it could exacerbate the existing trend towards monopolisation within the AI industry.

In my view, while strict liability is suitable for many consumer products, applying it to medical AI could present significant challenges. The inherent imperfections of this technology, even though it may outperform human doctors, make occasional errors unavoidable. Even the most advanced algorithms might result in substantial liability some of the time if subjected to strict

266 See, for instance, Lior, 2020: 1135 ff.
267 Shavell, 2018: 10–15.
268 Zech, 2021: 6.
269 Bathaee, 2018: 931.
270 Bathaee, 2018: 931–932.
271 Bathaee, 2018: 896.

liability regulations.[272] Implementing strict liability in this scenario will not yield positive results, as it could lead to a considerable slowdown or complete cessation of technology production. The fear of liability imposition could curtail development and production efforts, significantly impacting technological advancement.[273] Imposing strict liability for every misdiagnosis based on its accuracy rate would hinder the development of new – and much-needed – AI systems.[274]

Alternatively, socialising the risk (the so-called 'no-fault system') could be the solution.[275] This would not only provide victims of AI harm adequate compensation but would also not deter valuable technological innovation in this field. This is a system characterised by the distribution or spreading of risks across a larger group or society rather than confining it to a single individual or entity. The process involves the sharing or transfer of potential negative outcomes or hazards among members of a community, society, or a larger collective.[276] It is commonly advocated for medical liability[277] (and actually in place in some jurisdictions),[278] but it also works for manufacturer liability in some scenarios. Some jurisdictions have in place such a system for harm caused by pharmaceutical products,[279] and the same system could also work for harm caused by AI systems.

3.5.2 The Concept of 'Product' in the 1985 Directive on Defective Products

Originally conceived more than 40 years ago, the 1985 Directive on Defective Products was tailored to address conventional, tangible products, defining the 'product' as any movable good, along with electricity. Uncertainties surrounding the definition of a 'product' in light of the original version of the Directive left doubts about whether it encompasses software and other intangible entities.[280] An additional hurdle was its definition of products as movables, which posed uncertainty regarding the inclusion of AI software within its

272 Thomas, 2017.

273 Sullivan and Schweikart, 2019: 162.

274 Lai, 2021: 597–654.

275 Also in favour, Marchisio, 2021; Epstein, 2023: 85–89.

276 Mann, 1945: 43–57.

277 Mello, Studdert, Kachalia, et al., 2006: 459–492.

278 Raposo, 2016b: 57–69.

279 See, for instance, the Countermeasures Injury Compensation Program (CICP) Data, in the US (https://www.hrsa.gov/cicp/cicp-data).

280 Bertolini, 2020: 57.

scope.[281] While this definition worked when the AI is integrated into a movable object like a self-driving car, it raised questions about the coverage of purely software-based AI. Modern technologies often blend both software and hardware components or exist as intangible entities, such as cloud-based software. This posed challenges and it was generally claimed that the Directive did not account for liability concerning intangible products.[282]

The revised PLD, instead, holds manufacturers accountable for defects in both tangible and intangible components by expanding the definition of 'product' to encompass software and AI systems, whether they are embedded or standalone. Additionally, the new definition covers digital services when they are essential for the functionality of the interconnected or integrated product.[283]

3.5.3 The Concept of 'AI Defect' in the PLD

3.5.3.1 *AI Defective Product*

The crucial element triggering liability under the PLD is the defectiveness of the product, as defined in Article 7 of the PLD. A defect under this legal framework refers to a safety flaw where a product fails to meet the expected level of safety considering various circumstances,[284] including its presentation, intended use, and time of circulation.[285] Economic operators, such as manu-

281 Koch, Borghetti, Machnikowski, et al., 2022: 31–32.

282 Bertolini, 2020: 57.

Whittam illustrates this difficulty with the following example: a consumer sustains an injury due to a flaw in a connected pacemaker (Whittam, 2022: 253). The connected pacemaker serves as a medical device responsible for regulating a patient's heartbeat, encompassing both physical hardware and cloud-based software that together generates patient data. In this case, it seemed illogical that a consumer might be barred from seeking compensation under the PLD (the author was referring to the 1985 PLD) due to a flaw in the intangible software component, which could be viewed as a service rather than a product. Conversely, the consumer could indeed pursue a claim if the flaw stemmed from a defect in the pacemaker's physical hardware. This incongruity raised concerns about the PLD's adequacy in addressing the intricacies of contemporary technologies.

283 Montagnani, Najjar, Davola, 2024.

284 For example, imagine a medical device such as a pacemaker. If this pacemaker is prone to unexpected battery failures, which could lead to interruptions in its functioning, it would not meet the expected level of safety for patients and healthcare professionals, especially considering its vital function of regulating heart rhythm. This safety flaw, if discovered, can be considered a defect under the PLD, as the product does not meet the necessary safety expectations for an essential medical device. On the reasonable expectations of safety, see Grozdanovski, 2024: 74–78.

285 Borghetti, Silva, Fairgrieve, et al., 2021: 11–30.

facturers/developers, have the best ability to avoid defects that could lead to harm.[286] The producer is liable for the product's lack of safety, even in cases of reasonably expected misuse; in contrast, unforeseeable misuse falls outside the manufacturer's sphere of liability (Article 7(2)(b) PLD).[287]

Within the framework of the PLD, the determination of a defect hinges on the notion of reasonable safety expectations.[288] This concept encompasses a wide scope, thus requiring consideration of all circumstances of the case. The central inquiry revolves around defining what constitutes reasonable safety expectations for an automated product. This criterion must be an objective one, reflecting the perspective of an average individual's typical expectations. Given the recognised fact that automated products might not be flawless at all times, the public may not inherently anticipate a guarantee of absolute safety.

An additional difficulty arises due to different approaches to defining defects in AI systems, with anthropomorphic and system-oriented perspectives offering contrasting conclusions.[289] The anthropomorphic approach would require a judge to find a defect if an average human would not have made the same mistake (note that sometimes AI makes mistakes that would not occur in a purely human scenario).[290] On the other hand, the system-oriented approach evaluates the risk level of the entire AI system across all applicable devices, considering systemic malfunctions. Economic operators need to compare the AI model's performance with the *status quo* and with reasonable alternative AI model designs. Superhuman performance should not automatically imply non-defectiveness, as incentives for innovation and improvement are important. From both a technical perspective and from an injured person's perspective, it does not matter if the error would have been made by a human or not.

286 However, identifying defects in AI systems can be challenging, especially considering their potential ability to learn after being deployed, and thus the PLD now includes the assessment of this post-deployment learning ability (Article 7(2)(c) PLD).

287 Baram, 2007: 20.

288 Article 7(1) PLD – 'A product shall be considered defective where it does not provide the safety that a person is entitled to expect or that is required under Union or national law'. On the particular domain of AI, see Ulfbeck, 2020: 194–196; Polat, 2023b: 145–175; Buiten, 2024: 258–265.

289 Hacker, 2022: 22–25.

290 Hacker, Krestel, Grundmann, et al., 2020: 424, 428. If an AI model, on average, performs worse than humans in a given task, Hacker suggests that there should be a rebuttable presumption of defectiveness (Hacker, 2022: 24).

3.5.3.2 *The Challenge of the 'Defective Design'*

Imagine that a hospital implements an AI-driven diagnostic system to aid radiologists in interpreting medical imaging scans, such as X-rays or MRIs, to detect anomalies or signs of disease. The AI system is trained on a large dataset of images and has shown high accuracy in identifying common abnormalities. However, an unforeseen circumstance arises when the AI system encounters a rare or previously unseen condition that was not adequately represented in its training data. Within the framework of conventional product liability law, it would not be reasonable to hold the manufacturer accountable for not including such specific information in the software, as it cannot be considered a design flaw to account for every exceptional circumstance. This is, however, a design limitation,[291] and we can – and should – establish rules that encourage manufacturers to minimise design limitations through increased investments in research and development (R&D) and safety updates.[292]

On the other hand, it must be acknowledged that designers of AI technology cannot accurately predict its behaviour once deployed in real-world medical settings.[293] Unpredictability arises from several factors inherent to both AI technology and the complexity of real-world environments. Consider an AI system designed to assist radiologists in diagnosing lung conditions from X-ray images. During testing in controlled environments, the AI shows high accuracy. However, once deployed in a busy hospital setting, the AI encounters a higher prevalence of comorbidities and varying image qualities. Additionally, radiologists might use the system differently than anticipated, perhaps relying too heavily on the AI's recommendations or ignoring it in cases where they disagree with its findings. These factors lead to the AI making more frequent errors than predicted, illustrating the challenges of anticipating real-world behaviour. This, in turn, is related to the "responsibility gap', already referred to in this book.[294]

3.5.3.3 *The Presumption of Defectiveness*

Trying to address these issues, the PLD established a presumption of defectiveness in its Article 10(2).[295] A product is considered defective if any of the following situations occur: if the defendant fails to provide important evidence

291 Hacker, 2023.
292 Guerra, Parisi and Pi 2022: 333.
293 Leslie, 2019: 26.
294 Matthias, 2004: 175–183.
295 Hacker, 2023.

they are required to disclose under Article 9(1); if the claimant demonstrates that the product does not comply with safety regulations established by Union or national law that are intended to prevent the type of damage that occurred; or if the claimant proves that the damage was caused by an obvious malfunction of the product during normal use.[296]

The primary legal consequence of this presumption is the increased likelihood that the court will find the defendant liable for damages caused by the defective product. The defendant must provide evidence to rebut the presumption, which can be challenging if the product clearly does not meet safety standards or even if they failed to maintain adequate documentation.[297]

Regulatory authorities may become involved if the defect stems from non-compliance with safety standards, potentially leading to penalties such as fines, product recalls, or the withdrawal of the product from the market.[298] The manufacturer's reputation could suffer as a result, leading to a loss of consumer trust and potential financial consequences. In some cases, failure to rebut the presumption of defectiveness can also open the door to additional claims from other consumers or affected parties, increasing the defendant's legal and financial exposure.

3.5.3.4 *The Development Risk Defence*

Within the product liability framework, there exists an exemption known as the development risk defence, which allows the producer to avoid liability by demonstrating that the defect in the product, later revealed, was not discoverable in light of the scientific and technical knowledge available at the time it was marketed. It creates a situation where producers of such AI systems may avoid liability if the defect is impossible to detect during the product's introduction.[299] Consequently, this could hinder injured parties from seeking rightful compensation.[300] It has also been said that the introduction of the development risk defence alters the original commitment to a strict liability regime by incorporating a degree of negligence. This deviation from strict liability principles may complicate liability assessment and

296 Grozdanovski, 2024: 78–80.
297 Hacker, 2023.
298 Hacker, 2023.
299 This should not be confused with the later-defect defence, which exempts manufacturers from liability if it is likely that the defect causing the damage did not exist when the product was put into circulation or emerged afterwards (Mazzini, 2019: 22–23), which will be discussed in the next section.
300 Whittam, 2022: 254.

protection for affected individuals.[301] The PLD addresses this issue explicitly, maintaining this defence (Article 11(1)(e) PLD).

3.5.3.5 *The Later-Defect Defence*

The later-defect defence is based on the claim that the producer cannot be held liable where it is able to demonstrate that the defect did not exist at the time the product was put into circulation, but rather arose only at a subsequent stage. This argument becomes particularly salient in relation to self-evolving products, such as AI systems, whose behaviour and performance may change over time as a result of learning, updates, or interaction with their operational environment.

There have been debates surrounding the applicability of strict liability to producers of self-adapting AI systems, with some arguing that the causal link between system defects and resulting damages may be disrupted. Bashayreh et al., for instance, maintain that strict liability should only apply if the AI system's adaptation is predictable and directly attributable to the producer.[302]

This position has been contested, with the argument that even in the case of self-adapting AI, producers can still be held strictly liable.[303] Hacker, for instance, notes that ML models generally do not continue learning autonomously after deployment, unless explicitly programmed to do so. Even within the current legal framework, defects that appear to arise after the system is on the market can often be traced back to the initial decisions made by the manufacturer before deployment, meaning that continued adaptation does not break the causal chain. Therefore, the later-defect defence should not apply to such scenarios.[304]

If the AI system requires training, it is expected that the system undergoes sufficient training during its development phase before market release. Consequently, the producer's decision to equip the AI system with self-adapting capabilities becomes a crucial factor in the equation. When an AI system adapts, it does so within the parameters set by the producer, and the producer holds the responsibility to ensure its safety. Therefore, the producer can control what the AI system considers in its learning and decision-making processes, safeguarding against undesirable information. Moreover, the producer is best positioned to assess the AI system's market readiness and accepts the risk in hopes of earning substantial profits.

---

301 Whittam, 2022: 254.
302 Bashayreh, Sibai and Tabbara, 2020: 173–174.
303 Whittam, 2022: 256; Cauffman, 2018: 530.
304 Hacker, 2023.

The PLD follows this path. It emphasises that manufacturers should bear liability for damages stemming from defects in their products, even if the defect arises from modifications made to the product, as long as they are within the producer's control post-market release.[305] Article 7(1)(c)[306] of the PLD acknowledges the relevance of the system's continued learning ability in assessing defectiveness. This clarification reinforces the interpretation that residual control over the product and its associated risks should be linked to continued liability. In light of Article 11(2)[307] of the PLD, economic operators remain liable if they have control over the product and the defect is connected to a related service, the software (including updates or upgrades), or the absence of necessary updates or upgrades for safety maintenance.

Imagine a company develops an AI-powered diagnostic tool designed to assist dermatologists in identifying skin cancers from images of skin lesions. After the product's initial release, the manufacturer rolls out a software update to enhance the AI's diagnostic accuracy and expand its capabilities. However, this update introduces a bug that causes the AI to misclassify certain types of benign moles as malignant, leading to unnecessary biopsies and treatments for patients. In this scenario, the defect in the AI diagnostic tool arises from the modification made to the product post-market release, specifically the software update issued by the manufacturer. As a result, under the PLD, liability can extend to the period following a product's release to the market if the manufacturer continues to exercise control over the product, such as by implementing software updates or utilising ML algorithms.[308]

In parallel, the PLD clearly states that any economic operator who significantly alters a product outside of the manufacturer's control will be held liable for any resulting defect, effectively treating that party as the manufacturer. For example, imagine a hospital acquires refurbished MRI machines from a third-party company that modifies the devices to enhance performance outside of the original manufacturer's specifications. If a defect in the refurbished MRI machine leads to a misdiagnosis or harm to a patient, the hospital or

305 Buiten, 2024: 266.

306 About this norm, Hacker, 2023.

307 'By way of derogation from paragraph 1, point (c), an economic operator shall not be exempted from liability, where the defectiveness of the product is due to any of the following, provided that it is within the manufacturer's control:
(a) a related service;
(b) software, including software updates or upgrades; or
(c) the lack of software updates or upgrades necessary to maintain safety.'
About this norm, Koch, Borghetti, Machnikowski, et al., 2022: 31–33.

308 Buiten, De Streel and Peitz, 2023.

third-party refurbisher will be treated as the manufacturer under the PLD. They would be held liable for any defect resulting from the modifications, and the patient can seek compensation from the party responsible for altering the product, even though that party is not the original manufacturer.

3.5.3.6 *The Presumption of Causation*

According to Article 10(3) of the PLD, the connection between a product's defect and the resulting harm is presumed when it is proven that the product is defective, and the type of harm suffered aligns with what would typically result from that particular defect.[309]

The presumption of causation, once established, places a significant burden on the defendant, typically the manufacturer or seller, as it shifts the responsibility to them to disprove the link between the defect and the damage. This presumption simplifies the claimant's task in proving their case because they no longer need to demonstrate the causal relationship between the defect and the harm, which can often be complex and technically challenging. As a result, the defendant must provide sufficient evidence to show that the damage was not caused by the defect, despite its presence in the product.

If the defendant is unable to successfully challenge this presumption, they will likely be held liable for the damages. The presumption of causation can also expedite the legal process by reducing the evidentiary burden on the claimant, potentially leading to quicker settlements or judgments.

Furthermore, the inability to rebut the presumption can expose the defendant to broader regulatory scrutiny, product recalls, or even penalties, especially if the defective product poses ongoing risks to consumers. The defendant may also face reputational harm, leading to decreased consumer trust and potential loss of market share. This presumption, therefore, carries significant legal and commercial consequences for those responsible for ensuring product safety.

3.5.4 The Insufficiencies of Strict Liability

Not even the strict liability regime guarantees compensation for victims in every single situation. There are gaps in the regime, through which the producer may avoid liability.

It starts with the types of harm being covered. Strict liability typically applies to specific categories of harm.[310] According to Article 6 of the PLD, this involves material losses resulting from death or personal injury, including medically recognised harm to psychological health, damage or destruction

309 Hacker, 2023.

310 Buiten, De Streel and Peitz, 2023.

of property and destruction or corruption of data. If harm falls outside these narrowly defined categories, victims may not receive compensation under this Directive, even if the harm is severe.

If an AI diagnostic system fails to provide the correct diagnosis, it could be argued that this was not due to a system malfunction but rather the AI's inability to interpret the information accurately, which could fall within the normal error range, leaving the patient without compensation.[311] AI systems, like human practitioners, operate within certain error margins.[312] A malfunction implies a technical failure, such as a software bug or hardware issue, which would typically fall under product liability. However, diagnostic errors and other wrongful outcomes may not necessarily indicate a malfunction but rather a limitation in the AI's algorithmic capacity.[313] Every diagnostic tool, whether AI-based or human-based, has an inherent error rate. AI systems are trained on datasets that may not cover every possible medical scenario. Consequently, an AI's inability to diagnose correctly could be a reflection of these limitations rather than a malfunction. This normal error range can be a defence against liability claims, as errors are statistically expected.[314]

Alternatively, the producer may then claim that there is no causal connection between the damage suffered and the AI's error, as the doctor holds ultimate responsibility for the treatment decision that led to the damage.[315]

Even under strict liability, plaintiffs must establish a causal link between the product and the harm suffered. In complex cases, such as those involving multiple potential causes or delayed manifestation of harm (e.g., certain environmental or health issues), proving causation can be difficult and may result in the denial of compensation. The presumption of causation established in Article 10(3) of the PLD is intended to limit this outcome.

3.6 *The AILD*

3.6.1 A Fault-Based Regime

The AILD proposal was based on fault.[316] In the AILD, the notion of fault was linked to the AI system's adherence to EU or national regulations. Specifically, Article 4(1) emphasised that 'fault', under its regime, involved non-compliance with a duty of care established in EU or national law to prevent the damage in

311 Babushkina, 2023: 572.

312 Gillner, 2024.

313 Evans and Snead, 2024: 279–287.

314 Whittam, 2022: 262.

315 Bertolini, 2020: 114–116.

316 The European Consumer Organisation objects to the fault-based approach because it creates significant hurdles, if not impossible hurdles, for consumers to exercise their entitlement to compensation for damages (The European Consumer Organisation, 2023: 5).

question. Its definition of 'duty of care' (Article 2(9) AILD) invoked the required standard of conduct set by national or EU law to avoid harm to recognised legal interests. However, consumers generally lack the legal and technical expertise required to substantiate such claims, given that it requires the engagement of costly experts, assuming they can even locate such experts.[317] Consequently, this stipulation set an intimidating obstacle to pursuing compensation claims.

Crucially, this interpretation of fault did not acknowledge the possibility of harm occurring even when an AI system complies with EU law. Conversely, the AIA contemplates the potential for harm arising from AI systems that comply with regulations in Article 82, titled 'Compliant AI Systems which present a risk'. In this norm, an acknowledgement is made of the likelihood that even AI systems adhering to regulations may pose a threat to the health, safety, and fundamental rights of individuals,[318] but such a possibility seemed to be lacking in the AILD.

3.6.2 A Procedural Legal Framework

3.6.2.1 *The Relevance of a Uniform Procedure*

The AILD proposal encompassed liability claims at the national level, whether attributed to negligence or omissions of any party (be it providers, developers, or users).[319] These regulations aimed to address compensation for various forms of damage as delineated by national law, spanning areas such as life, health, property, and privacy, among others.

It incorporated procedural innovations, such as evidence disclosure obligations and an easing of the burden of proof under specific circumstances. It primarily aimed to eliminate procedural barriers that injured parties encounter when seeking compensation for damages caused by AI systems,[320] under the assumption that procedural aspects like the burden of proof contribute to the challenges of harmonising substantial issues involving liability for damages.[321] Additionally, there was the purpose of harmonising the basics of the applicable legal framework, based on the recognition that there are significant differences among Member States regarding key concepts like 'damage', 'fault' and 'causation',[322] which can lead to inconsistencies in the application of laws and hinder the effective enforcement of rights across borders.[323]

317 Hacker, 2023.
318 The European Consumer Organisation, 2023: 6.
319 A critical overview of this proposal in Galbois-Lehalle, 2021.
320 La Diega and Bezerra, 2024.
321 Li and Schütte, 2024: 144.
322 Ziosi, Mökander, Novelli, et al., 2023.
323 La Diega and Bezerra, 2024.

3.6.2.2 *Disclosure of Information and Presumption of Fault*

Proving defectiveness and establishing a causal connection between the defect and resulting damage often requires technical expertise beyond the claimant's capabilities, leading to an informational imbalance between the defendant and the injured party. To address this, the Expert Group on Liability and New Technologies proposed the concept of 'logging by design'.[324] Under this principle, AI producers are required to log relevant data that can assist claimants in identifying the causal link between the AI system and AI-caused damage. Claimants will have access to this information, or a presumption in their favour will arise if critical data are missing.

Nevertheless, the approach to substantiating liability through this means is conspicuously imbalanced. Typically, the victim, often lacking specialised expertise, will struggle to dissect the data effectively for evidentiary purposes. By implementing 'logging by design', the producer is only obliged to log and provide accessible data that could help the claimant establish the nexus between the AI defect and the suffered damage. However, even with access to extensive logged data, many consumers might lack the expertise to interpret and analyse it effectively for causal link demonstration.[325]

The ability of injured parties to access relevant evidence became a key feature of the AILD.[326] In cases involving high-risk AI systems suspected of causing damage – according to Article 1(1)(a) of the AILD, this mechanism only applied to high-risk AI systems[327] – claimants could ask national courts to order the disclosure or preservation of evidence from relevant parties. Non-compliance with these orders resulted in a presumption of such non-compliance, which

---

324 Expert Group on Liability and New Technologies – New Technologies Formation, 202: 47.

325 Bertolini, 2020: 84; Whittam, 2022: 258.

326 Madiega, 2023: 7.

327 Note that the very labelling as a high-risk AI system is problematic (La Diega and Bezerra, 2024), especially in light of Article 6(3) of the AIA:

'By derogation from paragraph 2, an AI system referred to in Annex III shall not be considered to be high-risk where it does not pose a significant risk of harm to the health, safety or fundamental rights of natural persons, including by not materially influencing the outcome of decision making.

The first subparagraph shall apply where any of the following conditions is fulfilled:

(a) the AI system is intended to perform a narrow procedural task;

(b) the AI system is intended to improve the result of a previously completed human activity;

(c) the AI system is intended to detect decision-making patterns or deviations from prior decision-making patterns and is not meant to replace or influence the previously completed human assessment, without proper human review; or

(d) the AI system is intended to perform a preparatory task to an assessment relevant for the purposes of the use cases listed in Annex III'.

simplified the claimant's procedural burden and encouraged parties to adhere to the orders.[328]

As a result of this obligation to disclose information, Article 3(5) of the AILD proposal established that failure to comply with court-issued disclosure orders resulted in a presumption of non-compliance with the relevant duty of care. This means that the court would presume the claimant had successfully proven the point they intended to establish with the requested evidence. However, the defendant had the opportunity to challenge this presumption by providing counterevidence.

The specific duties of care to which the presumption applied were exhaustively listed in Article 4(2) of the AILD. The limited scope of these duties raised concerns, especially considering that in tort law even unwritten rules of behaviour or professional standards can establish fault. The AILD Proposal should not be interpreted as implying that a provider could not be found at fault if they complied with the duties listed in the referred provision. Instead, Article 4(2) of the AILD narrowed the determination of fault specifically for the purpose of presumption. A provider could still be found liable if they violated unwritten norms, professional standards, or other protective rules that were not explicitly outlined in Article 4(2) of the AILD, even if they have adhered to the listed duties of care.[329] However, in such cases, the claimant would be required to establish a causal link without relying on the presumption of causation offered by Article 4(1) of the AILD.[330]

3.6.2.3 *The Presumption of Causation*

Article 4(1) of the AILD introduced a presumption of causation between a defendant's fault and the AI system's output,[331] applicable to all AI systems (Recital 28 AILD), under specific conditions:[332] (a) The claimant has shown, or the court has presumed, that the defendant is at fault, meaning they have failed to comply with a duty of care specified in EU or national law that is intended to protect against the damage that occurred; (b) It is reasonably likely, based on the circumstances of the case, that the defendant's fault has influenced the output generated by the AI system or its failure to produce an output; (c) The

328 Ziosi, Mökander, Novelli, et al., 2023.

329 Polat 2023a: 120.

330 Polat 2023a: 120.

331 Two causal connections can be established: (a) between the defendant's fault and the AI system's output or its failure to generate an output; (b) between the AI system's output (or lack thereof) and the resulting damages. The presumption outlined in the AILD Proposal focused specifically on the first causal link. Cf. Polat 2023a: 116.

332 Buiten, De Streel, Peitz, 2023; Li and Schütte, 2024: 148–149.

claimant has demonstrated that the output produced by the AI system or its failure to produce an output directly caused the damage. Recital 25 of the AILD clarified that a reasonable likelihood of a causal link between fault and damage did not exist if the fault did not influence the AI system's output and, consequently, did not cause the damage.[333] The CJEU would ultimately determine what constitutes a 'reasonable likelihood', depending on the specific circumstances of each case.

If all three conditions for presuming a causal link were met, the court's decision would vary based on the AI system's risk level. For high-risk AI systems – which encompass the majority of AI systems used in healthcare – the court would presume a causal link unless the defendant could prove that the claimant had reasonable access to sufficient evidence and expertise to establish the link independently.[334] This might occur, for instance, when the claimant could access the documentation and logs of a high-risk AI system. Regarding systems that are considered low-risk, the presumption would only come into play if national courts determine that it was 'excessively difficult for the claimant to prove the causal link' (Article 4(5) AILD). Therefore, for low-risk systems, the presumption of causation was not automatic and only applied when proving the link between the defect and the damage presented an unusually high level of difficulty for the claimant.[335]

The defendant, however, could challenge this presumption by proving that their fault could not have led to the damage or by showing that the victim had sufficient access to evidence and expertise to prove the causal connection independently. For example, imagine that a hospital uses an AI system to assist in diagnosing patients. A patient sues the hospital, claiming that the AI system's faulty diagnosis led to harm. Under the AILD, the patient could argue that the hospital breached its duty of care by relying on the AI system and that this non-compliance caused their injury. The hospital, as the defendant, could

333 For instance, consider an AI system used to assist in diagnosing medical conditions. If the AI system's developers fail to comply with AIA requirements related to data governance – such as not properly securing patient data or ensuring the quality of the data used – the system might produce inaccurate diagnoses. This non-compliance could reasonably lead to harm, such as misdiagnosis or delayed treatment, directly affecting patient outcomes. On the other hand, if the developers simply fail to register the AI system as a high-risk tool, this oversight does not directly influence how the system analyses patient data or generates diagnoses. As a result, this failure to register might not be considered sufficient to establish a causal link with any harm that occurs, since it does not directly impact the AI system's performance or the accuracy of its outputs.

334 Grozdanovski, 2024: 74.

335 Grozdanovski, 2024: 74.

challenge this presumption by showing that its reliance on the AI system was not the cause of the harm. For instance, they might present evidence that the patient's condition was misdiagnosed due to other factors, such as incomplete information provided by the patient. Alternatively, the hospital could argue that the patient had access to enough evidence, including expert evaluations or second opinions, to independently prove a link between the harm and the AI's diagnosis, meaning that the presumption of causation should not apply.

Under this provision, the presumption applied only to the link between fault and output, leaving room for certain relevant omissions to go unnoticed, such as the duty of human oversight from the professional deployer. Hacker points out that the breach of this duty pertains to an action (human oversight) that occurs after the AI output is generated. Consequently, the deployer could easily refute the presumption by arguing that the AI system's output would have been the same, regardless of any oversight.[336] According to this author, it would be far more useful to presume that the user's action or inaction, specifically the failure to comply with the human oversight obligation under Article 26 of the AIA, was the cause of the damage. The focus, therefore, should be on the corrective role of the human in overseeing the AI system's results, rather than on the inherent accuracy of the AI output.

3.6.3 Connection to the AIA

The AILD applied to AI systems, as per the definition of the AIA. This showed how the definition of AI, and in particular high-risk AI, is crucial, not only for the AIA but also for connected legal frameworks. This dependency on the AIA definition creates an ironic dilemma for lawmakers, as the precise definition of AI continues to pose challenges and unresolved issues during the finalisation

336 Hacker, 2022: 36–37. Consider a scenario where a doctor (the professional deployer) uses an AI system to recommend treatment plans for patients with heart conditions. The AI generates a suggestion based on the patient's data, recommending a specific medication. The doctor fails to exercise the necessary human oversight and automatically prescribes that drug. If the patient suffers harm as a result of this decision, prima facie one would say that the liability falls on the doctor, for failing to 'review' the AI's recommendation. However, the deployer can exclude the presumption of causation by arguing that their omission did not contribute to the harm, as even with human oversight the AI's recommendation would have been the same. Therefore, the doctor can argue that the harm resulted from the AI system itself, not from the independent medical decision. The key issue in establishing causation against the professional deployer should revolve around demonstrating that proper human oversight would have led to a correction of the AI's output during post-processing, rather than proving that the AI's output itself would have been different.

process of the AIA.[337] The implications of this entangled relationship highlight the significance of establishing a clear and comprehensive definition of AI.[338]

Other 'borrowed' concepts were those of 'AI provider' and 'user/deployer,' which were to be understood as in the AIA. In healthcare provision, the 'user/deployer' will be the healthcare provider, either the human doctor or the hospital. The requirements imposed on them by the AIA include obligations to use and monitor the AI system as instructed, requirements to assign human oversight to individuals with the necessary competence, training, authority and obligations to provide relevant input data aligned with the AI system's intended purpose (Article 26 AIA).[339]

3.6.4 Recital 15 of the AILD

3.6.4.1 *The Interpretation of Recital 15 of the AILD and Deployer Liability*

Under the AILD proposal, deployer liability was limited to cases involving fully automated decisions, where humans were not involved in the decision-making process from the point of generating the output. This restriction was explicitly stated in Recital 15,[340] which clarified that liability claims could not be addressed under its legal framework when damage resulted from a human assessment followed by human action or inaction, with the AI system merely providing information or advice.

This direct cause-and-effect relationship happens in cases where the AI operates autonomously, without requiring the user's direct control or supervision. However, many AI systems are not fully autonomous,[341] and tasks are often shared between the system and the user or operator, creating a 'human-on-the-loop' or even a 'human-in-the-loop' scenario (see Chapter 6, Section 3.5.2).

337 Ruschemeier, 2023: 361–376.

338 Pointing out this dependency, Hacker, 2022: 11.

339 Mennella, Maniscalco, Pietro, et al., 2024.

340 Recital 15 of the AILD: 'This Directive should only cover claims for damages when the damage is caused by an output or the failure to produce an output by an AI system through the fault of a person, for example the provider or the user under [the AI Act]. There is no need to cover liability claims when the damage is caused by a human assessment followed by a human act or omission, while the AI system only provided information or advice which was taken into account by the relevant human actor. In the latter case, it is possible to trace back the damage to a human act or omission, as the AI system output is not interposed between the human act or omission and the damage, and thereby establishing causality is not more difficult than in situations where an AI system is not involved.' About this Recital, Hacker, 2022: 19–20; Li and Schütte, 2024: 145–146.

341 No AI system operates at a fixed level of autonomy throughout its entire use; instead, the degree of autonomy can vary during different phases of its operation. See Merat, Seppelt, Louw, et al., 2019: 93.

Specifically, the legal remedies available to plaintiffs by the AILD were not applicable if a human agent intervened between the AI output and the resulting damage. In other words, the obligations to disclose evidence and the presumptions that alleviate the plaintiff's burden of proof were only granted when decisions were fully automated, not when AI operated as a recommender or decision-support system. In this instance, it was considered that the human intervention did not influence the system's output.[342] Even if the AILD was withdrawn, this exclusion of liability might still be on the table in future EU legislative interventions, and thus its analysis is essential.

In a healthcare setting, imagine a situation where a doctor relies on an AI system to help diagnose a patient. If the AI system suggests a diagnosis and the doctor, despite having the expertise to evaluate the recommendation, makes an incorrect decision that leads to patient harm, the fault would be attributed to the doctor's judgment, not the AI. In this case, the AILD's provisions that reduced the burden of proof would not be applicable, since the error stemmed from the human's decision rather than a malfunction or flaw within the AI system itself.

This segment introduced a significant restriction to its 'plaintiffs-friendly' regime. Additionally, it created an incentive for companies to bypass the rule by having AI-generated outputs approved by humans who blindly follow the AI's recommendations, thus avoiding its legal regime.

Furthermore, the caveat arising from Recital 15 of the AILD risked rendering parts of the AILD inapplicable and obsolete. The reason relates to the AIA. Articles 14 and 26 of the AIA require general human oversight of high-risk AI systems.[343] If compliance with these norms on human oversight takes place after the output is generated, in a way that involves an independent human assessment and decision that directly causes the harm, the specific high-risk AI system would fall outside the AILD proposal's scope, because causation can be traced to a human act or omission. Conversely, if human intervention, as per the AIA, occurs before or contemporaneously with the output generation, but the damage is still causally linked to the AI system's output, the AI system would fall under the purview of the AILD proposal.[344]

342 Li and Schütte, 2024: 145.

343 Enqvist, 2023: 508–535.

344 Recital 15 of the AILD resembles the limitation found in Article 22(1) of the GDPR, which scholars have warned can be circumvented by having human actors blindly follow machine advice, effectively rubberstamping the AI output (Wachter, Mittelstadt and Floridi, 2017b: 92).

The ironic note is that no AI systems – particularly high-risk ones – are supposed to work on a standalone basis. On the contrary, there must always be a human in the loop, prepared to revise or override 'AI decisions',[345] as many consequential decisions undergo human review, especially when the confidence level of the AI output is low.[346] If that is the case, Recital 15 of the AILD would apply every single time, thus systematically excluding the application of the AILD.[347]

3.6.4.2 *The Interconnection Human-Machine in Light of Recital 15 of the AILD*

In real-world scenarios, the intricate nature of interactions between humans and AI can make it challenging to determine whether a deployer's actions actually influenced the AI's output. Consider a situation where a radiologist uses an AI tool to help interpret medical images. During the time the system has been used by the radiologist, the latter systematically provides corrective feedback to the AI system, based on their own expertise and experience. Suppose that one day the AI system issues a diagnosis, later proven to be incorrect, leading to patient harm. In this case, it may be difficult to determine whether the AI's initial output or the radiologist's adjustments were responsible for the error. The complexity of such human-AI interactions makes it challenging to pinpoint whether the radiologist's actions influenced the AI's output or if the AI itself contributed to the mistake.

One major limitation of the AILD Proposal was that it did not account for cases where both the user and the AI system collectively contribute to the damage. For instance, in decision-support systems, the user can influence the AI's output, leading to damage. A professional using such a system could input incorrect data, such as misreporting a patient's health condition or a client's investment preferences. Even if the input is correct, the user may still cause damage by improperly applying the AI's recommendations in the real world. However, the AILD Proposal did not cover situations where both the user's error and the AI system's output were jointly responsible for the damage, leaving a significant gap in its application to shared-fault scenarios.

345 Tschider, 2024: 324–429.
346 Hacker, 2023: 19.
347 Hacker, 2023: 19.

4 Liability of the Authority That Approved the AI System

4.1 *Liability of Approval Bodies and Authorities*

The liability of the authorities and/or bodies that assess and approve products to reach the market is not common under EU law.[348] The assessment conducted by these authorities does not provide an absolute guarantee of the safety and effectiveness of the product. Regardless of the number of studies and assessments required, achieving such absolute certainty is impossible. Therefore, there is no legal basis for holding these authorities liable solely because the product caused harm unless it can be proven that the authority did not perform its duties in accordance with the law. In other words, the authority can only be held responsible if it granted market authorisation (in the case of AI medical devices, the CE mark of conformity) despite clear evidence that the product did not meet the required standards. Outside of these clear-cut scenarios, it would be difficult to build a solid case against the authority that approved the AI system.[349]

4.2 *The Specific Case of Notified Bodies under the MDR and the AIA*

Both the MDR and the AIA require a third-party assessment for some products,[350] to be granted by the Notified Bodies.[351] Suppose that an AI-based medical device designed to detect heart abnormalities was approved by a Notified Body; however, during the assessment, the Notified Body failed to thoroughly evaluate how the AI was trained and whether it would perform accurately across diverse patient populations (the AI system had been trained on data from a limited group of patients and performed poorly when used with other demographic groups). Additionally, the Notified Body did not adequately review the device's plan for post-market monitoring, meaning there was no clear system in place to track and correct errors after the device was in use.

---

348 Garcia-Micó, 2024.

349 Raposo, 2018: 136–139. Note that the CE mark allows a product to be sold and distributed across all EU Member States, regardless of the country where the Notified Body is based (see Chapter 3, Section 2). As a result, the Notified Body may find itself subject to legal action in several EU nations for essentially the same issue, yet facing the possibility of differing judgments in each jurisdiction (Danish Medical Agency, 2023).

350 The MDR requires third-party conformity assessments for some class I medical devices and for the ones in classes IIa, IIb, and III, and most AI devices will likely fall under one of these latter classifications (Zapata, Patil, Ward, et al., 2023: 19–34).

The AIA, in turn, requires third-party conformity assessments for AI systems listed in Annex I already subject to such assessments under their respective sectoral legislation, and for the ones in Annex III (Thelisson and Verma, 2024:113–121).

351 De Bruyne, 2019: 203–211.

Despite these issues, the device was approved, potentially putting patients at risk due to inaccurate diagnoses and lack of oversight. Will the Notified Body be held liable?

A landmark case in this regard is the Schmitt case,[352] where the CJEU examined the potential liability of TÜV Rheinland, a Notified Body, concerning women who received breast implants manufactured by the French company Poly Implant Prothèse SA (PIP). The CJEU determined that the Medical Devices Directive[353] should be interpreted to ensure that the involvement of a Notified Body in the EC declaration of conformity process was aimed at protecting the end users of medical devices. However, the CJEU also stated that the conditions under which a Notified Body's failure to meet its obligations might result in liability towards end users are governed by national law, as long as they adhere to the principles of equivalence and effectiveness. Thus, in theory, Notified Bodies can be held accountable for the injuries caused by the devices they recklessly approved.

The considerations made by the court regarding the Medical Devices Directive and the role of Notified Bodies in relation to medical devices can be transposed to the AIA and the role of Notified Bodies concerning AI systems that require their assessment. The CJEU recognised that the Medical Devices Directive's goal extended beyond just protecting health; it also encompassed the safety of individuals, including not only patients and users of medical devices but also 'third parties' or 'other persons' in general. While manufacturers had the primary responsibility to ensure compliance with the Medical Devices Directive, the legislation also imposed certain obligations on Member States and Notified Bodies related to safeguarding, vigilance, and health surveillance, which collectively ensure protection for health and safety.

Regarding the culpability of Notified Bodies for harm to end users, the CJEU referred to its previous judgment in Paul and Others,[354] noting that the imposition of surveillance obligations on certain bodies and the Directive's objective to protect injured parties did not necessarily create rights for those parties if the bodies fail in their duties, especially when the Directive did not explicitly provide for such rights. Moreover, the Directive did not specifically address how civil liability for Notified Bodies was incurred or how end users might seek compensation for injuries resulting from Notified Bodies' failures.

352 Elisabeth Schmitt v TÜV Rheinland LGA Products GmbH, C-291/15, ECLI:EU:C:2017:128.

353 At the time the incident took place the legal Framework for medical devices was the Council Directive 93/42/EEC of 14 June 1993 concerning medical devices.

354 Judgement of 12 October 2004, Peter Paul, Cornelia Sonnen-Lütte and Christel Mörkens v Bundesrepublik Deutschland, Case C-222/02, ECLI:EU:C:2004:606.

The conditions under which a Notified Body's failure to fulfil its obligations might lead to liability towards end users, as outlined in the Medical Devices Directive (and, arguably, in other EU directives and regulations), are a matter for national law.

5 Overview of Possible Scenarios of Medical Liability When Using AI

There are multiple liability scenarios in healthcare settings involving AI, and this section can only provide an overview of the most common ones. However, real life is always more complex than scholarly examples and various possible event combinations might arise.

5.1 *No AI Intervention in the Outcome*

Sometimes harm to the patient results purely from the negligence of human medical experts. This might happen either because there was no intervention of AI in the medical act or because, though AI had some intervention, the human intervention was the relevant one. Regarding the latter case, suppose the AI system recommended the correct medical prescription for a drug and the doctor accepted it, but then mistakenly registered the wrong dosage when issuing the actual prescription, leading to harm.

In such cases, only the human healthcare provider (or the institution they work for)[355] will be held accountable. It will be a case of medical liability and not manufacturer liability.

5.2 *Correct AI Outcome and Wrongful Deviation from the Human*

It might happen that the diagnosis and treatment provided by the AI system were correct, but the human expert decided not to follow the system's suggestion, opting for a different course of action. In such cases, the liability depends on who holds the decision-making authority. If AI is regarded as the decision-maker, human experts, like doctors, could be held liable for the mere fact that they deviate from the system's advice. However, considering the particularities of the medical profession, it is very unlikely that AI would be deemed the ultimate decision-maker.[356] Thus, assuming that the human was the one in charge of the decision, the outcome regarding the allocation

---

355 This could be the case if, for instance, such an error could be attributed to fatigue due to an excessively long working period imposed by the healthcare facility (Kachalia and Studdert, 2004: 1051–1056).

356 Lysaght, Lim, Xafis, et al., 2019: 300.

of liability will depend on the factors that justify (or not) the trust in the AI model, such as the level of accuracy recognised for that AI system, its level of explainability and the existence of sound reasons for the human to deviate from AI's guidance.[357]

If the AI system has proven to be extremely accurate through clinical trials and other studies, the doctor would be required to provide exceptionally solid evidence to justify why the AI recommendation was disregarded.[358] In such a scenario, the patient could argue that the AI's treatment suggestion was correct and aligned with best practice guidelines, holding the doctor liable for negligence due to their failure to adopt the AI's recommended treatment plan.[359] This situation is unfavourable as it could lead doctors to rely excessively on and conform to AI system recommendations to avoid the potential liability risk of not doing so.[360] This has led some authors to argue that medical malpractice regulations fail to adequately safeguard both claimants and medical professionals concerning the use of AI systems.[361]

5.3 *Wrongful AI Outcome and Excessive Trust from Humans*

In other situations, the AI outcome is incorrect, and the harm results from the fact that the doctor adhered to the AI system's suggestion. In such cases, the reliance on the AI's flawed output directly contributes to patient harm, raising concerns about over-reliance on technology in medical decision-making.[362]

Consider the case of a CPOE employed for prescribing medications. An alert denoted as a 'drug-drug interaction' could notify a doctor to potential side effects stemming from the concurrent consumption of two pills. This is a case of complementary automation. Should this drug-drug interaction alert prove to be inaccurate and lead to harm, the patient could potentially file a lawsuit against both the doctor and the CPOE system's provider.[363] Regarding the manufacturer's liability, the onus lies on the defendant (the manufacturer) to demonstrate that the harm was not due to the product, as strict liability applies. The doctor, in turn, might still bear significant responsibility, as per the doctrine of 'competent human intervention',[364] which emphasises that

357 Wang, Zhou, Li, et al., 2023: 209–220.
358 Kempt and Nagel, 2022: 226.
359 Bertolini, 2020: 114–115.
360 Froomkin, Kerr and Pineau, 2019: 38.
361 Bertolini, 2020: 114–115.
362 Quinn, Senadeera, Jacobs, et al., 2021: 892–893.
363 Pasquale, 2020.
364 The doctrine of 'competent human intervention' refers to the legal and ethical principle that, despite the use of advanced technologies such as AI, a competent human

human professionals must critically evaluate AI recommendations before acting on them.[365] Therefore, even though the system generated the faulty alert, the doctor's failure to intervene appropriately could be seen as a key factor in the harm caused.

The considerations mentioned in the previous scenario about the factors that can justify trust in AI are valid here, *mutatis mutandis*. That is, if there is a general consensus among experts that the accuracy of that particular AI system is low and the doctor still followed its recommendation without further assessment based on their own expertise, this will likely be considered a form of malpractice. Moreover, it is also likely that the hospital will be found negligent for failing to provide practitioners with tools (in this case, the AI) that have a reasonable level of accuracy.[366]

5.4 *Wrongful Use of the AI System*

In some cases, the AI outcome is incorrect because the AI system was used improperly by the human practitioner. For instance, the healthcare provider disregards specific instructions or used a robot for an unintended purpose.

A specific example could involve an AI-based medical imaging system designed to detect early signs of lung cancer, where the system analyses chest X-rays and highlights areas of concern for the radiologist to review. If the radiologist improperly uses the AI system by uploading a low-resolution scan that does not meet the system's input requirements, the AI may, due to the poor quality of the image, fail to flag a small tumour that would have been noticeable in a higher-quality scan. Suppose that relying solely on the AI's incorrect output, the radiologist dismisses the case as normal, and the patient is not diagnosed with lung cancer until the condition worsens. Who will be held liable?

Note that in the example, the AI outcome is incorrect because the human practitioner fails to follow the proper protocol for using the system, leading to an improper diagnosis. This scenario highlights how the misuse of an AI tool by the practitioner can result in errors that cause patient harm.[367]

Still, the assignment of liability will depend on the specific cause of the lapse in expertise. The hospital will be held liable if it has not provided

---

professional must oversee and intervene in the decision-making process to ensure safety, accuracy, and accountability. This doctrine is especially relevant in fields like healthcare, where the consequences of decisions can significantly impact human lives. Cf. Pasquale, 2022: 5, 13.

365 Sampson v. HeartWise Health Sys. Corp., No. SC-2022-0847 (Ala. May. 26, 2023).

366 Griffin, 2021: 102–103.

367 Volovici, Syn, Ercole, et al., 2022: 1996–1999.

adequate training to its employees on the use of AI.[368] Alternatively, the manufacturer may be held liable due to a lack of information on the product (possibly in parallel with the healthcare provider). There might also be a situation where the practitioner has all the necessary information and training but still used the AI incorrectly, leading to their own liability.

6 The Liability of AI

6.1 *The Direct Liability of the AI Entity*

6.1.1 The Refusal to Grant Legal Personality to AI

AI is not a legal person and, as such, cannot be held liable.[369]

The vast majority of legal systems recognise two types of legal persons: natural persons and legal entities. Natural persons are recognised as legal persons simply by virtue of being human.[370] Legal entities, on the other hand, are not human, which requires a more complex legal construction to establish their legal personality.[371] Legally, 'legal personality' is defined as the capacity to hold rights and obligations.[372] Despite the specificities of each legal system, this basic definition remains the common denominator.

Even though the EU has been legislating widely about digital issues, AI included, the regulation of legal personality does not fall within the competence of European law.[373] At most, the EU may encourage Member States, who alone are responsible for deciding who can hold rights and obligations, to act in one direction or another. However, given how important this topic is to the EU (particularly in light of the 'Digital Europe' project),[374] the position of European law will undoubtedly strongly influence national lawmakers.

In 2017, the European Parliament, in its resolution on 'Civil Law Rules on Robotics',[375] recognised that the more autonomous robots become, the less they can be viewed as mere tools in the hands of other actors, which raises the question of whether ordinary liability rules are sufficient, or if new principles

368 Maliha, Gerke, Cohen, et al., 2021: 631.
369 In detail see Raposo, 2024a: 1049–1072.
370 Kurki, 2023: 34.
371 Kurki, 2023: 46.
372 Kurki, 2023: 30–32.
373 Novelli, Bongiovanni and Sartor, 2022: 198.
374 European Commission, 2022.
375 European Parliament Resolution of 16 February 2017 with recommendations to the Commission on Civil Law Rules on Robotics (2015/2103(INL)), https://www.europarl.europa.eu/doceo/document/TA-8-2017-0051_EN.html.

and regulations are needed to clarify the legal responsibility of various actors for the actions and omissions of robots, especially when the cause cannot be attributed to a specific human and the harmful acts or omissions by the robots could have been avoided (Recital AB). Under the assumption that, in the future, an autonomous robot could be considered a legal person, the European Parliament urged the Commission to explore the implications of various legal solutions. One such solution would be the creation of a specific legal status for robots in the long term, allowing the most sophisticated autonomous robots to be recognised as 'electronic persons', responsible for compensating any damages they may cause (Paragraph 59.F).

However, this proposal generated numerous negative reactions, including a letter addressed to the European Parliament in which a significant group of experts in the field stated that 'From an ethical and legal perspective, creating a legal personality for a robot is inappropriate whatever the legal status model'.[376]

Perhaps due to the commotion caused by the 2017 Resolution, the European Parliament changed its position three years later, as is clearly reflected in two of its resolutions from October 2020: the Resolution 2020/2014(INL) (the 'Civil Liability Resolution')[377] and the Resolution 2020/2015(INI) (the 'Intellectual Property Resolution').[378] The Civil Liability Resolution begins by clarifying its fundamental premise that AI systems have neither legal personality nor human consciousness and their sole purpose is to serve humanity (Paragraph 6). It then explicitly states in Paragraph 7 that 'it is not necessary to give legal personality to AI-systems'. Similarly, in its Intellectual Property (IP) Resolution, the European Parliament recommended against granting legal personality to AI, arguing that doing so would reduce incentives for human creators.[379]

376 AAVVV, 2018.

377 European Parliament, Resolution of 20 October 2020 with recommendations to the Commission on a civil liability regime for artificial intelligence (2020/2014(INL)), https://www.europarl.europa.eu/doceo/document/TA-9-2020-0276_EN.html.

378 European Parliament, Resolution of 20 October 2020 on intellectual property rights for the development of artificial intelligence technologies (2020/2015(INI)), https://eur-lex.europa.eu/legal-content/EN/TXT/?uri=celex:52020IP0277.

379 Among the many arguments that could be made in this field, this one seems particularly inadequate. The refusal to grant intellectual property rights to AI based on this reasoning seems more like a protective measure for human inventors than a well-founded stance on AI's legal personality.

6.1.2 The Root of the Argument: AI Does Not Meet the Requirements to be a Legal Person

6.1.2.1 *Analogy with Legal Persons*

Compelling arguments are made by those who oppose granting legal personality to AI. This approach starts by identifying the characteristics that enable certain entities (both natural and legal) to be recognised as legal persons and then assesses whether it is possible to apply any of these characteristics to AI systems.

Regarding non-human legal persons, several theories have emerged to justify their legal personality.[380] The most well-known is perhaps the 'legal fiction theory', which acknowledges that the legal personality of these entities is a fiction, as these entities are not persons themselves, and the reason why the law treats them as such is to fulfil the purposes of the legal system. From this perspective, granting legal personality to AI would depend on whether such a legal fiction would help achieve the objectives of the legal system. The 'symbolic theory' instead recognises that it is easier to act under the law by treating groups of people as a single entity.[381] This reasoning works well for companies and foundations, but cannot be applied to AI, as AI lacks a human foundation. In contrast, the 'realist theory' rejects the idea that non-human entities are merely fictions or symbols and instead acknowledges their real existence as entities that operate in the legal world. According to this theory, such entities should be recognised regardless of the individuals that compose them and even independently of the legal status that the law might confer on them. However, none of these theories provides a clear answer to the question of AI's legal personality. The analogy between legal entities and AI encounters numerous inconsistencies, making it difficult to apply these traditional frameworks to AI.[382]

6.1.2.2 *Analogy with Natural Persons*

Moving the comparison to natural persons, the most common line of reasoning argues that AI lacks the human qualities deemed essential for the recognition of legal personality, such as consciousness, intentionality, emotions, desires, and interests. Even when AI exhibits behaviour that appears to demonstrate

380 Analysing these theories, see Banteka, 2020: 553 ff.

381 For example, instead of Mr. X having to deal with the 1,000 individuals that make up a company, he only needs to contract with the company itself, thus performing a single legal act. The legal personality of a company or foundation, in this sense, is a symbol representing all the people involved.

382 Raskulla, 2023.

such qualities, it would merely be an imitation of human behaviour, not an inherent action of the machine.[383]

The argument starts with the idea that both humans and AI can think, although AI does not yet think like a human.[384] Consciousness is another possible criterion, but it presents two challenges. First, consciousness is difficult to define. It involves self-awareness and awareness of the surrounding reality, but much about it remains beyond current understanding. Second, the legal system has not traditionally considered consciousness as a requirement for legal personality, but rather as something that can influence legal responsibility.[385] Autonomy is often cited as a key characteristic of human legal persons.[386] However, the concept is not universally agreed upon. Although AI has moved from automation to autonomy[387] (as it can now learn independently), this does not automatically make it autonomous in the legal sense. AI can act without human control, but it cannot make independent ethical choices like a human, at least not at its current stage.[388] Another possible criterion for legal personality is the capacity to have interests,[389] as legal personality allows for the protection of these interests and the responsibility for violating others' interests. However, AI currently lacks its own interests. Solaiman highlights AI's lack of intentionality, desires, and interests.[390] Intention is also suggested as a potential criterion for legal personality, in connection to the notion of 'liability', but it has its weaknesses. Intention is not always necessary for liability, as negligent actions are punishable, and civil liability often involves strict liability without fault.[391] On the other hand, AI might even exhibit a form of basic intention, like seeking more data to continue learning.[392] Finally, regarding the capacity to assume duties and obligations, the question is whether this capacity requires an understanding of those duties. Consciousness, intentionality, and the ability to understand the nature of rights and responsibilities are requirements.[393]

383 Solum, 1992: 1262.
384 Nehez, 2020: 467–492.
385 Patterson, 2015: 771–774.
386 Banteka, 2020: 654–565.
387 Pagallo, 2017: 18–19.
388 Baum, 2000: 174.
389 Shelton, 2015.
390 Solaiman, 2017: 173 (in his text, the author refers to industrial robots, but it is highly likely that these are robots powered by AI).
391 Brown, 2021: 221.
392 Brown, 2021: 221.
393 Nonetheless, some legal scholars argue that such understanding is not necessary. See Brown, 2021: 221.

AI lacks these attributes, operating instead based on algorithms and data-driven decision-making without awareness or understanding.

Moreover, many of these criteria do not apply to many human beings (such as young children, people in comatose states, or those with severe mental illnesses). Therefore, imposing any of these criteria as requirements for AI to be considered a legal person would demand more from AI than is required of human beings.

Perhaps the mistake in this debate lies in basing the discussion on possible similarities between non-human legal personality and human legal personality, thereby confining the reasoning to an exclusively anthropocentric view of the law.[394] In other words, the question of what – or who – a human person is should be separated from the question of what constitutes a person under the law, which should focus simply on identifying who or what can be the subject of legal rights and obligations.[395]

On the other hand, even if we consider these criteria inapplicable to the AI systems we currently have (no matter how advanced), this theory may no longer hold in the future when general AI finally becomes a reality. AI is a 'moving target', and the legal boundaries and limitations it currently faces may not remain stable in the near future.

6.2 *Responsibility of the Institution through the Accountability of AI*

It is common for a person (human or not) to be responsible for the actions of a third party that is accountable to them.[396] See, for example, the responsibility of the employer involving acts of his/her employee, or even the responsibility of the state for the acts of its agents or employees (see Chapter 5, Section 2.7.4). These scenarios can be transposed to medical responsibility.[397]

If it comes to be accepted that AI has legal personality, the next question is whether the AI can, in itself, be considered an employee or agent of the hospital in a way that leads to the liability of the latter. For this conclusion to be reached, we first need to assess whether the hospital has some kind of control or power over the agent, such as an employee. That is, can AI be considered an agent at the service and under the responsibility and control of the institution?[398]

394 Dyschkant, 2015: 215.
395 Brown, 2021: 215.
396 Giliker, 2010.
397 Shenoy, Shenoy and Shenoy, 2021.
398 On the role of AI as an agent, Aksoy, 2022: 146–160.

The relation of agency governs the relationship between agents and principals, involving situations where one party (the agent) is authorised to act on behalf of another party (the principal), usually to enter into contracts or perform specific legal tasks.[399] This legal framework is essential in business and personal transactions where someone might delegate authority to another to handle their affairs. A key feature of this relation is the concept of 'authority', that is, the power given to the agent by the principal, which can be actual (explicitly given) or apparent (based on how the principal represents the agent's authority to third parties). Conversely, the agent owes the principal a duty of loyalty and care, meaning the agent must act in the best interest of the principal and avoid conflicts of interest (fiduciary duty).

Even if the agency relation applies – which is far from clear – the responsibility of the hospital varies, depending on the nature of the contract in medical settings.[400] If the contract is considered a 'total contract', with non-delegable duties assigned to the hospital, all doctors – including the person-like AI system – who participate in the patient's treatment are considered assistants of the hospital. Consequently, the hospital will bear liability for any malpractice resulting from the use of AI, under the doctrine of vicarious liability. By contrast, in the case of a 'split contract' where the consulting doctor (in this scenario, the AI system) has a separate contract with the patient, the consulting doctor (the AI) will also assume legal responsibility through its owner or agent, but this last hypothesis is not legally possible under current law (see Chapter 5, Section 6.1).

Even if AI is not classified as a person under the law, courts may be inclined towards considering AI more akin to a doctor rather than mere equipment, given its remarkable capabilities and substantial involvement within the medical team. Should this analogy hold, hospitals may have the potential to secure separate insurance policies to cover the potential risks associated with AI-induced harm, mirroring their existing practices of insuring healthcare provider employees.[401]

6.3 *Alternative Liability Models*

An intriguing solution to address the intricate human-machine interactions lies in attributing legal personality to the association between humans and

399 Seavey, 1919–1920: 859–895.
400 Raposo, 2014: 193–197.
401 Cf. Allain, 2013: 1066.

machines.[402] By merging the two entities into a single legal identity, we could better manage the complexities of human-machine collaboration.

The doctrine of common enterprise liability appears particularly suitable for human-algorithm interactions.[403] In scenarios where individual actors unite for a shared goal and create a network, this approach holds the network accountable as a whole rather than holding individual actors accountable. This becomes crucial when attributing liability in cases where injuries or errors occur during medical procedures involving AI, as it allows the human-machine hybrid to be held accountable.

Alternatively, the legal network theory offers an approach or framework within legal studies that examines the interactions, relationships, and dynamics among various legal actors and institutions within a network or interconnected system. In various contexts, especially in the digital age, network liability often refers to the question of who holds legal responsibility for actions, content, or consequences within a networked environment. It considers human-machine interactions as interconnected networks, where algorithms and human actors co-produce a common purpose, forging a collective human-machine entity.[404] It views the legal system not just as individual components (such as laws, courts, lawyers, etc.) but as a network of interconnected elements that influence and shape legal outcomes, decisions, and the evolution of the law itself. This new entity could be subject to limited legal personality, giving rise to 'network liability' when joint actions lead to any harm or wrongful acts.

7 A Possible Way Ahead: the No-Fault Mechanism

7.1 *The Criticism of the Negligence-Based Models*

7.1.1 General Criticism of Negligence-Based Model

Legal scholars have pointed out that the current civil liability legislation poses a significant hindrance to the development and use of AI and the numerous benefits it offers. Multiple studies and expert groups have recognised this issue, underscoring its potential negative impact on the advancement of AI technology and the realisation of its advantages, especially in healthcare.[405]

Under the current paradigm of civil liability, compensation is typically granted only when a specific party is identified as responsible for the harm,

402 Mehta, 2023.
403 Schweikart, 2021: 19–21.
404 On this theory, see Lior, 2020: 1103–1166; Whalen, 2016: 539–565.
405 See Magrani, 2019: 1–19; Montagnani and Cavallo, 2021: 208–231; Pellegatta, 2019: 135–161.

either due to fault or under a strict liability rule. However, the use of AI can introduce unpredictability,[406] leading to negative consequences even when there is no flaw in design or implementation. Establishing a breach of duty can be particularly problematic, especially in cases where responsibility cannot be easily attributed to a single agent.[407] Additionally, demonstrating causation becomes complex, especially when errors involve multiple agents from different organisations.[408]

Another commonly argued set of objections refers to the very nature of any court proceeding and its detrimental effects in specific domains, such as in healthcare provision. Litigation can be time-consuming and resource-intensive, and the confrontational nature of clinical negligence cases hinders open discussions between patients and healthcare professionals about errors and their causes.[409] This adversarial approach can create a culture of blame, where the focus is on assigning fault rather than understanding the underlying causes of medical errors.[410] As a result, opportunities for learning and improvement within the healthcare system are missed. Healthcare professionals may become defensive and less willing to discuss mistakes openly, fearing legal repercussions. This can lead to a lack of transparency and hinder the implementation of necessary changes that could prevent future errors. Furthermore, the stress and burden of litigation can take a toll on the well-being of both patients and healthcare providers. Patients may experience prolonged emotional distress as they navigate the complexities of the legal system, seeking justice and resolution.[411] On the other hand, healthcare providers might suffer from professional and personal stress, which can impact their performance and the overall quality of care they provide.[412]

Given these complexities, it has been suggested that the approach to AI-related liability in healthcare (and potentially in other domains) should evolve from a focus on civil liability to one centred on the financial management of losses.[413] This shift could better accommodate the unique challenges posed by AI and foster the responsible development and use of AI technologies.

406 Issa, Jaber and Lakkis, 2024.
407 Buiten, De Streel and Peitz, 2023.
408 Kim, 2023: 162–163.
409 Raposo, 2016b: 57–69.
410 Weinberg, 2002: 93–113.
411 Wright, 2021: 645–658.
412 Madan, Das, Patley, et al. 2024: 317–325.
413 See, for instance, Marchisio, 2021.

7.1.2 Negligence, Deterrence and Effects on the Market

The role of deterrence in civil liability is not universally accepted.[414] Still, many see it as an integral part of civil liability and a form of indirect market regulation.[415] This is because the potential for liability in cases of harm encourages parties to prioritise investments in safety measures, reducing the likelihood of damages and promoting responsible practices.[416] Still, it remains a strong argument in favour of negligence-based models.

The issue is that when tort law is applied to AI harm, traditional fault-based deterrence or prevention might become ineffective.[417] In conventional tort law, deterrence works by holding individuals or entities responsible for negligent or wrongful actions, encouraging them to take steps to prevent harm. However, with AI, this model faces significant challenges, as already discussed (Chapter 5, Sections 1, 2 and 3).

Marchisio, for instance, argues that the current civil liability framework would not incentivise the AI industry to make robust investments in safety.[418] The application of traditional liability principles, particularly strict liability regimes, could lead to producers and programmers facing unpredictable and potentially unlimited claims. Instead of safety improvements, this scenario would deter them from entering the market and hinder technological progress, an effect often referred to as the 'technology chilling' risk.[419] For this author, mere deterrence-based investments would not effectively prevent the risks associated with AI behaviour, which can be unforeseeable. Producers and programmers, exposed to such liability claims without the possibility of mitigating risks through safety investments, would simply be discouraged from engaging in AI market activities. As a result, striking a balance in civil liability becomes vital to encourage responsible innovation and advancement in AI technology.

414 Van Rooij and Brownlee, 2021: 311–325.

415 Marchisio, 2021.

416 This holds across legal systems, although the extent and intensity of deterrence may vary significantly. For example, civil law systems typically limit deterrence to compensating for the actual damages suffered by the injured party, while US law may allow courts to award punitive damages.

417 Marchisio, 2021.

418 Marchisio, 2021.

419 Marchisio, 2021:. 53.

7.2 *The No-Fault System*

7.2.1 The Functioning of the No-Fault System

In the domain of traditional tort liability, some advocate applying fault-based rules,[420] while others favour strict liability regimes.[421] The proposals for a no-fault system aim instead to address the changing landscape of AI-related liabilities and to offer a compelling alternative.[422]

This system aims to provide compensation for unexpected harmful outcomes in healthcare without resorting to litigation or the need to attribute fault to any specific healthcare professional or part of the healthcare system.[423] The process is intentionally designed to be non-adversarial and more efficient than traditional litigation.

In our scenario, patients who could prove they were harmed by medical AI would be able to receive predetermined compensation without needing to prove fault.[424] This compensation could be managed by a government agency or another independent organisation.[425] The funding for such a system could come from a mix of taxes and fees, distributed based on the hospital's usage of medical AI or through contributions from AI developers.[426]

7.2.2 Pros and Cons of the No-Fault System

Supporters of the no-fault system argue that it benefits all parties involved, including patients, healthcare professionals, and the healthcare system. The non-adversarial nature of the no-fault compensation process fosters an environment where healthcare professionals and providers feel more comfortable reporting errors. It also facilitates reconciliation, with patients receiving explanations and apologies from healthcare professionals and providers involved.[427]

No-fault compensation systems require certain conditions to be met before compensation can be awarded. The most common requirements are

---

420 Abbott, 2018: 1–45.

421 Buonanno, 2019.

422 The term 'no-fault' pertains to a system that offers compensation through a specialised fund without requiring the establishment of fault by the responsible party. This concept differs from strict liability schemes, which also disregard 'fault' as a prerequisite for liability but function differently. Strict liability imposes the obligation to provide compensation on the responsible agents irrespective of their culpability, whereas in the no-fault model no legal responsibility is assigned to any party. See Gaylord, 1972: 589–93.

423 Holm, Stanton and Bartlett, 2021: 173–174.

424 Holm, Stanton and Bartlett, 2021: 184.

425 Mungwira, Guillard, Saldaña, et al., 2020.

426 Holm, Stanton and Bartlett, 2021: 184.

427 Raposo, 2016b: 61–67.

as follows.[428] First, the healthcare outcome must be avoidable. Secondly, the unexpected outcome resulted (in terms of causation) in harm. When an advisory AI system makes an error that could have been prevented if the system had worked better,[429] and its advice leads to harm, both prerequisites for compensation in a no-fault scheme are met. Consequently, the patient is entitled to compensation. The beauty of this approach lies in its simplicity, as it does not demand an exhaustive investigation into the specific causes of the AI malfunction, such as issues with inputs or internal models, the reasons behind the doctor's adherence to the advice, the actions of other doctors in similar situations, or whether another AI system might have performed better. Instead, we can readily conclude that compensation is warranted without having to pinpoint individual instances of responsibility or negligence.

However, no-fault systems also encounter challenges.[430] A major obstacle to implementing such systems is the projected cost. If all patients with unexpected harmful outcomes become eligible for compensation, costs are likely to rise, even considering potential savings on litigation expenses.[431] Additionally, determining what qualifies as an 'avoidable' event in this context poses difficulties, as it involves establishing something akin to negligence. Another issue is determining the threshold for the lowest amount of harm that warrants compensation. Moreover, critics of no-fault systems contend that, since there is no allocation of blame, they reduce the deterrent effect of clinical negligence litigation.[432] A reasonable solution, compensation-wise, could be a scheme funded by contributions from the State and relevant stakeholders, including AI system designers and manufacturers.[433] This fund could potentially be granted the authority to pursue recovery actions on behalf of patients, aiming to recover pay-outs in cases of AI-related harm. Additionally, a hybrid system might be considered to address patients' rights in this context.

428 Holm and Bartlett, 2021: 173–174.

429 'Avoidability' is a tricky concept, but it seems to refer to the possibility of preventing the harm had the patient been subjected to proper medical care (whatever that is). Cf. Raposo, 2016b: 62–64.

430 Raposo, 2016b: 61–67.

431 Howard, McWilliams, Hannant et al., 2019: 387–390.

432 Farrell, Devaney and Dar, 2010: 10.

433 Holm and Bartlett, 2021: 184.

8 Insurance

8.1 *The Added Benefit of AI Insurance Schemes*

Building a strong AI insurance system is a promising way to create a safer environment where medical AI tools can grow and improve.[434] Beyond the scope of national regulation, AI insurance can play a crucial role in certifying that technologies demonstrate limited and quantifiable risks. When insurers provide coverage for AI-related risks, it signals a level of trust and reliability in the technology.[435] This assurance encourages investors and stakeholders to allocate resources towards the development and deployment of medical AI tools, knowing that potential risks are mitigated by insurance policies.[436] By offering actuarially assessable solutions, insurers help create a potential market for insurance products, attracting investors to safer AI projects.[437]

Moreover, insurers possess the expertise to evaluate and quantify risks associated with AI technologies. They can use risk analysis to spot weaknesses and help AI developers create better safety measures, lowering the risk of failure or other problems.[438]

Ultimately, insurance coverage can reduce the financial burden associated with potential risks, allowing AI developers to focus more on innovation and less on risk avoidance. This fosters a more dynamic and experimental approach to developing advanced medical AI tools, accelerating progress and breakthroughs in the field.[439]

In turn, end-users, including healthcare providers and patients, are more likely to adopt medical AI tools if they are backed by insurance. This coverage reassures users that they are protected against unforeseen issues, thereby increasing acceptance and integration of AI technologies in clinical settings.[440]

By managing and distributing risks, insurers contribute to a balanced and sustainable market, where AI companies can thrive without being

---

434 See Stern, Goldfarb, Minssen, et al., 2022; Deloitte; Lior, 2022: 468–530; Faure and Li, 2022: 1–24.

435 Lior, 2022: 482–484.

436 Lior, 2022: 482–487.

437 The idea presented by the Expert Group on Liability and New Technologies involves mandating insurance obligations and contemplates the potential creation of compensation funds (European Commission, Directorate-General for Justice and Consumers, 2019: 61–62).

438 Lior, 2022: 511–512.

439 Lior, 2022: 523.

440 Deloitte, 2024.

overly exposed to catastrophic risks. This dynamic will influence insurance policy costs and drive wider adoption not only of AI technologies but also of liability insurance for their use, creating a positive cycle in the evidence-based development and deployment of AI products.[441]

8.2 *Insurance for Humans and Insurance for Software*

Traditional malpractice insurance typically covers medical practitioners exclusively. However, if AI were endowed with legal personhood, it would essentially become another entity akin to a medical professional within healthcare settings. This recognition would position AI at the same level as doctors and doctors in hospitals, clinics, or healthcare facilities.

A similar result could be achieved without granting legal personality to AI. By assigning liability to a legal person (the hospital, for instance), the system would ensure that medical malpractice insurance provides adequate compensation for patients adversely affected by AI-related incidents. While hospitals would be responsible for paying insurance premiums, the coverage would effectively mitigate the financial burden of legal proceedings in the event of a malpractice lawsuit involving the AI system. Moreover, instituting this insurance scheme would motivate healthcare providers to maintain the AI system's optimal functionality.

Similar to medical malpractice insurance, liability insurance for algorithms shares the characteristic of covering errors whose aggregate risk probabilities are assessed by insurance providers.[442] AI liability insurance differs from medical malpractice insurance due to the issue of scale, especially with software, where a single problem can quickly spread and affect a large number of users.[443] Unlike a single malpractice claim specific to one medical encounter, an AI liability claim can involve errors impacting hundreds of patients, such as a drug-dosing algorithm affecting an entire health system.[444]

8.3 *The Costs of AI Insurance*

The acceptance of insurance becomes advantageous when society collectively recognises the overall benefit of incorporating AI in clinical decision-making. Usually, the burden of insurance costs ultimately falls on those who pay for the healthcare provided. Thus, it is not the developer/manufacturer or doctor

441 Stern, Goldfarb, Minssen, et al., 2022.

442 Stern, Goldfarb, Minssen, et al., 2022.

443 Stern, Goldfarb, Minssen, et al., 2022.

444 Stern, Goldfarb, Minssen, et al., 2022.

directly bearing the cost, but rather the patients or those responsible for their care. From this perspective, the community genuinely carries the liability cost.[445]

Alternatively, manufacturers can incorporate liability insurance costs into contracts with healthcare providers, freeing the latter from liability related to AI tool usage. This approach creates more attractive AI products for smaller healthcare providers who may not have the capacity to research their own insurance policies, especially during the early stages of AI tool implementation.[446]

In cases where a product presents exceptionally high risk or its risk is not well-established, insurance may be expensive or not offered at all, acting as a deterrent to adopting or developing such products. This selective approach ensures that insurers play a vital role in credentialing, complementing regulatory review to ensure the safety and effectiveness of AI products.[447] Still, if market agents – either manufacturers or deployers – look for 'financial security',[448] insurance companies will eventually react.[449]

However, depending heavily on insurance systems, particularly those that are mandatory, could turn out to be insufficient under circumstances where the market for such coverage does not justify it. This situation might arise when the risks linked to a particular technology are exceptionally high, difficult to control or estimate, or when the number of users is insufficient in proportion to the identified risks.[450]

445 Smith and Fotheringham, 2020: 148.
446 Stern, Goldfarb, Minssen, et al., 2022.
447 Stern, Goldfarb, Minssen, et al., 2022.
448 Faure and Li. 2022: 6.
449 Faure and Li. 2022: 11–13.
450 Bertolini and Episcopo, 2021: 657–658.

CHAPTER 6

A Standard of Care for AI Medicine

1 Medical Standard of Care and Technological Innovation

1.1 *A Delicate Balance between Regulation and Innovation*

A rigid interpretation of the standard of care could impose significant barriers to the early adoption of new medical devices and technologies, as deviating from customary practices may be deemed unreasonable. While a demanding liability model in the medical AI context deals with accountability and recompense concerns, it could inadvertently hinder the advancement of AI innovation in the field of medicine.[1]

A more sophisticated approach acknowledges the evolution of clinical practices over time and the frequent utilisation of new technologies and techniques by doctors. Even when these devices are novel, their use by doctors might be considered reasonable and even desirable. In fact, a new device that offers clear benefits without additional risks would be seen as a highly desirable innovation that does not harm patients.[2] Still, a new device that carries additional risks presents more challenges under the malpractice standard of care; nevertheless, there may still be circumstances where the adoption and use of such a device would be considered reasonable under the law and the applicable standard of care.[3]

Overly strict regulations can unfairly slow down beneficial innovations. AI in healthcare is still developing, and too much regulation too soon could delay important advancements that could help patients and the healthcare system.[4]

However, it is equally vital to tread cautiously and avoid a regulatory framework that is overly permissive, as it could erode the trust between patients and doctors or compromise fundamental human rights. While fostering innovation through liberal regulations is crucial, it should never come at the expense of ethical considerations or violations of basic human rights.[5]

1 Jassar, Adams, Zarzeczny, et al., 2022: 187.
2 Greenberg, 2009: 432.
3 Greenberg, 2009: 431.
4 Schönberger, 2019: 182.
5 Muraille, 2019.

 | HTTPS://DOI.ORG/10.1163/9789004543980-008

1.2 *The Doctor-Machine Relationship*

The role of medical experts should not be reduced to that of mere executors of AI systems' advice or mere intermediaries between AI and patients.[6] A more effective approach involves integrating human and automated judgments. This allows doctors to review and possibly adjust the AI's suggestions to align with the unique goals and preferences of individual patients.[7]

When humans collaborate with AI systems to achieve a common goal, it is essential to view them not as separate entities but as part of a joint cognitive system, where performance is shared between humans and AI.[8] In this setup, the human maintains the ability to oversee the AI's overall activity, including its legal and ethical impact on patient care. The human should be empowered to decide whether and how to use the system and rely on its recommendations.[9] This shared model enables doctors to base their decisions not only on literature and clinical evidence but also on the individual patient's unique biology, values, and preferences. Additionally, it allows them to account for factors that the AI cannot adequately perceive, such as the patient's emotional state and beliefs.[10]

A significant portion of current medical AI does not replace human doctors but, rather, expedites and aids human diagnosis.[11] While AI provides data-driven insights, human doctors contribute their clinical expertise, intuition, and experience.[12] Despite AI's capabilities, the ultimate decision-making authority still resides with humans[13] (a modality of 'Augmented Intelligence').[14] This ensures that ethical considerations, patient preferences, and complex medical judgments are factored into care decisions.

---

6 Lagioia and Contissa, 2020: 272.

7 Lagioia and Contissa, 2020: 272.

8 Naikar, Brady, Moy, et al., 2023: 1669–1694.

9 Lagioia and Contissa, 2020: 278.

10 Lagioia and Contissa, 2020: 278.

11 Sezgin, 2023.

12 Asan, Bayrak and Choudhury, 2020.

13 Several studies are drawing our attention to the fact that AI is endangering our decision-making power and that over-reliance on AI are making humans 'lazy' (Ahmad, Han, Alam, et al., 2023). This is a growing concern, but we must make sure that the decision-making powers is always on our side.

14 'Augmented intelligence (AuI) integrates human intelligence (HI) and artificial intelligence (AI) to harness their strengths and mitigate their weaknesses. The combination of HI and AI has seen to improve both human and machine capabilities, and achieve a better performance compared to separate HI and AI approaches' (Yau, Lee, Chong, et al., 2021: 136744).

2 Medical AI Specific Patient's Rights

2.1 *Privacy Rights and Data Protection Rights*

Privacy threats and data breaches are a global concern, extending beyond the sphere of medical AI alone.[15] These are pervasive issues affecting all sectors of society, from finance to e-commerce, including healthcare delivery, where this is not a new problem. For instance, the safeguarding of patient records remains a significant concern irrespective of the presence or absence of AI use.[16]

The GDPR safeguards individuals' rights through several provisions.[17] One such right is the right to be informed, outlined in Articles 13 and 14, which ensures that patients are provided with clear information about how their personal data are used.[18] This includes cases where automated decision-making systems, such as AI, are involved in medical diagnoses or treatment recommendations. Healthcare providers must explain the logic behind these systems and the possible impacts on the patient's care when such decisions carry legal or significant health-related outcomes. Another critical right under the GDPR is the right of access, as described in Article 15.[19] This right allows patients to request and obtain access to their health data, including any data related to automated decision-making. Patients also have the right to object to the processing of their personal data under Article 21.[20] This means that, in certain circumstances, they can object to how their data are being used, especially if it involves profiling or automated decision-making. Additionally, there is an absolute right to object if their data are being used for direct marketing purposes, such as targeted pharmaceutical ads. As per Article 22 GDPR, if an AI system is used to assess a patient's condition or suggest a treatment plan, the patient has the right to understand how these decisions were made and how they might affect their health outcomes. These GDPR provisions are designed to ensure that patients maintain control over their personal (and sensitive) health data, particularly in the context of data-driven healthcare decisions involving automated systems like AI.

---

15 Farayola, Olorunfemi and Shoetan, 2024: 606–615.

16 Dalton-Brown, 2020: 116. Interestingly, a machine that does not genuinely remember patient details could provide better security than a doctor who retains specific patient information. Doctors might inadvertently share patient details in casual conversations or through forgetfulness, while machines do not have such lapses (Beltran-Aroca, Girela-Lopez, Collazo-Chao, et al., 2016).

17 Bharti and Aryal, 2022: 1391–1402.

18 Selbst and Powles, 2017: 233–242.

19 Selbst and Powles, 2017: 233–242.

20 Zanfir-Fortuna, 2020.

2.2 *The Right Not to Have AI Involved in Medical Care*

2.2.1 Sub-Dimensions: Information and Opt-Out

The right of patients to refuse AI involvement in their medical care is a crucial aspect of patient autonomy and informed consent.[21] This right requires patients, from the outset, to be fully informed about the use of AI in their medical care (informed consent), with a clear explanation of when and how AI tools are used, what the AI does, how it reaches decisions, and any limitations or risks associated with its use. [22] In parallel, patients must be informed of available non-AI alternatives for their care, including traditional diagnostic and treatment methods.[23]

In a second moment, clear opt-out mechanisms must be made available – possibly, within the informed consent form – for patients to opt out of AI involvement.[24] In addition, I suggest verbal confirmation, whereby the healthcare provider verbally confirms patients' choices and documents their preferences in medical records. EHR software must be designed in such a way that it can accommodate patient preferences. This will require, for instance, modifying EHR systems to record and flag patient preferences regarding AI involvement and create alerts and integrate within clinical workflows to remind healthcare providers of patients' choices.

This procedure involves an ongoing review, thus allowing patients to change their decision at any point during their care.[25] Patient counselling must be put in place to help patients understand their choices and the potential impacts on their care.

2.2.2 The Right Not to Be Subject to Automated Decision Making

This right can be connected to Article 22 of the GDPR, which encompasses rights related to automated decision-making and profiling, providing individuals with the right to avoid being subjected to decisions that are based solely on automated processes and have legal or similarly significant effects.[26] Prima facie, this is not allowed.

However, there are exceptions, which impose obligations on organisations to implement appropriate measures to protect individuals. This includes granting them the right to obtain human intervention, express their viewpoint, and

21 Iserson, 2024: 225–230; World Health Organization, 2021: 143.
22 Prakash, Balaji, Joshi, et al., 2022.
23 Park, 2024.
24 Ploug and Holm, 2020: 107–114.
25 Solaiman, 2024.
26 About this norm, Edwards and Veale, 2017: 44–51.

contest the decision.[27] Recital 71 offers interpretative guidance primarily on Article 22 rights, specifying that individuals have the right to request an explanation of a solely automated decision following its execution.

Article 22 of the GDPR comes into play solely when processing occurs exclusively through automated means. AI systems, however, are typically not entirely automated. They function more as decision-support tools and are often inappropriate or premature for full automation.[28]

The ambiguity around this norm lies in discerning the point at which 'nominal' human involvement ceases to qualify as involvement.[29] Several European data protection authorities are currently grappling with this question. Merely nominal human participation can happen either because there is insufficient human intervention or because of 'automation bias', a psychological tendency for humans to overly rely on or underutilise decision support systems.

Another complexity emerges from Article 22's requirement for a 'decision' that either 'produces legal effects, concerning him or her, or significantly affects him or her'. The GDPR's limited explanation of a 'decision' raises two issues. First, does an AI system's output qualify as a 'decision'?[30] AI experts contend that an algorithm's output serves as a precursor to a decision, whether made by another system or a human (like a judge). When queried, AI models generally output classifications or estimations, often accompanied by uncertainties. Independently, they lack the capacity to translate these into actionable decisions. Second, even if we assume that algorithmic 'output' is a 'decision', the question remains: when does an AI 'decision' produce 'legal effects concerning him or her or similarly significantly affects him or her' (Article 22(1) GDPR)?[31] For instance, if an AI system classifies individual X as 75% more likely than average to belong to group A, where group A is associated with an undesirable attribute (such as a tendency not to comply with medical guidance or a high likelihood of therapeutic failure due to excessive weight), is this a decision about ('concerning')' X? In multiple cases involving AI systems, including in healthcare delivery, the distinction is not clear.

2.2.3 The Right to be Treated Equally

Healthcare providers must respect patients' choices (to accept or not accept AI involvement) without prejudice, ensuring that patients who opt out of AI

27 Vale and Zanfir-Fortuna, 2022.
28 Gulley and Hilliard, 2024.
29 Lazcoz and De Hert, 2023.
30 This discussion can be found in Langer and Landers, 2021.
31 Hintze, 2020.

involvement receive the same quality of care and attention as those who do not. Healthcare providers must guarantee that opting out of AI does not result in any form of discrimination or reduced quality of care.[32] Patients who choose traditional methods should receive the same level of attention, resources, and medical expertise.

Suppose a case where the patient asks to opt out of AI involvement in their care and prefers to rely solely on the doctor's expertise. If, over the course of the appointment, the doctor spends less time with that patient compared to other patients who accepted AI involvement, performs only basic checks and provides a generic advice sheet rather than a detailed, personalised treatment plan, this would be an example of discriminatory treatment.

2.2.4 Difficulties in Exercising the Opt-Out

There might be situations where a healthcare facility cannot provide the required care without using AI. The staff in charge must clearly communicate the facility's limitations, the reasons why AI is integral to the care provided and discuss potential alternatives.

In such cases, an alternative path must be in place to respect the patient's right to opt out of AI's involvement while still ensuring they receive appropriate medical care. Healthcare units must develop a customised care plan that minimises or excludes the use of AI, leveraging available resources and more classic methods.

Healthcare staff must work with the patient to find acceptable compromises that align with their preferences while ensuring they receive necessary care. If the situation is complex, involve an ethics committee to review the case and provide recommendations that balance patient autonomy with the need for effective medical care.[33]

Ultimately, the patient might be referred to other facilities, which can offer the required healthcare without the use of AI. The transfer from the original healthcare facility must be facilitated, ensuring a smooth transition and continuity of care. If another facility is not easily accessible, consider using telemedicine services to consult with healthcare providers who can offer non-AI-based

---

32 De Miguel Beriain, 2020: 251.

33 Ploug and Holm use the case of Jehovah's Witnesses to illustrate cases in which the healthcare system addresses and accommodates the fears and concerns of some patients, even when these fears are not shared by the majority or considered rational by the standard definition, and claim the same could work for patient refusing to use AI. See Ploug and Holm, 2020: 108.

advice or treatment plans. Engage specialists who can provide input remotely, allowing the local facility to follow non-AI-based treatment protocols.

It must be acknowledged that such procedures might become complex, lengthy (while often there is urgency medical care) and costly. To avoid additional hurdles, healthcare facilities must develop policies within the healthcare facility to handle cases where AI use is integral to care, ensuring that patient rights are respected, and alternative solutions are considered.

Nonetheless, if AI becomes recognised as the best standard of care and the remaining alternatives become substandard medicine, to offer such an alternative could be seen as a form of medical malpractice. Therefore, there might be situations in which the patient would not be able to opt out simply because there is no alternative.[34]

2.2.5 Balancing Patients' Autonomy with (Other) Medical Duties

In any case, the right of patients to refuse AI involvement in their treatment decisions should not compel doctors to violate widely accepted medical ethical standards.[35] Doctors are bound by the principles of non-maleficence and beneficence, which requires them to act in the best interest of their patients.[36] If refusing AI involvement means that a doctor would be forced to provide suboptimal care, they are ethically obliged to seek the best alternatives while ensuring the patient's decision is respected.[37]

Doctors are professionals entrusted with the responsibility to adhere to the best practices in medicine. Medical ethics requires that decisions should not only respect patient preferences but also align with evidence-based standards of care. Forcing a doctor to ignore AI when it is scientifically proven to aid in decision-making or diagnosis would be asking them to disregard established medical protocols, potentially placing them in an ethically compromising position. Therefore, doctors should not be compelled to offer treatment options that are inferior or outdated simply because a patient refuses AI assistance. Ethical practice demands that doctors inform patients about the risks

34 Mourby et al. suggest that if no alternative care is available, then using consent as the legal basis for processing data under the GDPR becomes problematic. The patient's consent cannot be considered 'freely given' if refusing AI-enhanced treatment results in the denial of care, thus violating the condition for valid consent under the GDPR (Mourby, Ó Cathaoir and Collin, 2021).

35 De Miguel Beriain, 2020: 250.

36 Moulton and King, 2010: 85–97; Varkey, 2021: 18–20.

37 Varkey, 2021: 21.

and benefits of refusing such technologies, but they are not obligated to violate clinical guidelines by providing lesser care.[38]

For instance, if an AI system offers enhanced diagnostic precision or treatment planning that significantly improves patient outcomes, a doctor must weigh the refusal of AI against the ethical duty to offer the most effective care. Denying the use of AI may result in lower-quality outcomes, putting the patient at risk of harm. Legally, doctors cannot be compelled to administer care that they believe contradicts their professional judgment, even by court order.[39]

2.2.6 The Doctor's Perspective: the Right Not to Be Compelled to Refrain from Using AI

No healthcare provider can be compelled to perform medical acts that fall below the standard of care, regardless of patient requests or institutional pressures.[40] This principle is grounded in several key considerations.

First, there is the unavoidable risk of liability. Doctors bear legal and professional responsibility for their actions. Performing substandard procedures exposes them to malpractice claims, disciplinary action by medical boards, and reputational harm. The duty of care they owe to their patients cannot be waived or diluted by external demands.

Second, while patients have rights, including autonomy and informed decision-making, doctors also have rights, notably the right to exercise their clinical judgment free from coercion.[41]

Third, medicine is traditionally conceived as a 'liberal profession',[42] meaning that it combines technical expertise with moral responsibility.[43] Doctors are not mere service providers; they are entrusted with protecting health and life. This professional ethos obliges them to refuse interventions that are unsafe, futile, or harmful, even under patient insistence or administrative pressure.

Finally, patients are not customers in the conventional market sense.[44] Healthcare cannot be equated with consumer goods or services that can be demanded and delivered on request. It operates under a different normative framework: one where decisions must be clinically justified, ethically

38 Rubin and Prager, 2018: 826–829.

39 Pruski, 2024.

40 'In no circumstance should a physician be compelled to provide substandard care to a patient, nor be forced to choose between a legal imperative or their ethical obligations to each patient' (Resneck, 2023).

41 World Medical Association, 2025.

42 European Economic and Social Committee, 2013.

43 European Economic and Social Committee, 2013: 109.

44 Heubel, 2000: 240–253.

defensible, and aligned with professional standards of care. Medicine is not a purely transactional activity where professional expertise is subordinated to client instructions; it is a fiduciary relationship structured around standards of science, ethics, and law.

Thus, compelling a doctor to perform a substandard act undermines not only patient safety but also the integrity of medical practice as a profession governed by science, ethics, and law. The doctor-patient relationship is one of trust and expertise, not of contractual obedience.

3 AI Standard of Care: When to Use AI / When to Rely on AI

3.1 *Trust in Technology*

When examining trust in technology, three key factors come into play: human characteristics, environmental characteristics, and technology characteristics.[45]

Human characteristics encompass various aspects such as the personality of the individual placing their trust (trustor), their inherent disposition to trust, and the capabilities of the party being trusted (trustee) to manage risks. The trustor's personality and disposition to trust are influenced by their experiences, personality traits, and cultural background. Meanwhile, the trustee's ability refers to their competence and skill set in fulfilling tasks within a specific domain.[46]

Environmental characteristics include factors in the surrounding context, such as the nature of tasks, cultural backgrounds, and institutional influences. The nature of tasks can vary in terms of importance, ranging from highly significant to trivial. Cultural background encompasses elements such as ethnicity/race, religion, socioeconomic status, as well as associations with specific countries or regions.[47]

---

45 Siau, Wang, 2018: 50.

46 For example, a manager may trust an employee with negotiation tasks if they demonstrate a high level of competence in that area. Likewise, a doctor may rely on AI for diagnostic decisions if the AI consistently demonstrates a high level of accuracy and reliability in identifying medical conditions.

47 An example of trust in AI in medicine influenced by cultural background can be seen in the use of AI-powered telemedicine platforms in regions with lower socioeconomic status or distinct cultural practices, such as rural areas in India. In many of these regions, patients may be hesitant to trust healthcare technologies due to various factors, including religious beliefs, traditional medical practices, or mistrust in Western medicine. However, the success of these AI tools in building trust with local populations heavily depends on how well they account for local cultural practices, languages, and common health issues specific to that community (Nagarathinam, Chellasamy, Elangovan, et al., 2023).

Technology characteristics focus on the attributes and features of the technology itself that contribute to establishing trust.[48] This includes factors such as reliability, security, transparency, and ease of use. Each case is a different case. For instance, the challenges faced by CDSS vary depending on the specific decision-making task.[49] While CDSS for clinical diagnosis can be built on established links between clinical data and gold standards for accuracy, the same cannot be said for therapy planning. In complex therapeutic scenarios, where no gold standard exists and expert opinions may differ, the formulation of a therapeutic plan becomes more subjective.[50] Evaluative studies should not aim to determine the 'correct approach' but rather compare the performance of the CDSS against other doctors, typically recognised experts, in similar tasks. If a therapeutic advisory system and an expert doctor offer differing conclusions for managing a specific case, neither may be definitively 'correct', and both plans, or neither, could be reasonable. Thus, the focus should be on assessing how well the CDSS performs compared to other doctors in the given task, rather than seeking a definitive correct answer.[51]

3.2 *Trust in AI Medicine*

3.2.1 Trust within the Doctor-Patient Relationship

Trust involves several key aspects that shape human interactions.[52] These dimensions are particularly relevant when analysing trust in various contexts. The first dimension, often referred to as 'behavioural trust', is built upon predictability derived from past experiences. For instance, individuals may trust that their alarm clock will ring each morning based on its consistent performance in the past. This form of trust is limited in its applicability to known scenarios and lacks the capacity for generalisation to novel situations. In

48 Hosseini Shoabareh, Ghasri, Roberts, et al., 2024.

49 Shortliffe and Sepúlveda, 2018: 2199–2200.

50 For instance, a study focusing on antibiotic therapy selection for meningitis patients prior to laboratory identification of the infecting organism found that even infectious disease experts did not consistently suggest acceptable regimens (Tunkel, Hartman, Kaplan, et al., 2004: 1277). This underscores the absence of a single 'right answer' in many therapeutic decisions.

51 Some trust assessment systems are already being developed. The Aletheia Framework, a ground-breaking ethics toolkit and trustworthiness process developed by Rolls-Royce, empowers AI developers to address pressing concerns related to automated systems (Rolls-Royce, 2020). It encompasses a wide range of crucial aspects, such as preventing the perpetuation of social biases, preserving job opportunities and human skills, ensuring accountability, and building trust in algorithmic outputs. About this tool, see Hallows, Glazier, Katz, et al., 2022.

52 I closely follow LaRosa and Danks, 2018: 211.

contrast, 'understanding trust' delves into the mechanisms that drive actions. Trust based on understanding arises from comprehending the reasons and motivations behind a trustee's behaviour. An illustration of this is the trust one places in a close friend, not solely due to the ability to predict their actions, but due to an awareness of their beliefs, values, and intentions. This type of trust offers the advantage of extending beyond familiar situations to encompass unfamiliar contexts. Another dimension, known as 'role-based trust', acts as a bridge between behavioural and understanding trust. It is not an independent form of trust but serves as a conduit for either behavioural or understanding trust to flourish, contingent upon the nature of the role itself. When roles are defined by consistent behaviour, such as the reliability of a postman delivering the mail on specific days, role-based trust aligns with behavioural trust. Conversely, roles grounded in skills or functions engender understanding and trust. The role of a doctor shows this, as it relies on expertise and knowledge. Patients entrust doctors not merely based on predictability but on their understanding of the doctor's qualifications, beliefs, and intentions.[53] Within medical teams, role-based trust takes on a leading role. In complex patient care scenarios involving diverse professionals, role-based trust serves as the adhesive that binds the team's combined expertise. Patients place their trust in the collective competence of the medical team, relying on their individual roles to contribute to the overall quality of care.[54]

The rapport a patient forges with their doctor serves as a formidable catalyst, capable of yielding both positive and negative consequences for the foundation of trust within the patient-doctor relationship. Through a series of recurrent interactions, the patient conditions the doctor's appraisal of their own medical condition,[55] and also the foundation of trust.[56] A clear example is when a patient maintains open communication with their doctor, discussing treatment options and care plans. This ongoing dialogue gradually builds trust and understanding. Conversely, if a doctor consistently ignores the patient's preferences, it can weaken the trust between them. These interactions create a blend of experiences and social connections that reinforce trust over time.[57]

53 Chipidza, Wallwork and Stern, 2015.
54 Langlois, 2020: 71–73.
55 Roter, 2000: 5–15.
56 Thom, Kravitz, Bell, et al., 2002: 476–483.
57 LaRosa and Danks, 2018: 213.

3.2.2 Patient's Trust in AI

The advent of AI-powered clinical tools has the potential to transform the way doctors and patients interact. Some patients may be hesitant to embrace AI devices due to their desire for human social interaction when seeking healthcare services.[58] Namely, patients may express concerns about the potential loss of in-person cues and personal interactions with their healthcare providers.[59]

Users' perception of trustworthiness in AI-based medicine is also influenced by the level of autonomy exhibited by AI systems.[60] There are no fully autonomous AI systems in healthcare currently, but some of them are more autonomous than others. Systems that operate with more human oversight often foster greater trust,[61] as users feel more secure knowing that a human expert is involved in interpreting and acting on AI recommendations.[62]

A major factor in building trust in AI is the level of scrutiny applied to the approval, market entry, and post-market surveillance of AI systems. Patients (and doctors alike) are more likely to trust technology that has undergone rigorous regulatory approval or other validation and certification processes, especially for high-risk technologies. Therefore, a strong regulatory framework for medical AI would help to establish and build that trust.[63]

It has been argued that if AI manages to 'gain' patients' trust, it can bring more openness and transparency to the doctor-patient relationship. Imagine a situation where a smart AI device in healthcare gives important information to a doctor. Interestingly, people tend to share more information with an AI than with a human, especially when the information might make them look bad socially, like not taking prescribed medicine or abusing illicit substances.[64] As a result, the AI device can gather a lot of unbiased information, which can improve the patient's experience and outcomes. This could help build trust between the patient and the healthcare provider. By showing how well the patient follows their treatment plan and how they respond to treatments, this AI helps the doctor find the best ways to treat the patient, possibly avoiding the need for lots of expensive medical tests. Still, this is merely a forecast, as

---

58 Gursoy, Chi, Lu, et al., 2019: 159.

59 Esmaeilzadeh, Mirzaei and Dharanikota, 2021.

60 Kühne, Jacobsen, Legewie et al., 2025.

61 However, as Holzinger, Zatloukal and Müller (2025: 59–62) highlight, the increasing complexity of contemporary AI architectures – including large-scale neural networks and generative AI systems – erodes human interpretability and constrains meaningful oversight and decision-making capacity.

62 Witkowski, Okhai and Neely, 2024.

63 Meskó and Topol, 2023: 120.

64 Jiang, 2024.

up until now we do not have enough clinical evidence to attest to the veracity of this prediction.

3.2.3 Doctors' Trust in AI

Doctors' trust in AI is crucial for building patients' trust. However, doctors must be candid about their own level of trust in the technology. By being transparent about their trust in the AI, doctors protect themselves from potential liability and ensure that they do not mislead patients into believing the AI is infallible.

The central issue is whether healthcare providers should be held accountable for relying on AI technology,[65] especially when AI systems are designed to assist rather than replace human judgment. Under frameworks like the AIA and the liability directives, providers may still be expected to exercise their professional judgment and not blindly follow AI recommendations. Failing to do so could result in negligence claims if harm occurs.

However, if liability rules impose penalties on healthcare providers for relying on AI technology that yields incorrect conclusions, they might opt to use the technology merely to confirm their judgments. Although this approach could shield them from liability, it would discourage the full utilisation of AI's potential, which lies in complementing and enhancing human judgment rather than merely validating it.[66]

3.3 *Disagreements between the Doctor and the AI Systems*

When the machine and the human converge in the same scenario, it might be the case that the data, when interpreted by the machine, could diverge from the thought process of the human doctor regarding the significance of some elements (the data) within the context. To illustrate, a doctor might deliberately disregard a profoundly unusual blood test outcome, potentially resulting from a mishap during blood collection, test analysis, or the transcription of results, while the AI system might give particular importance to the blood test results.[67] What to do when the doctor and the AI system have different 'opinions' in the medical acts, for instance, regarding the diagnosis or the therapeutic measures to be carried out, that is, when there is a 'peer disagreement' between a doctor and an AI system?

When these equally competent peers hold conflicting views on a particular proposition within their domain of expertise, it creates a dilemma for the doctor. Both the doctor and the algorithm can be seen as experts in their respective

65 Herington, McCradden, Creel, et al., 2023: 1512.
66 World Health Organization, 2021: 76.
67 Habli, Lawton and Porter, 2020: 253.

fields, but their training and reasoning processes differ significantly.[68] This scenario shows the importance of balancing the differences in how human doctors and AI systems make decisions to achieve the best and most effective results.

A mechanism to address conflicts between humans and AI systems can be found in the GDPR, which prohibits decisions made entirely by automated systems that have a significant impact on individuals (Article 22(1) GDPR). Likewise, the AIA imposes human oversight for high-risk AI systems (Articles 14 and 26 AIA) to ensure that decisions with substantial consequences, such as those in healthcare, involve human judgment as the 'last word'. Thus, in case of conflict, the human decision will always prevail.

Another possible approach to mitigate issues resulting from disagreement scenarios is to implement an adjudicator algorithm that can resolve conflicts between AI systems and doctors.[69] However, building an effective adjudicator algorithm for conflicts between AI systems and doctors is not an easy endeavour. It requires a multidisciplinary approach involving experts in medicine, AI, ethics, and law. It also implies a commitment to transparency and ethical decision-making, as well as an ongoing refinement to ensure the best possible patient care.

An alternative approach is to involve the patient in the decision-making process through informed consent or shared decision-making, leaving the final decision – whether to follow the doctor's opinion or the AI's recommendation – to the patient.[70] Although some doctors may view this as a sign of professional weakness and fear that it might cause patient anxiety, from an ethical standpoint, it is an acceptable strategy. In fact, many patients may appreciate their doctor's honesty, trustworthiness, and willingness to engage them in the decision-making process. Moreover, patients can provide valuable insights into their perspectives on the risks and benefits of different treatment options.[71] From a liability perspective, discussing the clinical uncertainties arising from a disagreement between the doctor and the algorithm and ensuring the patient's understanding can prevent the patient from later alleging negligent nondisclosure.

---

68 Kempt and Nagel, 2022: 222–229.

69 Suggestion made by Banja, Hollstein and Bruno, 2022: 818.

70 As proposed by of Banja, Hollstein and Bruno, 2022: 818.

71 I am referring to cases where the doctor negligently failed to disclose important clinical information that could have influenced the patient's decision. By actively involving the patient and promoting transparency, the risk of such allegations can be mitigated.

While involving the patient in decision-making through informed consent or shared decision-making seems ethical and empowering,[72] this approach has potential weaknesses.[73]

First, increased anxiety and burden on patients. Although technology promises various benefits, there are worries that it shifts too much responsibility onto patients. What is often framed as 'empowerment' may actually lead to an overburdening of individuals, requiring them to take on more responsibility for managing their own healthcare.[74] While some patients may appreciate the transparency, others may feel overwhelmed by having to choose between their doctor's opinion and an AI's recommendation. The responsibility of making a complex medical decision can induce anxiety, especially if patients feel they lack the expertise to fully understand the implications of their choice.

Secondly, shifting the decision-making burden onto patients could dilute the doctor's role as an expert guide. If the final decision rests heavily on the patient, there is a risk that the doctor's professional judgment is undervalued. This, in turn, may lead to suboptimal outcomes, especially if the patient lacks sufficient understanding, since this shift could potentially degrade the quality of care and result in lower healthcare standards overall.

Last but not least, it should be noted that although shared decision-making can reduce liability for nondisclosure, it does not eliminate the risk of litigation.[75] If a patient chooses an AI's recommendation that results in harm, they might still blame the doctor for not persuading them to take the doctor's advice or for insufficiently explaining the AI's limitations.

3.4 *Liability Considerations in the Use of AI in Healthcare Delivery*

3.4.1 Risk Mitigation

When doctors use new technology in their practice, it can increase the risk of malpractice claims, especially if it involves unfamiliar risks or changes how care is usually delivered.[76] Considering the increased potential for liability, it becomes essential to explore measures that doctors can undertake to mitigate their malpractice risks while remaining aligned with their legal obligations of providing care. Some assessments may be conducted before the policy or project launch to pre-empt potential risks (ex-ante assessments), while others are

72 Whitney, McGuire and McCullough, 2004: 54–9; Coulter, 1997.
73 Coulter, 1997.
74 Solaiman and Malik, 2024.
75 Schoenfeld, Mader, Houghton et al., 2019: 126–136.
76 Greenberg, 2009: 436.

carried out once the project is operational to evaluate its performance (ex-post assessments).[77]

This is particularly true for AI, as 'AI cannot be deployed in healthcare without sufficient infrastructure and risk mitigation'.[78] Although there is no universal solution, institutions may contemplate various practices to reduce AI risks.[79] These practices could involve implementing oversight and continuous monitoring, improving explainability and interpretability, and delving into emerging risk-reduction methods such as differential privacy (a technique that adds controlled noise to data to protect individual privacy) and watermarking (embedding hidden signals in AI-generated content to verify its origin), among other strategies.

Algorithmic impact assessments serve as a mechanism to evaluate the potential societal consequences of an AI system before its deployment, often accompanied by continuous monitoring.[80] These assessments find their origins in the field of policy evaluation and have been widely used in various fields like finance, cybersecurity, and environmental studies. Closely aligned assessments include data protection impact assessments (DPIAs), which measure the impact of data processing on data protection rights,[81] as well as human rights impact assessments (HRIAs).[82]

The AIA brought with it the Fundamental Rights Impact Assessment (FRIA), as outlined in Article 27 of the AIA. The FRIA's key objective is to pinpoint the specific risks that AI systems may pose to the rights of individuals or groups and outline the necessary measures to address these risks if they occur.[83] Certain high-risk AI systems, detailed in Annex III of the AIA, must undergo FRIA. These systems include those used in the following fields: biometrics, education and vocational training, employment and workforce management, access to key private and public services, law enforcement, migration, asylum, border control, justice systems and democratic processes. It is important to note that the generality of AI systems in healthcare is not currently included among the areas subject to FRIA requirements. Article 27 of the AIA only applies to the AI systems referred to in Article 6(2) of the AIA (and even then, with many exceptions), so medical AI will hardly fit the scope of the norm. Exceptions include

77 Ada Lovelace Institute, 2022: 15.
78 Renda, Arroyo, Fanni, et al., 2021: 64.
79 Mello and Guha, 2024: 276–277.
80 Ada Lovelace Institute, 2022.
81 Demetzou, 2019.
82 Walker, 2014.
83 Mantelero, 2024.

AI used in healthcare that is specifically mentioned in Annex III, such as systems for determining eligibility for healthcare services, emergency healthcare patient triage or dispatch systems, and AI systems used for risk assessment and pricing in life and health insurance.

In the context of AI, these tools and methodologies are still in their nascent stages, lacking a consensus on their application, timing, and objectives, and few instances of practical implementation exist.

3.4.2 Clinical Trials and Evidence-Based AI Medicine

A significant barrier to widespread adoption of AI in healthcare delivery is the lack of robust, evidence-based medicine validating its effectiveness and reliability.[84] This challenge involves multiple factors, from proving clinical effectiveness to overcoming regulatory obstacles, all of which are key to gaining the trust of healthcare professionals and patients. Unlike drugs, which undergo validation through clinical trials despite their inherent complexity,[85] AI systems have yet to fully adopt this model, creating uncertainty about their performance rates.

Despite this, there are some ongoing AI clinical trials.[86] Recent studies confirm the likelihood of an increase in clinical trials assessing AI for diverse applications as it continues to advance and become more integrated into healthcare.[87] Should this trajectory gain traction, it will showcase adequate safety and efficacy levels, mirroring the standards applied to drugs.

However, the reasoning behind AI systems will not follow simple, easily articulated patterns that can be tested through standard clinical trials. Instead, it might depend on complex connections that are hard to control or on different types of knowledge. Given these concerns, traditional clinical trials are insufficient to establish the efficacy of AI devices,[88] giving preference to randomised clinical trials.[89] Departing from large premarket pivotal trials, reliance on post-market data and post-market surveillance is a strategy already employed for more 'traditional' medical devices that can also be used for AI medical devices.[90]

84 Tsopra, Fernandez, Luchinat, et al., 2021.
85 Friedman, Furberg, DeMets, et al., 2015.
86 Angus, 2020: 1043–1045.
87 Pearce, Cruz Rivera, Liu, et al., 2023.
88 Senger and O'Leary, 2018: 292.
89 Han, Acosta, Shakeri, et al., 2024: e367–e373.
90 Badnjević, Pokvić, Deumić, et al., 2022: 1–15.

3.5 *Human Oversight*

3.5.1 Forms of Human Intervention

What is expected from doctors when using AI in healthcare is that they apply their expertise to interpret the AI's outcomes, considering the unique circumstances of each patient's case.[91] This collaborative approach recognises the importance of human judgment and expertise in conjunction with AI-generated insights. In such cases, humans evaluate the output generated by the AI model alongside other available information before making a final decision.

In the domain of medicine, numerous scenarios arise where a binary 'yes/no' response is inadequate, and the most suitable reply often falls into the category of 'perhaps'.[92] Even when AI is a suitable option for a concrete case, that does not mean that its results do not require additional human interpretation. A case that illustrates the need for 'human adaptation' (which is also a form of human oversight) relates to Sepsis Watch, a tool powered by ML designed to aid healthcare professionals in promptly diagnosing and treating sepsis.[93] The findings of Sendak et al. revealed that in order to contextualise the information presented in Sepsis Watch, the Rapid Response Team nurses acquired expertise and established practices over time.[94] For example, the nurses developed a practice of accessing a patient's EHR chart before contacting the emergency department doctor. By familiarising themselves with the patient's complete clinical history, the nurses were able to provide a comprehensive clinical overview and demonstrate 'due diligence' in anticipation of potential inquiries from doctors. This practice was not initially outlined by the design team but emerged as an ad hoc approach that facilitated effective integration. These observations show that the nurses and the doctors retained their professional discretion in diagnosing sepsis, which ultimately enhanced the use of the ML-powered tool.

---

91 This involves drawing upon their experience and a profound comprehension of the illness to form a integrated understanding. To truly encompass all facets, doctors must take into consideration not only the patient's physical state but also their psychological, emotional, social, and even religious dimensions. Cf. Astromskė, Peičius and Astromskis, 2021: 513.

92 Johnson, 2019: 434.

93 Wu, Du, Gu, et al., 2021. Sepsis is a life-threatening infection that, if left untreated, can result in organ failure or even death. It is worth noting that sepsis is a major contributor to inpatient fatalities within hospitals.

94 Sendak, Elish, Gao, et al., 2019.

The requirement of human supervision might take the form of an 'algorithmic audit',[95] which may involve regular updates of the AI system. These updates become necessary to keep pace with advances in medical knowledge and access to new datasets. Suppose an AI system in the field of oncology, which is used to assist doctors in diagnosing various types of cancer based on imaging data. This algorithm, initially trained on a large dataset of historical cancer images and clinical outcomes, has proven to be highly accurate in identifying early signs of different cancers, thus significantly aiding in early diagnosis and treatment planning. However, medical knowledge and technology are constantly evolving. New types of imaging techniques, such as advanced MRI or PET (positron emission tomography) scans, are developed, providing more detailed and accurate images of tumours. If the AI algorithm is not regularly updated to incorporate the latest imaging techniques, genetic markers, and new research findings, its diagnostic accuracy will start to decline. For instance, the algorithm may miss early signs of newly identified cancer types or fail to recognise updated patterns in existing cancer types. Failure to update the algorithms risks malpractice.[96]

Humans are required to do much more than merely supervise the machine. Presently, despite the increasing autonomy of surgical robots, human intervention remains crucial in executing numerous tasks and influencing the robot's trajectory (for example, in selecting process parameters or positioning the patient).[97] Moreover, in the collaboration between human experts and AI systems, the human's role is indispensable because it allows consideration of the patient's legal and ethical values, personal preferences, moral considerations, and other information that may not be accessible to or within the purview of the AI systems. Suppose that an older patient, with advanced lung cancer, receives an AI recommendation for aggressive treatment involving chemotherapy, radiation, and an experimental drug. However, during the consultation, their oncologist learns that the patient values quality of life over aggressive treatment and has an advance directive favouring palliative care. Additionally, the patient's cultural and religious beliefs prioritise natural healing. A human doctor will be able to consider these personal preferences and moral considerations, which the AI system cannot fully account for, and recommend a treatment plan aligned with the patient's holistic needs, ensuring their values and

95 Dueno, 2020: 337–358.

96 Couture, Roy, Ganache, et al., 2021.

97 Denecke and Baudoin, 2022.

wishes are respected. This unique human touch ensures a comprehensive and holistic approach to patient care.[98]

3.5.2 Human in the Loop and on the Loop

The distinction between a 'human in the loop' and a 'human on the loop' lies primarily in the nature and timing of human intervention. A 'human in the loop' implies a model of continuous, active involvement, wherein human agents are integral to each stage of the system's operation, making or validating decisions in real time. By contrast, a 'human on the loop' assumes a supervisory role, whereby human oversight is maintained, but intervention is occasional and typically reactive, activated only when the system's performance deviates from expected parameters or poses potential risks. This distinction bears significant implications for accountability, as the former presupposes sustained human agency, while the latter may dilute the immediacy of human responsibility.[99]

Consider an AI system used to analyse radiological images for the early detection of breast cancer. Under a human-in-the-loop model, the system processes mammography images and highlights areas of potential concern, but the assessment has no clinical effect unless it is reviewed and confirmed by a doctor. The clinician examines the AI output alongside the original images, the patient's medical history and other diagnostic information, and decides whether further tests or treatment are warranted. At each stage, the AI operates as a decision-support tool, and the human professional remains actively involved in validating and contextualising the system's recommendations.

By contrast, the same system could be deployed under a human-on-the-loop model. In this configuration, the AI automatically analyses incoming images and classifies cases according to predefined risk categories, triggering follow-up procedures or referrals without immediate human review. Clinical staff intervene only when the system flags an atypical case, when performance metrics indicate anomalous behaviour, or during periodic audits of system outputs. Human oversight thus exists, but it is supervisory and largely ex post, rather than embedded in each individual decision.

3.5.3 Requirements of Oversight

True human oversight requires that the interaction between the system and its user be transparent and traceable. This ensures that if any action is called

98 Lagioia and Contissa, 2020: 245–289.

99 Methnani, Tubella, Dignum et al., 2021; Mosqueira-Rey, Hernández-Pereira, Alonso-Ríos, et al., 2023: 3005–3054.

into question or needs review, there is a clear chain of responsibility, allowing for the identification of a specific individual who can be held accountable.[100]

Oversight has to be more than a proxy. Simply having a human authorise actions does not equate to meaningful control.[101] Human presence alone does not guarantee meaningful control from a responsibility standpoint. For genuine human control to exist, the system must be able to both account for relevant ethical considerations and trace decisions back to a person who is fully aware of, and accountable for, the outcomes.[102]

This is a dynamic task. To put the concept of meaningful human control into practice, Methnani et al. propose the use of variable autonomy, defined as the capability of a system to dynamically adjust its level of autonomy, allowing it to shift anywhere along the continuum from complete autonomy to full teleoperation.[103]

3.5.4 Drawbacks of Human Oversight

However, human oversight does not come without concerns, especially within a 'human in the loop approach'. It has been argued that imposing legislative or regulatory mandates for healthcare practitioners to thoroughly review the observations and recommendations provided by AI platforms could introduce inefficiencies in healthcare decision-making, depending on how it is carried out.[104] For instance, a doctor could be required to approve every recommendation made by the system, such as adjusting medication dosages, administering fluids, or changing ventilation settings. While human oversight is essential for ensuring patient safety and applying expert knowledge, it can also slow down decision-making. In critical situations where rapid response is required, waiting for human input at every step could delay necessary interventions, leading to potential harm.

Moreover, such a requirement may entail substantial costs for healthcare practitioners and their patients. This could result in foregoing numerous beneficial outcomes simply because the combination of AI and doctor review becomes prohibitively expensive. The potential consequence of this situation is that some individuals may be unable to afford the care they could have otherwise received.[105]

---

100 Methnani, Tubella, Dignum et al., 2021.
101 Van der Stappen and Funk, 2021: 514–519.
102 Santoni de Sio and Van den Hoven, 2018): 15.
103 Methnani, Tubella, Dignum et al., 2021.
104 Methnani, Tubella, Dignum et al., 2021.
105 Alshanteer, 2020: 54.

In addition, the human operator may not always have sufficient information or viable options to meaningfully impact the system at every stage of the decision-making process.[106] For instance, if the AI flags a subtle abnormality in a radiology scan that the human expert cannot clearly interpret or if sufficient supporting clinical data are lacking, the human may struggle to provide meaningful input, especially when the human is too much 'inside' the loop, they will lose objectivity, as excessive confidence on technology might make the human biased.

3.5.5 Human Out of the Loop

The challenge of human error and slow reactions in time-sensitive tasks highlights the need for full autonomy, where the human is completely removed from the control loop, allowing the system to operate independently, referred to as 'human out of the loop' or 'outside the loop'.[107]

Despite removing the human from direct control, the idea of some kind of meaningful oversight can be maintained if the system's behaviour is transparent and known, ensuring that human involvement remains critical in defining the parameters for governance, operation, and evaluation. This oversight includes determining how the system is monitored and how failures are handled, ensuring that human authority is not entirely lost.

Feedback loops – also present in 'human out of the loop' settings – play a critical role in the continuous improvement of AI systems.[108] By analysing successes, failures, and unexpected outcomes, these loops provide valuable insights into areas in which AI needs adjustment. Unlike humans directly involved in the loop, those observing and interacting afterwards offer essential input that drives the refinement of AI over time.[109] Unlike 'human on the loop' oversight, which operates during system use and allows for intervention, this form of human involvement occurs only after the fact and serves exclusively to inform subsequent system refinement.

---

106 Methnani, Tubella, Dignum et al., 2021.

107 Methnani, Tubella, Dignum et al., 2021.

108 Suppose an AI system is used to analyse radiological images to detect signs of cancer. Over time, data are collected not only on the system's accuracy but also on instances where it missed critical details or provided false positives. Feedback from healthcare professionals and outcomes of patient treatments provide valuable insights after the fact ('humans outside the loop'). These insights inform developers of the AI system, leading to adjustments in its algorithms to improve future performance, making it more reliable and accurate in its diagnoses over time.

109 Tschider, 2024: 419.

Alongside feedback, internal monitoring and inspection processes play a crucial role, as does adversarial testing. This type of testing is conducted by parties with interests contrary to those of the organisation, such as competitors, and takes place outside the feedback loop.[110]

3.5.6 Digital Human Skills and Accreditation

Doctors who engage with AI systems and derive insights from them should undergo educational training.[111] This training must be meticulously overseen, evaluated, and endorsed by an impartial external entity. Regulatory bodies or even insurance entities could potentially enforce specific educational requirements or knowledge prerequisites prior to the use of healthcare AI or robotic technologies. This is not unique to AI. To ensure appropriate usage of any new and complex medical device, doctors must possess knowledge regarding its safety and effectiveness.[112]

If a doctor lacks information about the device, including its safety and comparative advantages over conventional treatments, using it would subject patients to unknown risks. Conversely, when doctors possess a clear understanding of the device's safety, effectiveness, and supporting scientific evidence, they are better equipped to determine its suitability for clinical use. Doctors can protect themselves from device-related malpractice liability by acquiring proper training and expertise before utilising a new device.[113]

In fulfilling their duty of care, doctors must seek to comprehend the scientific evidence supporting a new device before employing it in practice.[114] By following this principle, we protect the doctor's respected role from being reduced to merely being 'button-pushers'.[115] This also strengthens the foundation of mutual understanding and trust between doctors and patients.

Doctors must also have the capacity to work without AI. When AI fails, doctors must have the skills to continue the medical act without it. The trust in AI cannot be so great that its absence or malfunction paralyses the doctor.[116] Thus,

---

110 Tschider, 2024: 419–420.

111 Smith, 2024.

112 Greenberg, 2009: 435.

113 Greenberg, 2009: 435.

114 Alattal, Azar, Myles, et al., 2025.

115 LaRosa and Danks, 2018: 213.

116 Suppose a hospital in which an AI system is extensively used to assist radiologists in interpreting medical imaging, such as X-rays, CT scans, and MRIs. If one day, due to a critical software bug or a cyberattack, the AI system goes offline unexpectedly, human radiologists are expected to step in. Radiologists must not become so overly reliant on the AI system that they lose their skills in manual image interpretation; otherwise, this could lead

doctors must continue to be trained in tasks that are normally performed by AI, from the most complex to the most basic ones, in case there is a failure.[117]

The responsibility of ensuring adecuate knowledge is not solely incumbent upon the doctor; the patient also bears a role. A wide-ranging comprehension of the capabilities and limitations of healthcare AI is essential for patients. This understanding equips them to make informed decisions about the technology's use, its impact on the doctor's role, and subsequently, the patient's own experiences.[118]

3.6 *The Use of AI as Part of the Medical Standard of Care*

In the future, when AI becomes an integral part of the medical standard of care, the safest course of action for a doctor (in terms of liability) will be to use AI and follow its recommendations. Conversely, doctors who choose not to use AI will increase the likelihood of a negative ruling.[119]

As AI systems prove their value, medical boards, insurers, and courts can increasingly expect doctors to integrate these tools into their practice. Medical guidelines and protocols are likely to evolve to include AI as part of routine care in fields where it demonstrates clear advantages.[120] Insurers and professional boards will likely establish protocols that define AI as a core diagnostic or therapeutic tool, pushing doctors to justify non-use in particular cases.[121] In this scenario, failure to follow these guidelines might not only affect liability but might also influence malpractice insurance coverage and reimbursement, making adherence to AI advice both legally and financially prudent.

However, currently, as AI is not generally part of the existing standard of care, excluding the specific domains where it can be considered so.[122] Many medical protocols still omit AI, making its use in medical practice potentially reckless.

---

to delays in diagnosis, missed diagnoses, or incorrect diagnoses, potentially jeopardising patient care.

117 Griffin, 2021: 100.

118 LaRosa and Danks, 2018: 214.

119 This same consideration in Griffin, 2021: 99.

120 Arefin, 2024: 1–17.

121 As it happened with other technologies, for instance, computers (Fisk, 2024).

122 This could be the case of radiology, considering that the American College of Radiology has already established the ARCH-AI quality assurance programme to standardise and ensure the safe use of AI in imaging, emphasising its role in clinical practice (American College of Radiology, 2024). Likewise, in cardiovascular care, AI is used extensively, and according to the American College of Cardiology, AI-based digital biomarkers and predictive algorithms enhance diagnostics and optimise patient care by providing timely, individualised insights (American College of Cardiology, 2024).

Adding to this, courts tend to be very conservative in their assessments, avoiding radical changes.[123] In the domain of medical malpractice, courts assessing the reasonable standard of care may lean toward traditional, human-based practices in the absence of concrete guidelines endorsing AI-based decisions. Thus, in cases of adverse outcomes, AI reliance could be argued as a deviation from accepted medical standards, potentially exposing doctors and institutions to liability. Historically, courts have ruled against professionals who fail to adopt widely accepted tools that improve outcomes, viewing such failures as a breach of the duty of care. Liability standards could evolve to consider whether the AI was available, validated, and suitable for the case. As with other medical technologies, if AI demonstrates a significant reduction in error rates, failure to use it could be deemed negligent. However, we are not at that stage yet. Therefore, in the current scenario, healthcare providers must carefully consider whether or not to use AI, as the decision could significantly impact their legal liability and the quality of care provided.

4 AI Standard of Care: Informed Consent in AI Medicine

4.1 *The Ratio of Patients' Informed Consent*

Informed consent is the cornerstone of healthcare provision, and its omission may result in civil[124] and/or criminal liability.[125] It aims to facilitate treatment decision-making in the context of the relationship between a doctor with special knowledge in the field and a lay patient (as a rule) in a way that extends patients' comprehension of their medical conditions and concerns.[126]

The reason for instituting informed consent in AI medicine is to strengthen the trust between patients and doctors,[127] not just to increase trust in AI.[128] This educational effort will likely change the doctor-patient relationship, making it more like a team.

Doctors need to ensure that patients understand how AI is used in their diagnosis and treatment. This includes explaining the benefits, limitations, and uncertainties of AI-generated recommendations. Clear communication

123 Yeung, 2019.

124 Pereira, 2015.

125 Raposo, 2012: 379–390.

126 Hoffman, Rosenthal, Colditz, et al., 2018: 246.

127 LaRosa and Danks, 2018: 214.

128 In addition to the consent required for any medical procedure, medical acts involving personal data processing require additional requirements, namely a legal base on Articles 6(1) and possibly 9(2) of the GDPR.

and shared knowledge between doctors and patients will help support better and more informed healthcare decisions.[129]

The primary doctor using AI as a consultant should be responsible for providing the patient with the information needed to ensure informed consent. This responsibility includes informing the patient about alternative treatments or diagnoses proposed by the AI, as well as providing a clear understanding of the potential risks and benefits associated with such treatment.[130] This disclosure is deemed necessary for the patient to make an informed decision about their healthcare.

Another possibility is to have the AI system directly provide the patient with the necessary information. A significant difference is that when consulting a human doctor, there is an opportunity for dialogue. Patients can ask questions, seek clarifications, or present concerns, and the doctor can adjust recommendations based on human reasoning and patient input. In contrast, 'traditional AI systems' provide fixed recommendations without the ability to engage in discussions or adapt based on feedback.[131]

However, this scenario is evolving. Advances in AI are enabling systems to interact with patients more dynamically. Some AI systems are being designed to answer patient questions.[132] With improvements in natural language processing, AI could soon offer more conversational interactions, clarify uncertainties and provide explanations much like a human doctor would. This development will bring AI closer to fulfilling a more interactive, patient-centred role in healthcare.

4.2 *Content of Informed Consent in AI Medicine*

4.2.1 General Content of Patient's Information in AI Medicine

The range of information to be provided to the patient has always been a contentious issue. Most jurisdictions in Europe follow the patient-based approach,[133] that is, the criterion is what information a reasonable patient would require for an informed decision.[134] Still, in some situations, the specific

---

129 LaRosa and Danks, 2018: 214.

130 Lupton, 2018: 5.

131 Froomkin, Kerr and Pineau, 2019: 50.

132 Trofimiuk, 2024. Still, this section will only address the challenges of human doctors performing the legal and ethical obligation of patients' informed consent, not the cases – likely to happen in the future – of having the AI system itself providing the information and collecting the consent.

133 European Society for Emergency Medicine, 2019: 11–12.

134 In the United States, there are two different standards for informed consent, which vary depending on the state: one grounded in the reasonable doctor and the other in

needs and circumstances of the individual patient should also be taken into account.

What a reasonable patient wants/needs might change over time. For instance, we may assume that at the present moment it is relevant for most patients to be told about the AI intervention, due to the still disruptive nature of this technology and its novelty (though, with the caveats to be developed in the next section) However, we might consider that if the majority of individuals get used to AI and it becomes a routine medical tool, the expectation of a 'reasonable person' will change. People will expect AI to be integrated into medical care and to be informed if that is not the case.

The doctor may not always realise that a patient places importance on certain details, and some patients who are uneasy about AI might recognise their discomfort as somewhat irrational.[135] These patients might prefer not to know about AI's role in their treatment to avoid unnecessary anxiety, while still benefiting from the technology. This creates a challenge, as a doctor cannot anticipate that a patient would prefer not to be informed about AI unless the patient expresses this before the healthcare provider brings up the AI's involvement.[136] One possible approach is to invoke therapeutic privilege – meaning the doctor withholds information for the patient's own well-being – regarding the use of AI in their care.[137] However, this should remain an exception rather than standard practice.

The patient must grasp the overall process and the critical factors that contribute to it. The explanation should highlight the essential inputs, outputs, and the causal connections between them. Notably, fulfilling these requirements does not necessitate disclosing the system's internal workings or the specifics behind its decision-making process.[138]

4.2.2 Information about AI's Involvement

Still, even under current conditions, the involvement of AI does not necessarily trigger a duty to inform, just as the use of other technologies often goes unnoticed by the patient and remains outside the scope of clinical disclosure. In my view, doctors are not obliged to inform patients about the involvement

---

the reasonable patient. The doctor-based standard obliges doctors to disclose the risks that a reasonable doctor would typically inform a patient about. On the other hand, the patient-based standard requires doctors to reveal risks that a reasonable patient would expect to be informed of in a similar situation. See Braithwaite and Caplan, 2014: 558–559.

135 Pruski, 2024.

136 Pruski, 2024.

137 Shalak, Shariff, Doddapaneni, et al., 2022: 152–155.

138 Astromskė, Peičius and Astromskis, 2021: 516.

of AI in their decisions.[139] Most jurisdictions do not have norms in this regard, but I follow the thesis of Glenn Cohen, one of the most reputable experts in the matter.[140]

However, Cohen recognises that there are exceptions to the lack of obligation to provide information, which also seem reasonable to me: (a) when patients ask about AI involvement; (b) when medical AI is a black box; (c) when AI is given a disproportionate role in final decision-making; (d) when AI is used to reduce costs and not to improve patient health.[141] To this set of exceptions, I would add yet another: situations where the use of AI entails special and significant risks in the specific circumstances of the case. Even though the norm is the *prima facie* non-disclosure of AI's intervention, the exceptions might become so relevant and so frequent that they will likely surpass the rule.

There might be situations in which patients ask doctors: 'is this what the AI recommended?' (exception (a)).[142] If the doctor says no, lying, some authors claim that they may be facing a violation of informed consent, analogous to cases in which doctors lie about their qualifications for the medical procedure or about the identity of the person who is going to perform the medical procedure.[143] The analogy is stronger for me, the more the AI recommendation is adopted automatically,[144] i.e., when it becomes more of a true substitute, and weaker when the AI only provides recommendations.

The duty to disclose the intervention of black box AI (exception (b)) is justified by its opacity and the consequent lack of information about its reasoning. This feature is not enough to exclude black box AI from healthcare delivery (see Chapter 3, Section 3.4.9), but patients should be given the chance to refuse it.

At the current state of the art, we still do not have fully autonomous AI systems in healthcare delivery, and eventually we will never have. However, there are cases of complementary AI where the systems play a pivotal role and thus this should be communicated to the patient (exception (c)).

The cost-related exception (exception (d)) might be difficult to envisage nowadays, considering the novelty and thus the price of AI medicine.

139 On the contrary, Iserson (2024: 225) claims that 'To respect patient autonomy and informed decision-making, hospitals should explicitly mention in their general consent to treatment forms whether they use'.

140 Cohen, 2020: 1425–1469.

141 Cohen, 2020: 1429.

142 Currently, few patients will be aware of AI and thus will ask about it, although this may change in a future where the technology is more ubiquitous and well-known.

143 Cohen, 2020: 1444.

144 Cohen, 2020: 1445.

Still, in a scenario when the medical use of AI becomes the norm, it will be cheaper to use it instead of other technologies (or even instead of humans).

Cohen has put forth a fifth speculative scenario that involves the non-disclosure of AI participation in decision-making or procedures when AI is intended to elevate a regular doctor's capabilities to those of a specialist.[145] This non-disclosure could be perceived as an indication of the doctor's inadequate experience or qualifications, which should have been disclosed, potentially undermining the patients' trust in their healthcare provider. In my view, however, the legal and ethical outcome should be different. If a doctor relies on AI to conceal their lack of knowledge or proficiency in performing a medical act, the issue is not merely about disclosing this to the patient but about preventing such doctors from practising altogether, as AI should not be used to mask incompetence. Conversely, if the doctor possesses the knowledge and skills of a competent professional but uses AI to enhance their performance beyond the average standard, I see no reason why this should be communicated to the patient.

Cohen's thesis equates AI with a human doctor and works well for state-owned hospitals, where there is no contract between the hospital and the patient. In these cases, patients cannot choose their doctors or know who will be on their medical team, including whether AI is involved. However, in private hospitals, where patients usually choose their doctors, it can be argued that they have the right to know if AI is part of the medical team.

However, this solution does not work well if AI is, instead, equated to a medical tool.[146] Surgeons typically do not seek patient consent for the use of particular instruments during a procedure, as patients place trust in the surgeon's expertise to select the tools that will best facilitate the operation, and the same should happen for AI.

4.2.3 Information about the AI Risks

For the consent to be valid, patients must receive comprehensive information beforehand regarding all essential aspects. This encompasses not only the treatment process, potential consequences, and risks of the medical intervention but also, particularly, the extent and associated risks of employing AI.

To determine the necessity of disclosure, two key criteria are often considered: the 'nature' and the 'likelihood' of a risk.[147]

145 Cohen, 2020: 1435–1436.
146 As for instance in Topol, 2019: 44–56.
147 Kiener, 2021: 706.

The nature of a risk entails evaluating the severity of potential consequences arising from a medical procedure.[148] For instance, the risk of death is deemed more significant than the risk of finger amputation. However, assessing the nature of a risk requires considering not only objective physical or psychological harm but also the subjective relevance of certain harms to individual patients (for example, a musician might be particularly concerned with the amputation of a finger).

The likelihood of the risk is another crucial factor: the more probable the risk, the more compelling the reason to include it in the information provided to the patient. The likelihood of a risk is the probability of the harm occurring. Risks become more significant the more likely they are to materialise.[149] However, following the influential legal case of Montgomery v Lanarkshire Health Board,[150] in the UK, legal scholars and philosophers alike have claimed that even very low risks can require disclosure if a reasonable person would consider the risk material to their decision, e.g. a risk of paralysis should be disclosed even if it is considerably below 1%.

The interesting feature of using AI is that the level of risk involved might actually diminish. When looking at most robotic surgeries, for instance, the risk associated with modern robotic surgery appears to be comparable, if not marginally less, than that of traditional techniques.[151] As such, the involvement of a robotic system might not inherently necessitate a distinct patient disclosure based on the level of risk.

The novelty of the technology used does not exempt doctors from providing the necessary information. On the contrary, it is essential to explain the potential risks associated with new medical technologies to patients.[152] According to the case law of the German Federal Court of Justice, when a treatment method is relatively new and not widely accepted, the doctor must clearly inform the patient that this new method may carry unknown risks. While employing new treatment methods can be vital for medical advancement, this cannot justify failing to disclose the novel nature of the method and its potential risks to the patient.[153]

Comprehensibility – both intellectual and emotional –information about the risks is a crucial aspect of communication. For instance, when an AI

148 Paterick, Carson, Allen, et al., 2008: 314–315.

149 Beauchamp and Childress, 2013: 73–81.

150 Montgomery v Lanarkshire Health Board [2015] UKSC 11.

151 Crew, 2020: S5–S5.

152 Douglas, Lacey and Howard, 2023: 1127.

153 Federal Court of Justice (BGH), judgment of 13 June 2006, VI ZR 323/04. A comment on this case in Buchner, 2020: 228–229.

diagnosis reveals a 50% chance of a severe illness requiring substantial intervention, there are various methods for conveying this information to the patient effectively. Firstly, the information should be presented clearly and in straightforward language, avoiding medical jargon that may confuse the patient. The doctor should break down the diagnosis into understandable terms, explaining what the 50% probability means in practical terms and what the potential illness entails.[154] Secondly, the presentation should take into account the patient's emotional state and preferences.[155] For some patients, visual aids such as charts or diagrams can help them better grasp the likelihood and implications of the diagnosis. Others might benefit from a detailed verbal explanation, including examples of how similar cases have been managed.

4.2.4 Information about Alternative Options

4.2.4.1 *The Relevance of Knowing Alternative Options*

Informed consent in AI-driven medicine requires that patients are not only made aware of the risks and benefits of AI-based treatments but also made aware about alternative options.[156] It is essential for healthcare providers to explain any non-AI alternatives, including traditional methods or human-only interventions, so patients can make a fully informed decision. Presenting these alternatives ensures that patients understand their choices and the implications of each, fostering trust and respecting their autonomy in medical decision-making.

The importance of discussing alternative options cannot be overstated.[157] AI in medicine, while advancing rapidly, may come with uncertainties or limitations in certain contexts. Some patients may be more comfortable with established methods, while others may prefer AI's potential for greater accuracy or speed.[158] The same patient might also have different preferences along the way. For example, a patient might choose an AI-based diagnostic tool for its enhanced detection capabilities but might prefer a human-led approach in cases where they prioritise the doctor's personal judgment or experience.

Patients need to feel that they are not being pressured into accepting AI-driven interventions simply because they are the latest trend.[159] Instead, they should understand that their healthcare provider has considered all

---

154 Allen, Charpentier, Hendrickson, et al., 2023.
155 Sheeran, Jones, Pines, et al., 2022: 186–196.
156 Pruski, 2024.
157 Pallocci, Treglia, Passalacqua, et al., 2023: 2118.
158 Gerke, Minssen and Cohen, 2020: 295–336.
159 Lambert, Madi, Sopka, et al., 2023.

possible options and is recommending the most appropriate course of action based on their individual circumstances.

Informed consent in the context of AI involves a dynamic conversation between the healthcare provider and the patient.[160] It is not just about presenting a list of options but about engaging in a dialogue where the patient's concerns, preferences, and questions are addressed. This ensures that patients are not only informed but also truly understand the implications of their decisions.

4.2.4.2 *The Concept of 'Alternative Treatment'*

What can be discussed is what counts as an 'alternative treatment' in the sense of a different treatment. For example, should laparoscopic surgery performed by a robotic AI be considered an 'alternative treatment' compared to one performed by a human surgeon? If they pose different levels of risk, we could answer that they are different treatments, but what if they pose exactly the same risk?[161]

From the patient's perspective, the use of a robotic AI in surgery may introduce a psychological component that influences their decision-making process. Many patients may have reservations about placing their trust in an AI system due to unfamiliarity, concerns about machine errors, or the absence of human intuition during critical moments of the surgery. Even if the risks are statistically the same, the lack of direct human involvement might be seen as a significant difference by the patient, affecting their sense of control and comfort in the decision-making process.

Moreover, while both options may involve the same risks, robotic AI systems might operate under different skill sets compared to human surgeons. Robots excel in precision, endurance, and consistency, while human surgeons bring experience, adaptability, and problem-solving skills in unpredictable situations. A human surgeon's expertise, accumulated through years of practice and real-time decision-making, might offer advantages in certain scenarios that cannot yet be replicated by AI. Conversely, in procedures that require high precision and minimal margin for error, an AI could outperform a human surgeon. Thus, even with similar risk profiles, the potential differences in performance, adaptability, and error management could make them distinct treatment options.

160 Bekbolatova, Mayer, Ong, et al., 2024.

161 The same question is analysed in Pruski, 2024. The author concludes that in many cases the intervention of AI does not change the nature of the medical act and thus it does not make it an 'alternative treatment'.

On an ethical level, the introduction of robotic AI systems in surgery raises questions about accountability. In cases of adverse outcomes, determining responsibility becomes more complex with AI. While human surgeons are legally and ethically accountable for their decisions and actions, the accountability for a machine's performance is less clear (see Chapter 5, Section 2). This ambiguity in legal responsibility could further distinguish AI-assisted surgeries from traditional ones in terms of how the medical and legal systems view the two options.

As robotic AI technology continues to evolve, the distinction between the two options could narrow further. With advancements in ML and AI's ability to simulate human decision-making processes, AI could one day reach a point where its judgment and adaptability mirror that of human surgeons. However, society's trust in AI systems will need to evolve alongside these technological improvements. Until patients and medical professionals are equally confident in AI-assisted procedures, the perception of them as a separate treatment option will likely persist.

4.3 *Challenges Raised by Black-Box AI*

4.3.1 Lack of Relevant Information

It is widely acknowledged that understanding an AI's assessment criteria is crucial for empowering patients to make better-informed decisions:[162] 'Doctors have a moral responsibility to provide their patients with medical care that is effective, respects their dignity, and honors their autonomy. To uphold these values, it is essential to be able to explain diagnoses and treatment recommendations'.[163]

However, this may be infeasible within a black-box AI system.[164] Imagine an AI system recommending a specific treatment plan for a patient with a history of hypertension and diabetes. The AI might suggest a medication regimen based on data from thousands of similar cases, taking into account the patient's medical history, age, and other health factors. However, for the patient to trust and follow the recommendation, they need to understand the reasoning behind it; this is where explainability comes into play.[165] Without this information, patients may struggle to make informed decisions about

---

162 Schweikart, 2021: 301.

163 Chan, 2023: 286–287.

164 Price II, 2017: 430.

165 For example, in light of this perspective, if the AI suggests a specific chemotherapy regimen for cancer treatment, the doctor should clarify why the AI made this suggestion, including how it analysed the patient's previous treatments, current condition, and relevant medical research.

their treatment.[166] In a similar example, suppose a black-box AI system makes healthcare recommendations based on hidden assumptions, such as prioritising the extension of a patient's life over improving their quality of life, without taking into account the patient's personal preferences. Patients could make more informed decisions if they understood the criteria and underlying assumptions the AI system uses. For instance, if a patient knew that the AI prioritised life extension but did not consider quality of life, they could better weigh whether the recommended treatment aligns with their own values and preferences.[167]

4.3.2 Doctors' Inability to Comply with Informed Consent

A major challenge is the intrinsic complexity of AI, especially black box AI (See Chapter 3, Section 3.4). At first glance, it seems extremely difficult to provide the necessary information when a black-box system is involved.

Still, it is not required the technical comprehension of every single step.[168] Understanding an explanation does not necessitate comprehending the intricate pathways of data within an AI system, just as comprehending human explanations does not require understanding neural signal transmissions in the human brain. Moreover, many of the issues raised in AI-based medicine can also be found in non-AI medicine (e.g., doctors use an X-ray machine as a diagnostic tool, and they need not be well-versed in its intricate engineering).

However, while technical expertise is unnecessary, understanding the machine's performance characteristics and limitations is essential for making accurate diagnoses. In a similar vein, clinical decision support systems serve as aids to healthcare professionals, who may not be experts in the system's inner workings but must comprehend its capabilities and constraints to arrive at informed diagnoses and treatment decisions.

Doctors need to disclose that while they can assess the recommendations made by the AI, they may not have a comprehensive understanding of how the

---

In a similar example, the patient might ask why a particular drug was chosen over others. The doctor could explain that the recommended medication has been shown to be more effective in reducing blood pressure for patients with diabetes, as demonstrated by the analysis of vast datasets. The doctor might also outline potential side effects, drug interactions, and success rates based on the patient's individual health profile.

166 Quinn, Jacobs, Senadeera, et al., 2022.

167 Quinn, Jacobs, Senadeera, et al., 2022.

168 Andreotta, Kirkham and Rizzi, 2022: 1719.

AI system arrives at those recommendations.[169] This disclosure acknowledges the limitations in the doctor's ability to fully interpret the AI's reasoning.[170]

The process of generating information demands significant time and effort from doctors. This raises concerns not only about the direct financial costs but also about the missed opportunities to allocate that time and effort towards delivering healthcare services to other patients. As a result, the value of providing explanations needs to be weighed against the resources required to generate them.[171] An alternative is to allocate the task of generating information for the patient to an AI system, which would not only relieve the doctor but also make it quicker and potentially provide more comprehensive information, adapted to each patient.[172]

4.4 *Too Much Information versus Reasonable Information*

4.4.1 The Scope of Information

Potentially, the informed consent for medical AI could encompass a wide array of information, including the way these systems function; the degree to which they are comprehensible and trustworthy; the errors and limitations they may present; the methods for resolving disagreements between the AI and the healthcare professional; the security of the patient's identifiable information (personal data) within the AI system; the reliability and validation of the AI's performance; and the presence of any biases in the AI program.[173]

The disclosure of all this information, however, may not always be beneficial for patients. Excessive disclosure makes it difficult for patients to distinguish significant information from trivial information.[174] When patients are bombarded with excessive information – particularly highly technical or complex details – they may become confused, anxious, or even paralysed by the decision-making process.[175] This phenomenon, often referred to as

169 Federation of State Medical Boards, 2024.

170 The extent of disclosure depends on various factors, such as the specific AI system being used, its level of complexity, and the impact of its recommendations on patient care. Cf. Morley, Machado, Burr, et al., 2020.

171 Astromskė, Peičius and Astromskis, 2021: 517.

172 Technically, this would involve designing AI systems that furnish analogous explanations to those presently demanded from humans in accordance with legal requirements. It would amount to transferring the responsibility for elucidating the workings of the AI system from the doctor to the developers of the technology (Astromskė, Peičius and Astromskis, 2021: 517).

173 I am closely following the suggestion of Iserson, 2024: 225–226.

174 Cohen, 2020: 1453.

175 Benbassat, 2023.

'information overload',[176] can lead to decision fatigue, where the patient is unable to process the information adequately and may either defer to the doctor's judgment without truly understanding the implications or make decisions based on incomplete understanding. Information overload can also erode trust, as patients may feel overwhelmed or perceive the process as unnecessarily complicated.

For example, if the AI diagnosis is based on a combination of factors resulting in a probability range of 40–60% for a specific outcome, should the patient be informed about all the distinct factors contributing to that potential outcome? Not all factors that contribute to an AI's probability estimate may be equally relevant to the patient.[177] If some factors are routine clinical indicators that the patient is already aware of, it might be more beneficial to focus on explaining new or significant factors that have a larger impact on the outcome. This balanced approach can help the patient focus on the most critical information when making decisions about their treatment or care.

The challenge of balancing 'too much information' versus 'reasonable information' is a critical aspect of effective communication between healthcare providers and patients.[178] The complexity and volume of information that can be conveyed present a unique challenge: providing sufficient detail to enable informed decision-making without overwhelming the patient.

4.4.2 The Definition of Relevant Information in AI-Based Medical Care

Providing reasonable information involves distilling the necessary details into a format that is both comprehensible and sufficient for the patient to make an informed decision.[179] This approach focuses on clarity, relevance, and the patient's ability to understand and use the information in the context of their own health and treatment preferences. Reasonable information should include key risks and benefits, alternatives and the role of AI and other technologies.

Other types of information may be provided, depending on the circumstances.[180] Rather than explaining every detail of an AI system's diagnostic process, the doctor might simply state that the AI analyses a large amount of medical data to help identify potential diagnoses and treatment options, but that the final decision is made by the doctor based on their expertise and the patient's specific situation.

176 Bester, Cole and Kodish, 2016: 869–886.
177 Dalton-Brown, 2020: 117.
178 Murray, 2012: 563–566.
179 Koplin, Gyngell, Savulescu, et al., 2022: 655–665.
180 Lühnen, Mühlhauser and Steckelberg, 2018: 377.

There is some debate on the criteria to assess when the information is reasonable. It has been understood that doctors should convey to the patient information that a reasonable doctor (in some jurisdictions) or a reasonable patient (in others) need to know.[181]

Note that the 'reasonable' person – be it a doctor or a patient – might not coincide with that concrete individual. For instance, in medical AI, the prevailing focus revolves around maximising lifespan. However, this approach can clash with the desires of concrete patients who prioritise minimising suffering over longevity.[182] This illustrates how the notion of 'reasonableness' is context-dependent: what seems reasonable to a hypothetical patient or doctor may differ significantly from the preferences of real individuals.

It is entirely possible that in most scenarios involving AI, neither the reasonable doctor would be able to provide information nor the reasonable patient would be able to understand it. Strictly speaking, this incapacity is observed in other scenarios. Let's take the case of medicines:[183] neither does the doctor has information about how the medicine was developed and about the vicissitudes of clinical trials, nor do patients need to be informed about this.[184] In conclusion, being 'lost in information' is not new in healthcare.

4.5 *From Informed Consent to Patient Participation in Decision Making*

Beyond standard informed consent, the concept of educated consent emerges as a more robust approach,[185] which involves a meaningful dialogue between patients and healthcare providers. This educational process could manifest in various ways, spanning from the passive conveyance of information to direct interactions between patients and AI systems before their involvement in patient care.[186]

This approach advocates for a more radical form of informed consent, where patients are actively involved in their own medical decision-making process.[187] Traditionally, many doctors – especially those from older generations – may view this practice as an admission of their own lack of expertise, potentially

---

181 Sarela, 2024: 47–62.

182 Instead of customising treatment based on a patient's personal goals and values, the computer is often seen as having the authority to decide the 'correct answer' objectively. Whether this occurs, however, ultimately depends on how the AI system is developed and integrated into clinical decision-making. See Friedrich, Mason and Malone, 2022: 7.

183 Van Norman, 2016: 170–179.

184 Babic, Gerke, Evgeniou et al., 2019: 1202–1204.

185 Brenner, Brenner and Horowitz, 2009: 348–351.

186 LaRosa and Danks, 2018: 214.

187 Galletta, Piazza, Meloni, et al., 2022.

undermining their authority in the eyes of the patient. As a result, they might hesitate to fully engage patients in discussions about treatment options, preferring instead to guide the decision-making process themselves.[188]

However, in today's medical landscape, where AI systems are increasingly integrated into clinical practice, involving patients more deeply in their own care decisions is becoming not only beneficial but necessary. For instance, AI systems often present a range of therapeutic alternatives, some of which may differ from the doctor's preferred approach. In such scenarios, the patient's active participation in the decision-making process can be crucial. Consider a situation where both the doctor and the AI system offer different treatment plans. The doctor might lean towards a traditional approach based on their clinical experience, while the AI might recommend a newer, data-driven alternative. Rather than choosing one path unilaterally, the doctor could involve the patient in a detailed discussion, presenting the pros and cons of each option, including the reasoning behind the AI's recommendation. This collaborative approach ensures that the patient's values, preferences, and understanding are integral to the final decision.

Engaging patients in this way not only respects their autonomy but also provides a safeguard for both the doctor and the patient. If the chosen treatment outcome is not as expected and the case ends up in court, the fact that the patient was fully informed and involved in the decision-making process can be a powerful defence.[189] It demonstrates that the patient was aware of the various options and associated risks, and that their decision was made in collaboration with the healthcare provider. Moreover, this approach fosters greater trust between the patient and the doctor.[190] Patients who feel heard and involved are more likely to trust their healthcare providers and adhere to the treatment plan, even when outcomes are uncertain.[191] In a healthcare environment where AI is increasingly influencing decisions, this level of transparency and patient involvement can bridge the gap between traditional medical expertise and cutting-edge technological tools.

188 Driever, Stiggelbout and Brand, 2020: 77–82.
189 Banja, Hollstein and Bruno, 2022: 818.
190 Deniz, Akbolat, Çimen, et al., 2021.
191 Jiang, Wu, Zhang, et al., 2024.

5 Empathy in AI-Based Medicine

5.1 *The Relevance of Empathy in Healthcare*

Acknowledged as a cornerstone in the therapeutic alliance between caregivers and patients, empathy has been demonstrated to play a crucial role in improving health outcomes.[192] By enabling healthcare providers to discern and understand users' experiences, concerns, and perspectives, empathy contributes to the fortification and advancement of the therapeutic relationship.[193]

The empathetic bond between health professionals and patients improves collaboration in creating a personalised treatment plan, leading to greater patient satisfaction. Notably, the empathetic relationship cultivated during the care process reinforces therapeutic outcomes, as users exhibit improved adherence to the prescribed therapeutic course.[194] Consequently, this enhances the quality of care, reduces errors and results in a higher percentage of positive therapeutic experiences for healthcare recipients.

A practical example of this is in the management of chronic conditions like diabetes. When a healthcare provider builds a strong, empathetic relationship with a patient, the patient is more likely to follow the recommended treatment plan, such as adhering to medication schedules, making dietary changes, and attending regular check-ups. This improved adherence leads to better blood sugar control, reducing the risk of complications like nerve damage or cardiovascular issues. As a result, the overall quality of care improves, medical errors decrease, and the patient experiences more positive health outcomes, contributing to a higher satisfaction with the care they receive.

Some claim that empathy is overrated in healthcare. Dalton-Brown, for instance, argues that not all doctors are highly empathetic, and in some cases, empathy might even detract from care since it focuses on the caregiver rather than the altruistic nature of the care itself.[195] Still, a substantial body of evidence indicates that empathy plays a crucial role in healthcare, as it is associated with reduced patient anxiety, which can lead to physiological improvements across various medical conditions. Empathetic doctors are often more successful in gathering essential information from their patients, which positively impacts the quality of care and overall medical outcomes.[196]

192 Moudatsou, Stavropoulou, Philalithis, et al., 2020.
193 Spiro, 2009: 1177–1179.
194 Moudatsou, Stavropoulou, Philalithis, et al., 2020.
195 Dalton-Brown, 2020: 115–121.
196 Hirmiz, 2023.

5.2 *Changes in Doctor-Patient Relationship*

Over time, the doctor-patient relationship has undergone a significant transformation, evolving from a traditionally paternalistic model to a more patient-centric dynamic.[197] In the past, the paternalistic approach positioned the doctor as the primary decision-maker, with the patient playing a more passive role.[198] The doctor's authority was largely unquestioned, and the patient was expected to follow medical advice with little input or discussion. The traditional paternalistic model views the doctor as a 'benevolent father', making decisions for the patient's best interests. However, this model has gradually given way to a more collaborative approach, where doctors and patients work together as equal partners in making healthcare decisions. The model of 'doctor as partner' promotes equal power and open communication between doctor and patient. Meanwhile, new models emerged. One is the model of the 'doctor as service provider,' which reflects the rise of consumerism in healthcare, with increased private sector involvement. Another one is characterised by the disempowerment of doctors and patients, as the power of external market forces, driven by neoliberalism, reduces the influence of both, leading to increased social and medical inequality. Lastly, there may be a fifth 'technology' model, where technology reshapes the patient-doctor relationship.[199]

The most significant twist in this evolution is the transformation towards an 'active patient' archetype,[200] which has precipitated changes not only in patients but also in doctors.[201] It has redefined the essence of the doctor-patient relationship, leading to a re-evaluation of the foundations that support our understanding of trust.[202]

The shift towards a patient-decider model has been driven by several factors, with one of the most influential being the advent of easy and widespread access to information through computers and the Internet.[203] The Internet has made medical knowledge more accessible to the general public than ever before. Patients can research their symptoms, learn about treatment options, and even read about others' experiences with similar conditions. This increased access to information has levelled the playing field, allowing patients to engage in more informed discussions with their doctors.[204] As a result, patients are no

197 Truog, 2012: 581–585; Solaiman and Malik, 2024.

198 Lorenzini, Ossa, Shaw, et al., 2023: 425–426.

199 These models rely on a classification proposed by Solaiman and Malik, 2024.

200 Phillips and Scheffmann-Petersen, 2020: 1419–1430.

201 LaRosa and Danks, 2018: 212.

202 LaRosa and Danks, 2018: 212.

203 Fitzpatrick, 2023.

204 However, this also comes with challenges. Not all information found online is accurate or reliable, and patients may come to consultations with misconceptions or fears

longer passive recipients of care but active participants who bring their own knowledge, concerns, and preferences to the table.

In this evolved relationship, the dynamic between doctor and patient is characterised by mutual respect and shared decision-making.[205] Doctors, with their extensive clinical knowledge and expertise, provide the necessary medical insights, diagnosis, and treatment options. However, they now recognise that patients hold authority over their own bodies and personal lives.

While the shift to a patient-centric and patient-decider model offers many benefits, including greater patient satisfaction and better alignment of treatment with personal values, it also presents challenges. Doctors must balance their professional judgment with respect for patient autonomy, even when patients choose options that the doctor may not fully endorse. This can be particularly challenging in cases where a patient's choice might lead to less favourable health outcomes.[206] The stress of witnessing a patient's condition worsen due to their (the patient's) chosen course of action might lead to frustration or emotional distance, potentially reducing the perceived empathy if not managed effectively. Maintaining empathy and open communication is crucial to preserving trust and ensuring that patient autonomy does not come at the expense of the therapeutic relationship.

5.3 *The Progressive Erosion of Medical Empathy due to Technology*

The traditional concept of healthcare originated from the familiar and trusted family doctor who builds personal connections with patients. However, it can be questioned to what extent health professionals can still be empathetic with their patients in this time of technology.[207]

Medical care via information technologies has created a gap between doctors and patients. Examples include alert fatigue caused by pop-up warnings, the growing doctor dependence on EHRs, and the tendency for doctors to be more focused on the computer interface.[208] For instance, a doctor treating a patient with multiple chronic conditions might receive a barrage of alerts when prescribing a new medication. Instead of engaging with the patient to discuss the potential benefits and risks of the treatment, the doctor may be

based on faulty information. Here, the doctor's role includes helping patients navigate the vast amount of information available, correcting misunderstandings, and providing evidence-based guidance. See Lu and Schulz, 2023: 983–988.

205 Montori, Ruissen, Hargraves, et al., 2023: 213–217.

206 Hardavella, Aamli-Gaagnat, Frille, et al., 2017: 129–135.

207 Kerasidou, 2020: 245–250.

208 Sulmasy and Bledsoe, 2019: S1–S32.

preoccupied with managing the flood of alerts. This can leave the patient feeling unheard or neglected, weakening the relationship.

These technological shifts have, in many cases, eroded the doctor-patient relationship by reducing the time and quality of direct, personal interaction. The traditional model of care, where doctors and patients engage in meaningful dialogue, has been compromised by the demands of digital systems that, while designed to enhance care, often divert attention away from the human aspects of medicine.

5.4 *Empathy in Doctor-Patient-AI Relationship*

5.4.1 Can Doctors Be Empathic Using AI?

The doctor-patient relationship already changes substantially along the way, and if we gained some benefits (mostly, the vindication of patients' autonomy instead of medical paternalism), we also lost some, in particular medical empathy. In our existing healthcare system, providers rarely offer deeply personal care. The primary reason is the overwhelming patient-to-caregiver ratio, which leaves healthcare professionals with limited time to connect with patients on a more personal level. As a result, the opportunity for meaningful and individualised interactions is significantly constrained.[209]

The integration of AI in healthcare substantially changes the doctor-patient relationship, and it comes with challenges. Healthcare practitioners must recognise that the ultimate legal and ethical responsibility for decisions lies with them. To ensure successful implementation, it is essential to adequately train doctors on the advantages and drawbacks of AI and encourage them to use it as a complement rather than a replacement for their clinical decision-making abilities and duties towards patients.[210] Balancing the benefits of AI while preserving the essence of human care is key to delivering optimal healthcare outcomes.[211]

The inclusion of AI also offers substantial potential added value for the doctor-patient relationship. Under the right conditions, AI can enable doctors to allocate more time and attention to each patient, thereby enhancing the quality of care and increasing focus on individual patients.[212] This new paradigm could potentially enable healthcare providers to devote more attention to patients' needs and actively listen to their concerns, instead of being

209 Drossman, Chang, Deutsch, et al., 2021: 1670–1688.

210 Mittermaier, Raza and Kvedar, 2023: 1–2; Visweswaran, King and Cooper, 2022.

211 Smith, 2021: 539–540.

212 Sauerbrei, Kerasidou, Lucivero, et al., 2023.

preoccupied with documenting information in the EHR.[213] For instance, during a medical appointment, in a room equipped with a microphone and speaker capable of recording the visit, the technologies for passive data collection and analysis reduce the doctor's workload. This system operates by actively listening to the conversation, distinguishing between voices, and tracking the consultation, providing real-time suggestions to the doctor through an earpiece. It further transcribes the entire interaction, generating a comprehensive record for all parties involved and subsequently completes and navigates the EHR. This will enable doctors to allocate more time for personal interactions with patients, which fosters trust, a fundamental element in the doctor-patient relationship.[214]

5.4.2 AI-Patient Relationship

An essential aspect of fostering and sustaining the therapeutic relationship is 'empathetic understanding'.[215] Building a therapeutic bond depends on understanding the patient's emotions and experiences. Empathetic understanding conveys to the patient that the care provider truly comprehends and cares about their emotions and experiences. When the care provider reflects their own emotions back to the patient, it is termed 'reflection',[216] a term that expresses the condition where the care provider communicates that they are familiar and responsive to the patient's feelings.

Remarkably, even when patients are cognizant that their interaction is with an AI system, it can still evoke profound emotions, an aspect that holds therapeutic relevance in certain contexts. This phenomenon, known as 'transference', entails the redirection of the patient's emotions toward an inanimate entity. It is not uncommon to have situations in which patients unconsciously redirect feelings towards a person (e.g., a significant person from their past or current life) to another (e.g., the therapist).[217] Eventually, AI can also become the target of this 'transference'. While this transference can have therapeutic benefits, it is essential to be mindful of instances where patients might develop excessive attachment or 'attraction' to the AI system in a non-therapeutic manner.

213 Aminololama-Shakeri and López, 2019: 309.
214 Smith, 2021: 304.
215 The expression is borrowed from Luxton, 2014: 4.
216 Lupton, 2018: 3.
217 Luxton, 2014: 4.

In the literature, the concept of 'Turing Deceptions' has been suggested to refer to scenarios where individuals cannot discern whether they are interacting with a human or a machine.[218] In extreme cases, they might attribute a semblance of 'life' to the machine.[219] It is claimed that an increasing perception of the AI's autonomy and emotional attributes by users corresponds to a heightened sense of attachment towards it.[220] This connection to care robots does not always manifest, but when it does, it remains consistent over time.[221] The ramifications of such relationships can range from simple enjoyment of the company of an AI system to a potential emotional dependency. In extreme cases, patients' autonomy might be compromised, as their emotional bond with the AI influences their decision-making process. This situation could limit their freedom of choice, especially concerning the AI itself. For instance, due to attachment, patients might resist separating from the AI, regardless of its benefits to their health or the potential privacy concerns it poses. This establishment of a parasocial relationship also intertwines with the concept of emotional trust. This seemingly irrational trust holds a risk when questions of accountability arise.[222]

5.4.3 Empathic AI

The challenge in ascribing human-like intelligence to AI and all the correlated human features stems from its lack of sentience and the inability to internalise or express moral values.[223] Emotional intelligence has been regarded as essential aspects of overall intelligence and the capacity to make sound decisions.[224] This is why doctors are evaluated not only based on their analytical skills but also their 'bedside manner' and their ability to consider and communicate complex treatment options that impact the well-being of the patient.[225]

---

218 Miller, 2010: 51–52.

219 Lupton, 2018: 3.
Particular populations, such as the elderly (Xie, Xie, Wang, et al., 2025: 1687–1699) and children (especially those on the autism spectrum) (Langer, Marshall and Levy-Tzedek, 2023) are more susceptible to this phenomenon, although it can develop across all demographics.

220 Boch, Lucaj and Corrigan, 2020: 4.

221 Van Maris, Zook, Caleb-Solly, et al., 2020.

222 Boch, Ryan, Kriebitz, et al., 2023: 9.

223 This separation of morality from AI's cognitive capabilities is a fundamental obstacle, as traditionally, we have associated morality with the complete enjoyment of rights and legal responsibilities granted to natural persons. Cf. Chung and Zink, 2018: 57.

224 Chauhan and Chauhan, 2007: 217–38.

225 Chung and Zink, 2018: 57.

For machines to demonstrate what we consider a truly caring attitude, they would need to show the ability to recognise, understand, and respond appropriately to human emotions.[226] Arguably, one of the key issues with AI's inability to provide genuine care lies in its lack of understanding of the human experience, including our shared mortality and suffering, which are fundamental to authentic caregiving.[227]

However, advancements in affective sensing and processing have equipped AI systems with the ability to detect, interpret, and express emotions, as well as recognise other behavioural signals from human interactions.[228] Some AI systems possess an appearance of empathetic concern, making the interactions incredibly interactive and human-like.[229]

Indeed, current AI products are beginning to show some features that emulate rudimentary forms of empathy. Amazon's Alexa has a 'whisper mode' to communicate in whispers when people also whisper. This is, clearly, not empathy, but simply the compliance with a human command. Still, it can be extremely relevant for reassuring patients or when a baby is nursing to sleep.[230] ELIZA is another good example. Its basic text interface and limited computer code demonstrated how easily it could mirror a person's emotions, display empathy, and interactively engage the individual.[231]

As technology evolves, we might have AI systems able to recognise our state of mind, namely by analysing biometric data (facial features, tone of voice, the words used)[232] and automatically adapt to the counterpart.[233] For instance, an AI system could use the most adequate words and the most adequate tone of voice for a patient who is anxious, fearing a diagnosis, or who is afraid of a given medical act, or even someone who lost a loved one in the hospital. We might even have caretaker bots and end-of-life comfort bots.[234] Even if a patient is aware that they are interacting with a machine, they can still

226 Hirmiz, 2023.

227 Chen, 2020: 260.

228 Mengoni, Ceccacci and Generosi, 2024: 325–354.

229 Colombo, Rampino and Zambrelli, 2021: pp. 547–559. Presenting robots with human-like characteristics can significantly influence how people perceive and interact with them. For instance, when robots are designed to appear more anthropomorphic, individuals tend to show empathy towards them and may even hesitate to harm or strike them. See Davis, Armstrong, Groggel, et al., 2023.

230 Powell, 2019.

231 O'Brien, 2023: 111–126.

232 Bustamante, Alama-Maruta, Ng, et al., 2022: 112–121.

233 Guingrich and Graziano, 2024.

234 Terry, 2019: 185.

experience intense emotions during the encounter with the simulation.[235] Emotional engagement, and even profound emotional interactions, may be beneficial in appropriate therapeutic contexts.

The question arises as to whether it is ethically and feasibly possible to recreate all that with AI.

As for the feasibility, one might question how robots, incapable of mortality, can provide care for humans. However, this line of reasoning proves irrelevant if we agree that empathy does not necessarily require personal experience or the ability to undergo certain concepts.[236]

For example, while palliative care workers cannot reverse the course of illness or bring patients back to life, they play a crucial role in improving the quality of life for their patients by providing physical, emotional, and psychological support. This raises an interesting question: could an AI doctor do the same? Can AI, even with sophisticated algorithms, truly understand and respond to the emotional and spiritual needs of patients facing end-of-life decisions? Currently, AI can assist by predicting pain levels, suggesting optimal medication doses, or analysing patient data in real time. However, when it comes to sitting with a patient in their most vulnerable moments, offering a reassuring presence, or tailoring care to the psychological and emotional state of the patient, this is more dubious. The obvious answer is that human touch remains irreplaceable.[237] I agree that considering the current stage of the art, it would be challenging for AI to replicate the warmth, presence, and deeply personal care that palliative professionals provide. However, it is not excluded that in the near future AI will be able to emulate this human aspect of caregiving.[238]

Ethically, it is often argued that endowing AI with human-like traits, particularly empathy, is problematic. AI does not feel; it merely performs empathy. Framing a machine as empathetic risks misleading users – especially those in vulnerable situations – into placing undue trust in it, disclosing sensitive personal information, or even developing unhealthy emotional attachments to a system that is, at its core, indifferent (see Section 6.4.2 in this Chapter).

5.4.4 The Human Touch

AI systems must adhere to human values to ensure their acceptance in society and the specific environments they are designed to operate in. This underscores the importance of aligning AI systems' behaviour with human

235 Yun, Lee and Kim, 2021: 610–625.

236 Distinguishing between ethics of care and ethics of empathy see, Dalton-Brown, 2020: 118.

237 Meskó, 2023.

238 Editorial, 2024.

values, facilitating their integration and effectiveness within their intended contexts.[239] Examples of the 'human touch' in healthcare can be seen when a doctor patiently answers all of a patient's questions and concerns, no matter how many there are, ensuring the patient feels heard and understood; or when a doctor makes a follow-up phone call to check on a patient's progress after a procedure, providing a personal touch and ensuring they're recovering well.[240]

Within AI, it is imperative to analyse to what extent this technology affects the traditional human touch in the patient-doctor relationship.[241] It is commonly argued that AI lacks the capacity to form such an intimate relationship with individuals.[242] Many fear that AI could take technological intrusion to the other extreme, so that the only 'people' in the room will be the patient and a robot, which may be unacceptable to many.[243] As tasks are increasingly delegated to intelligent machines, the dehumanisation of healthcare becomes more evident, as machines often lack the ability to fully understand and respond to the needs of human patients.

There is also the concern that if machines entirely dominate the healthcare sector, the quality of care we receive might diminish. We risk losing the profound, human-centred care that goes beyond just completing medical tasks.[244] The reason for such fear is that we often connect care not just, or even primarily, with the provision of services, but also with the deeper aspects of 'caring for' and 'caring about' individuals.[245] Arguably, true care can only arise from a mutual understanding of human nature, including our inherent limitations and vulnerabilities.[246] Establishing and nurturing a therapeutic relationship is based on an empathetic understanding of the patient's emotions and experiences. Effective communication of the healthcare provider's own feelings, known as reflection, plays a significant role in this process.

239 Rossi, 2016.

240 Epstein, 2006: 272–278.

241 Jorstad, 2020.

242 For instance, a report from a European Parliament Committee has emphasised the paramount importance of human interaction in the realm of caregiving (European Parliament, 2017). The report highlights that human contact constitutes a fundamental element of providing care, and the substitution of humans with robots may lead to the dehumanisation of caregiving practices.

243 Michael Mittelman and colleagues argue that '[patients] need to be cared for by people, especially when we are sick and most vulnerable' (Mittelman, Markham and Taylor, 2018).

244 Hirmiz, 2023.

245 Coeckelbergh, 2014: 39–40.

246 Sparrow and Sparrow, 2006: 151–153.

In contrast, some argue that AI will not dehumanise medicine.[247] With a vision of restoring the human touch in medicine, proponents of AI in healthcare envision a future where doctors can devote more time to fostering meaningful patient relationships (see Chapter 6, Section 5.4.1). They aim to liberate healthcare professionals from the burdens of extensive documentation in EHRs and the arduous task of sifting through vast amounts of literature to determine the optimal treatment plan.[248] As Topol suggests, AI has the potential to reshape the healthcare landscape, allowing medicine to regain its human essence.[249]

247 Even when patients are aware of interacting with a machine, the interaction can evoke intense emotions, which, in a suitable therapeutic context, can be desirable. The redirection of a patient's emotions toward a machine (transference) can lead to some patients forming excessive attachments or attraction to an AI clinical provider system, and in some cases, patients may even perceive the machine as 'alive'. Cf. Lupton, 2018: 5.
See also Song, Xu and Zhao, 2022.

248 Bekbolatova, Mayer, Ong, et al., 2024.

249 Topol, 2019. LaRosa and Danks, 2018: 212–213.

Conclusive Notes

1 Will AI Replace Human Doctors?

As people's views of doctors change, so does their role in healthcare. If doctors are relegated to the status of mere users of AI, the distinct repository of knowledge and skill that they traditionally embody might be undervalued, weakening trust. On the other hand, when doctors are positioned as skilled operators of AI systems, the narrative shifts to enhance their perceived expertise, thus strengthening the foundation of role-based trust.[1]

In my view, AI's aim is – or must be – to boost human capabilities and facilitate better decision-making, leading to what we can call AI-enhanced medicine. By leveraging the unique strengths of AI and humans, organisations can achieve optimal results. AI is not a threat that will replace human involvement. Instead, it should be embraced for its potential to enhance and extend human lives.[2]

Still, I am far from the mantra 'computer knows best'.[3] Humans, assisted by AI, know best.[4] 'Collaboration between AI and human intelligence has the potential to lead to extraordinary improvements in human skills and the resolution of complicated issues. AI augmentation, in which humans and machines collaborate, has potential in a variety of fields, ranging from healthcare to scientific study'.[5]

While AI shows promising abilities in calculating probabilities similar to humans, only human healthcare providers employ critical thinking and clinical judgment to interpret diagnostic information, evaluate treatment options, and make informed decisions. Human healthcare providers are equipped to handle unforeseen circumstances, adjusting treatment plans based on real-time assessments.[6] Human providers excel at managing this complexity by considering multiple factors in a holistic way, which AI, as of now, cannot replicate.

1 Nickel, 2022.
2 Wagner, 2019: 489–501.
3 McDougall, 2019: 156–160. In a similar sense, Goodman, 2007: 132.
4 Jackson (2019: 42) argues that the potential for increased fairness in most decision-making processes through the implementation of AI is quite feasible, even if some degree of error is present.
5 Rawas, 2024.
6 Haraldseid-Driftland, Billett, Guise, et al, 2022.

 | HTTPS://DOI.ORG/10.1163/9789004543980-009

On the other hand, prescribing and treatment decisions require meaningful conversations with patients to ensure that the chosen approach aligns with their personal values, goals, and preferences. Humans are naturally skilled at having these personal discussions and making value-based decisions; something AI currently struggles with because it cannot fully understand context or social cues. Healthcare is complex, involving diverse patient needs, medical histories, and changing conditions.[7] The competence of a medical professional extends beyond technical knowledge and judgment. It encompasses the capacity to establish a comforting and effective rapport with patients, engaging in sensitive conversations to understand their symptoms and concerns. This ability to offer responsive and meaningful feedback is crucial in building trust with patients.[8] Maybe in the future, AI systems will develop in a way that they manage to mimic some of these human features; still, AI will never be able to be human, just as we cannot be machines.

I also believe that the goal of AI is not to 'become human'. Neither humans nor machines would benefit from such a development. Instead, AI is destined to enhance us, serving as a form of digital augmentation. Doctors will become better at their jobs by leveraging AI's capabilities. They will also become more human, not only because AI will free up their time for deeper patient interactions, but also because it will prompt them to emphasise the uniquely human qualities that distinguish them from machines.

2 Closing Observations

AI holds extraordinary promise in the healthcare sector, from streamlining diagnostic processes to enabling personalised treatment plans and improving overall system efficiency. Its capacity to process vast datasets, detect subtle patterns, and support complex clinical decisions is already reshaping expectations around medical practice. However, this transformation is not without friction. The integration of AI into clinical workflows challenges not only existing protocols but also the legal and ethical frameworks that underpin healthcare delivery.

One critical area of concern is the misalignment between AI-assisted care and the current standard of care. Medical liability frameworks are still largely built around human decision-making, professional discretion, and well-established clinical pathways. AI systems, by contrast, introduce novel

7 Smith, 2021: 303–304.
8 Lagioia and Contissa, 2020: 277.

forms of risk that are not easily accommodated by existing norms. If we fail to revisit and adapt the legal definition of what constitutes acceptable medical practice in the age of AI, we may inadvertently penalise both doctors who rely on such tools in good faith and institutions that deploy them responsibly.

This legal lag creates the risk of conflicting expectations. Healthcare providers may face increased exposure to malpractice claims where AI advice leads to poor outcomes, even if their reliance on the system was reasonable. At the same time, if courts continue to assess clinical decisions exclusively by reference to pre-AI standards, this could discourage the adoption of technologies that might otherwise improve patient safety and outcomes. More broadly, failing to update the standard of care could send a mixed message to patients, undermining trust in both AI technologies and the professionals who use them.

For the potential of AI in healthcare to be fully realised, the standard of care must evolve in tandem with technological innovation. This means establishing clear guidelines for the responsible use of AI, updating liability regimes to reflect the realities of human-machine collaboration, and ensuring that both doctors and patients can rely on the system with confidence.

References

AAVV (2018) Open Letter to the European Commission Artificial Intelligence and Robotics, Politico, https://perma.cc/2UXV-Z3YF.

Aaron DG, CT Robertson, LP King and WM. Sage (2025) 'A New Legal Standard For Medical Malpractice', *JAMA*, 333(13): 1161–1165.

Aboy M, T Minssen and E Vayena (2024) 'Navigating the EU AI Act: Implications for Regulated Digital Medical Products', *npj Digit. Med.* 7, 237, https://doi.org/10.1038/s41746-024-01232-3.

Ada Lovelace Institute (2022) Algorithmic Impact Assessment: A Case Study in Healthcare, https://www.adalovelaceinstitute.org/report/algorithmic-impact-assessment-case-study-healthcare/.

Adhamov S (2023) 'Robotic Surgery in World Medicine', *Science and Innovation*, 2(D2): 196–199, https://doi.org/10.5281/zenodo.7671218.

Aggarwal R, OT Mytton, M Derbrew, D Hananel, M Heydenburg, B Issenberg, C MacAulay, M Mancini, T Morimoto, N Soper, A Ziv and R Reznick (2010) 'Training and Simulation for Patient Safety', *Quality & Safety in Health Care*, 19(2): i34–i43, https://doi.org/10.1136/qshc.2009.038562.

Ahmad I, HF Sheikh and A Aved (2020) 'Benchmarking the Task Scheduling Algorithms for Performance, Energy, and Temperature Optimization', *Sustainable Computing: Informatics and Systems*, 25, 100339, https://doi.org/10.1016/j.suscom.2019.07.002.

Ahmad SF, H Han, MM Alam, MK Rehmat and M Irshad (2023) 'Impact of Artificial Intelligence on Human Loss in Decision Making, Laziness and Safety in Education', *Humanities and Social Sciences Communications*, 10(1): 1–14, https://doi.org/10.1057/s41599-023-01787-8.

Ahuja AS (2019) 'The Impact of Artificial Intelligence in Medicine on the Future Role of the Physician', *PeerJ* 4(7): e7702, https://doi.org/10.7717/peerj.7702.

Aksoy PÇ (2022) 'AI as Agents: Agency Law'. In *The Cambridge Handbook of Artificial Intelligence: Global Perspectives on Law and Ethics* (LA DiMatteo, C Poncibò, M Cannarsa, (Eds.), 146–160. Cambridge University Press, Cambridge.

Alattal D, AK Azar, P Myles, R Branson, H Abdulhussein and A Tucker (2025) 'Integrating Explainable AI in Medical Devices: Technical, Clinical and Regulatory Insights and Recommendations', ArXiv, https://arxiv.org/abs/2505.06620.

Alamgir A, O Mousa and Z Shah (2021) 'Artificial Intelligence in Predicting Cardiac Arrest: Scoping Review', *JMIR Med Inform*, 17;9(12): e30798, https://doi:10.2196/30798.

Aldoseri A, KN Al-Khalifa and AM Hamouda (2024) 'AI-Powered Innovation in Digital Transformation: Key Pillars and Industry Impact', *Sustainability*, 16, https://doi.org/10.3390/su16051790.

 | HTTPS://DOI.ORG/10.1163/9789004543980-010

Ali O, W Abdelbaki, A Shrestha, E Elbasi, MAA Alryalat and YK Dwivedi (2023) 'A Systematic Literature Review of Artificial Intelligence in the Healthcare Sector: Benefits, Challenges, Methodologies, and Functionalities', *Journal of Innovation & Knowledge*, 8(1), 100333.

Ali S, T Abuhmed, S El-Sappagh, K Muhammad, JM Alonso-Moral, R Confalonieri, R Guidotti, J Del Ser, N Díaz-Rodríguez and F Herrera (2023) 'Explainable Artificial Intelligence (XAI): What we Know and what is left to Attain Trustworthy Artificial Intelligence', *Information Fusion*, 99, 101805, https://doi.org/10.1016/j.inffus.2023.101805.

Allain, JS (2013) 'From Jeopardy! to Jaundice: The Medical Liability Implications of Dr. Watson and Other Artificial Intelligence Systems', *Louisiana Law Review*, 73: 1049–1079.

Allen C, MK Hendrickson, R Gotlieb, J Marmet, E Hause, C Praska, S Lunos and MB Pitt (2023) 'Jargon Be Gone – Patient Preference in Doctor Communication', *Journal of Patient Experience*, 10, 23743735231158942.

Alowais SA, SS Alghamdi, N Alsuhebany, T Alqahtani, AI Alshaya, SN Almohareb, A Aldairem, M Alrashed, KB Saleh, HA Badreldin, MS Al Yami, S Al Harbi and AM Albekairy (2023) 'Revolutionizing Healthcare: The Role of Artificial Intelligence in Clinical Practice', *BMC Med Educ* 23, 689, https://doi.org/10.1186/s12909-023-04698-z.

Alshanteer M (2020) 'A Current Regime of Uncertainty: Improving Assessments of Liability for Damages Caused by Artificial Intelligence', *North Carolina Journal of Law & Technology*, 21(4): 27–58.

AMA (2018) AMA Passes First Policy Recommendations on Augmented Intelligence, https://www.ama-assn.org/press-center/ama-press-releases/ama-passes-first-policy-recommendations-augmented-intelligence.

AMA Board of Trustees Policy Summary (2019) 'Augmented Intelligence in Healthcare', https://www.ama-assn.org/system/files/2019-08/ai-2018-board-policy-summary.pdf.

Amann J, A Blasimme, E Vayena, D Frey, VI Madai on behalf of the Precise4Q consortium (2020) 'Explainability for Artificial Intelligence in Healthcare: A Multidisciplinary Perspective', *BMC Med Inform Decis Mak*, 20, 310, https://doi.org/10.1186/s12911-020-01332-6.

American College of Cardiology (2024) 'The Potential of AI in Cardiovascular Care Focus of Review', https://www.acc.org/latest-in-cardiology/articles/2024/06/24/16/51/the-potential-of-ai-in-cardiovascular-care-focus-of-review.

American College of Radiology (2024) 'ARCH-AI: Setting the Standard for Responsible Artificial Intelligence in Radiology', https://www.acr.org/Member-Resources/Medical-Student/MESO-News/Responsible-Artificial-Intelligence-in-Radiology.

American Medical Association (2019) 'AMA Adopt Policy, Integrate Augmented Intelligence in Doctor Training [News Release]', https://www.ama-assn.org/press

-center/ama-press-releases/ama-adopt-policy-integrate-augmented-intelligence-physician.

American Medical Association (2018) 'Augmented Intelligence in Health Care', https://www.amaassn.org/system/files/2019-01/augmented-intelligence-policy-report.pdf.

Aminololama-Shakeri S and JE López (2019) 'The Doctor-Patient Relationship with Artificial Intelligence', *American Journal of Roentgenology*, 212(2): 308–310, https://doi.org/10.2214/AJR.18.20509.

Anagnostaki M (2023) Synthetic Data: Anonymized Data or Pseudonymized Data?, Datenshutzt, https://www.datenschutz-notizen.de/synthetic-data-anonymized-data-or-pseudonymized-data-3541386/.

Anastasia S, M Madonnab and L Monica (2021) 'Implications of Embedded Artificial Intelligence – Machine Learning on Safety of Machinery', *Procedia Computer Science*, 180: 338–343.

Anaya JM, C Duarte-Rey, JC Sarmiento-Monroy, D Bardey, J Castiblanco and A Rojas-Villarraga (2016), 'Personalized Medicine. Closing the Gap Between Knowledge and Clinical Practice', *Autoimmunity Reviews*, 15(8): 833–842.

Anderson SL (2008) 'Asimov's "Three Laws of Robotics" and Machine Metaethics', *AI & Society*, 22: 477–493.

Andreotta AJ, N Kirkham and M Rizzi (2022) 'AI, Big Data, and the Future of Consent', *AI & Soc*, 37: 1715–1728, https://doi.org/10.1007/s00146-021-01262-5.

Angus DC (2020) 'Randomized Clinical Trials of Artificial Intelligence', *JAMA*, 323(11): 1043–1045.

Antich-Isern P, J Caro-Barri and J Aparicio-Blanco (2021) 'The Combination of Medical Devices and Medicinal Products Revisited from the New European Legal Framework', *International Journal of Pharmaceutics*, 607, 120992, https://doi.org/10.1016/j.ijpharm.2021.120992.

Anvari M, C McKinley and H Stein (2005) 'Establishment of the World's First Telerobotic Remote Surgical Service: For Provision of Advanced Laparoscopic Surgery in a Rural Community', *Ann Surg*, 241(3): 460–464, https://doi:10.1097/01.sla.0000154456.69815.ee.

Arefin S (2024) 'AI Revolutionizing Healthcare: Innovations, Challenges, and Ethical Considerations', *MZ Journal of Artificial Intelligence*, 1(2): 1–17.

Arnold MH (2021) 'Teasing out Artificial Intelligence in Medicine: An Ethical Critique of Artificial Intelligence and Machine Learning in Medicine', *Bioethical Inquiry*, 18: 121–139, https://doi.org/10.1007/s11673-020-10080-1.

Arogundade OR (2023) 'Structuring Knowledge Bases with AI and Machine Learning', *Information and Knowledge Management*, 8: 31–39.

Artetxe A, A Beristain and M Grana (2018) 'Predictive Models for Hospital Readmission Risk: A Systematic Review of Methods', *Computer Methods and Programs in Biomedicine*, 164: 49–64.

Asan O, A Bayrak and A Choudhury (2020) 'Artificial Intelligence and Human Trust in Healthcare: Focus on Clinicians', *J Med Internet Res*, 22(6): e15154, https://doi.org/10.2196/15154.

Asaro PM (2012) 'A Body to Kick, but Still No Soul to Damn: Legal Perspectives on Robotics', In *Robot Ethics: The Ethical and Social Implications of Robotics*, Patrick Lin et al. (Eds.), 169–186. MIT Press, Cambridge.

Asimov I (1950) 'Runaround. I, Robot'. In *The Isaac Asimov Collection*. Doubleday, New York City.

Askitopoulou H and AN Vgontzas (2018) 'The Relevance of the Hippocratic Oath to the Ethical and Moral Values of Contemporary Medicine. Part I: The Hippocratic Oath from Antiquity to Modern Times', *European Spine Journal*, 27(7): 1481–1490.

Association for Computing Machinery (2018) ACM Code of Ethics and Professional Conduct. https ://www.acm.org/code-of-ethics.

Astromskė K, E Peičius and P Astromskis (2021) 'Ethical and Legal Challenges of Informed Consent Applying Artificial Intelligence in Medical Diagnostic Consultations', *AI & Soc*, 36: 509–520, https://doi.org/10.1007/s00146-020-01008-9.

Ataíde RP and IS Craveiro (2020) 'Critérios de Averiguação da Culpa Negligente', *O Direito*, 152(2): 343–355.

Athey S (2017) 'Beyond Prediction: Using Big Data for Policy Problems', *Science*, 355(6324): 483–85.

Avati A, K Jung, S Harman, L Downing, A Ng and NH Shah (2018) 'Improving Palliative Care with Deep Learning', *BMC Med Inform Decis Mak*, 18(4), https://doi.org/10.1186/s12911-018-0677-8.

Babic B, S Gerke, T Evgeniou and IG Cohen (2019) 'Algorithms on Regulatory Lockdown in Medicine', *Science*, 366(6470): 1202–1204, https://doi.org/10.1126/science.aay9547.

Babushkina D (2023) 'Are we Justified Attributing a Mistake in Diagnosis to an AI Diagnostic System?', *AI Ethics*, 3: 567–584, https://doi.org/10.1007/s43681-022-00189-x.

Badnjević A, LG Pokvić, A Deumić and LS Bećirović (2022) 'Post-Market Surveillance of Medical Devices: A Review', *Technol Health Care*, 30(6): 1315–1329, https://doi.org/10.3233/THC-220284.

Baig, MA, MA Almuhaizea, J Alshehri, MS Bazarbashi and F Al-Shagathrh (2020) 'Urgent Need for Developing a Framework for the Governance of AI in Healthcare', *Stud Health Technol Inform*, 272: 253–256, https://doi.org/10.3233/SHTI200542.

Bajwa J, U Munir, A Nori and B Williams (2021) 'Artificial Intelligence in Healthcare: Transforming the Practice of Medicine', *Future Healthc J*, 8(2): e188–e194, https://doi.org/10.7861/fhj.2021-0095.

Baker A, Y Perov, K Middleton, J Baxter, D Mullarkey, D Sangar, M Butt, A DoRosario and S Johri (2020) 'A Comparison of Artificial Intelligence and Human Doctors for the Purpose of Triage and Diagnosis', *Frontiers in Artificial Intelligence*, 3, 543405, https://doi.org/10.3389/frai.2020.543405.

Balasubramaniam N, M Kauppinen, A Rannisto, K Hiekkanen and S Kujala (2023) 'Transparency and Explainability of AI Systems: From ethical Guidelines to Requirements', *Information and Software Technology*, 159, 107197, https://doi.org/10.1016/j.infsof.2023.107197.

Baldwin R and HJ Arata III (2021) 'Causal Reasoning with Autonomous Systems and Intelligent Machine Applications', *The Cyber Defense Review*, 6(4): 95–115.

Banja J, R Hollstein and MA Bruno (2023) 'Defensive Medicine in the AI Era', *European Journal of Radiology*, 168, 111096, https://doi.org/10.1016/j.ejrad.2023.111096.

Banja J, R Hollstein and MA Bruno (2022) 'When Artificial Intelligence Models Surpass Physician Performance: Medical Malpractice Liability in an Era of Advanced Artificial Intelligence', *JACR – Journal of the American College of Radiology*, 19(7): 816–820, https://doi.org/10.1016/j.jacr.2021.11.014.

Banteka N (2020) 'Artificially Intelligent Persons', *Houston Law Review*, 58: 537–596.

Baram M (2007) 'Liability and its Influence on Designing for Product and Process Safety', *Safety Science*, 45(1–2): 11–30, https://doi.org/10.1016/j.ssci.2006.08.022.

Barnsteiner JH (2008) 'Medication Reconciliation', in *Patient Safety and Quality: An Evidence-Based Handbook for Nurses*, RG Hughes (ed.), Rockville (MD): Agency for Healthcare Research and Quality (US), https://www.ncbi.nlm.nih.gov/books/NBK2648/.

Bartlett B (2023) 'The Possibility of AI-Induced Medical Manslaughter: Unexplainable Decisions, Epistemic Vices, and a New Dimension of Moral Luck', *Medical Law International*, 23(3): 241–270, https://doi.org/10.1177/09685332231193944.

Bartlett B (2024) 'Clinical Negligence in an Age of Machine Learning: Res Ipsa Loquitur to the Rescue?', *Journal of European Tort Law*, 15(3): 295–326.

Bashayreh M, FN Sibai and A Tabbara (2020) 'Artificial Intelligence and Legal Liability: Towards an International Approach of Proportional Liability Based on Risk Sharing', *Information and Communications Technology Law*, 30(2), 169–192.

Bashir D, GD Montañez, S Sehra, PS Segura and J Lauw (2020) 'An Information-Theoretic Perspective on Overfitting and Underfitting', in *AI 2020: Advances in Artificial Intelligence: 33rd Australasian Joint Conference*, AI 2020, Canberra, ACT, Australia, November 29–30, 2020, Proceedings 33, 347–358. Springer International Publishing, Cham.

Bates DW, D Levine, A Syrowatka, M Kuznetsova, KJ Craig, A Rui, GP Jackson and K Rhee (2021) 'The Potential of Artificial Intelligence to Improve Patient Safety: A Scoping Review', *Npj Digital Medicine*, 4(1): 1–8, https://doi.org/10.1038/s41746-021-00423-6.

Bathae Y (2018) 'The Artificial Intelligence Black Box and the Failure of Intent and Causation', *Harv J Law Technol*, 31(2): 889–938.

Bathae Y (2020) 'Artificial Intelligence Opinion Liability', *Berkeley Tech. LJ*, 35: 113–170.

Batko K and A Ślęzak (2022) 'The Use of Big Data Analytics in Healthcare', *Journal of Big Data*, 9(1), https://doi.org/10.1186/s40537-021-00553-4.

Baum, SD (2020) 'Social Choice Ethics in Artificial Intelligence', *AI & Soc*, 35, 165–176, https://doi.org/10.1007/s00146-017-0760-1.

Beauchamp TL (2019) 'The Principle of Beneficence in Applied Ethics', in *The Stanford Encyclopedia of Philosophy*, Zalta (ed.), https://plato.stanford.edu/archives/spr2019/entries/principle-beneficence/.

Beauchamp TL and Childress JF (2013) *Principles of Biomedical Ethics* 7th ed., Oxford University Press, New York.

Beck AH, AR Sangoi, S Leung, RJ Marinelli, TO Nielsen, MJ Van de Vijver, RB West, M Van de Rijn and D Koller (2011) 'Systematic Analysis of Breast Cancer Morphology Uncovers Stromal Features Associated with Survival', *Sci Transl Med*; 3(108): 108ra113, https://doi.org/10.1126/scitranslmed.3002564.

Beck J and T Burri (2024) 'From "Human Control" in International Law to "Human Oversight" in the New EU Act on Artificial Intelligence', in *Research Handbook on Meaningful Human Control of Artificial Intelligence Systems*, G Mecacci, D Amoroso, LC Siebert, D Abbink, J Van den Hoven and FS De Sio (Eds.), 104–130. Edward Elgar Publishing, Cheltenham.

Beede E, E Baylor, F Hersch, A Iurchenko, L Wilcox, P Ruamviboonsuk and LM Vardoulakis (2020) 'A Human-Centered Evaluation of a Deep Learning System Deployed in Clinics for the Detection of Diabetic Retinopathy', in *Proceedings of the 2020 CHI Conference on Human Factors in Computing Systems*, 1–12. Association for Computing Machinery, New York, https://doi.org/10.1145/3313831.3376718.

Bekbolatova M, J Mayer, CW Ong and M Toma (2024) 'Transformative Potential of AI in Healthcare: Definitions, Applications, and Navigating the Ethical Landscape and Public Perspectives', *Healthcare*, 12(2), 125, https://doi.org/10.3390/healthcare12020125.

Beltran-Aroca CM, E Girela-Lopez, E Collazo-Chao, M Montero-Pérez-Barquero and MC Muñoz-Villanueva (2016) 'Confidentiality Breaches in Clinical Practice: What Happens in Hospitals?', *BMC Medical Ethics*, 17(1), https://doi.org/10.1186/s12910-016-0136-y.

Benbassat J (2023) *Teaching Professional Attitudes and Basic Clinical Skills to Medical Students*. Springer, Cham, https://doi.org/10.1007/978-3-031-26542-6_2.

Bernal J and C Mazo (2022) 'Transparency of Artificial Intelligence in Healthcare: Insights from Professionals in Computing and Healthcare Worldwide', *Appl. Sci.*, 12, 10228, https://doi.org/10.3390/app122010228.

Bernstein MH, MK Atalay, EH Dibble, A Maxwell, AR Karam, S Agarwal, RC Ward, TT Healey and GL Baird (2023) 'Can Incorrect Artificial Intelligence (AI) Results Impact Radiologists, and if So, What Can We Do About It? A Multi-Reader Pilot Study of Lung Cancer Detection with Chest Radiography', *Eur Radiol*, 33: 8263–8269, https://doi.org/10.1007/s00330-023-09747-1.

Bertoli P and V Grembi (2017) 'Medical Malpractice: How Legal Liability Affects Medical Decisions', WP 17/14, HEDG – Health, Econometrics And Data Group, http://www.york.ac.uk/economics/postgrad/herc/hedg/wps/.

Bertolini A (2020) *Artificial Intelligence and Civil Liability.* Bruxelles: European Parliament – Committee on Legal Affairs.

Bertolini A and F Episcopo (2021) 'The Expert Group's Report on Liability for Artificial Intelligence and Other Emerging Digital Technologies: A Critical Assessment', *European Journal of Risk Regulation*, 12(3): 644–659, https://doi.org/10.1017/err.2021.30.

Bester J, CM Cole and E Kodish (2016) 'The Limits of Informed Consent for an Overwhelmed Patient: Clinicians' Role in Protecting Patients and Preventing Overwhelm', *AMA J Ethics*, 18(9): 869–886, https://doi.org/10.1001/journalofethics.2016.18.9.peer2-1609.

Bhagat SV and D Kanyal (2024) 'Navigating the Future: The Transformative Impact of Artificial Intelligence on Hospital Management – A Comprehensive Review', *Cureus*, 16(2), https://doi.org/10.7759/cureus.54518.

Bharti SS and SK Aryal (2022) 'The Right to Privacy and an Implication of the EU General Data Protection Regulation (GDPR) in Europe: Challenges to the Companies', *Journal of Contemporary European Studies*, 31(4): 1391–1402, https://doi.org/10.1080/14782804.2022.2130193.

Bignami E, J Montomoli, V Bellini and M Cascella (2023) 'Uncovering the Power of Synergy: A Hybrid Human–Machine Model for Maximizing AI Properties and Human Expertise', *Crit Care*, 27, 330, https://doi.org/10.1186/s13054-023-04598-0.

Birmingham A (2017) 'MD Anderson: What Can we Learn from the Biggest IA Failure in the World So Far', https://www.linkedin.com/pulse/md-anderson-what-can-we-learn-from-biggest-ia-failure-birmingham.

Blake V (2012) 'When is a Physician-Doctor Relationship Established?', *AMA Journal of Ethics*, 14(5): 403–406.

Blasimme A and E Vayena (2020) 'The Ethics of AI in Biomedical Research, Patient Care, and Public Health', in *The Oxford Handbook of Ethics of AI*, MD Dubber, F Pasquale and S Das (Eds.), 703–718. Oxford Academic, Oxford, https://di.org/10.1093/oxfordhb/9780190067397.013.45.

Bleher H and M Braun (2022) 'Diffused Responsibility: Attributions of Responsibility in the Use of AI-Driven Clinical Decision Support Systems', *AI and Ethics*, 2(4): 747–761, https://doi.org/10.1007/s43681-022-00135-x.

Boch A, S Ryan, A Kriebitz, LM Amugongo and C Lütge (2023) 'Beyond the Metal Flesh: Understanding the Intersection between Bio- and AI Ethics for Robotics in Healthcare', *Robotics*, 12(4), 110, https://doi.org/10.3390/robotics12040110.

Boch A, L Lucaj and C Corrigan (2020) 'A Robotic New Hope: Opportunities, Challenges, and Ethical Considerations of Social Robots', Tech. Univ. Munich, https://

ieai.sot.tum.de/wpcontent/uploads/2021/05/ResearchBrief_April2021_SocialRobots_FinalV2.pdf.

Bohr A and K Memarzadeh (2020) 'The Rise of Artificial Intelligence in Healthcare Applications', *Artificial Intelligence in Healthcare*, 25–60, https://doi.org/10.1016/B978-0-12-818438-7.00002-2.

Borghetti JS (2019) 'Civil Liability for Artificial Intelligence: What Should its Basis Be?', *La Revue des Juristes de Sciences Po*, 17: 94–102.

Borghetti JS, MS Silva, D Fairgrieve, P García, A Keirse, P Machnikowski, E Rajneri, C Schmon, V Ulbeck, V Vallone and H Zech (2021) 'Relevance of Risk-Benefit for Assessing Defectiveness of a Product: A Comparative Study of Thirteen European Legal Systems', *European Review of Private Law*, 29(1): 91–132.

Borrego-Díaz J and J Galán Páez (2022) 'Knowledge Representation for Explainable Artificial Intelligence: Modeling Foundations from Complex Systems', *Complex & Intelligent Systems*, 8(2): 1579–1601.

Bouderhem R (2024) 'Shaping the Future of AI in Healthcare Through Ethics and Governance', *Humanit Soc Sci Commun*, 11, 416, https://doi.org/10.1057/s41599-024-02894-w.

Božić V (2023) 'Integrated Risk Management and Artificial Intelligence in Hospital', *Journal of AI*, 7(1): 63–80, https://doi.org/10.61969/jai.1329224.

Braithwaite RS and A Caplan (2014) 'Does Patient-Centered Care Mean that Informed Consent is Necessary for Clinical Performance Measures?', *J Gen Intern Med*, 29: 558–559, https://doi.org/10.1007/s11606-013-2645-4.

Braun M, P Hummel, S Beck S and P Dabrock (2021) 'Primer on an Ethics of AI-Based Decision Support Systems in the Clinic', *Journal of Medical Ethics*, 47(12): e3, https://doi.org/10.1136/medethics-2019-105860.

Brazier M and J Miola (2000) 'Bye-bye Bolam: A Medical Litigation Revolution?', *Medical Law Review*, 8(1): 85–114, https://doi.org/10.1093/medlaw/8.1.85.

Brennan PA and RS Oeppen (2022) 'The Role of Human Factors in Improving Patient Safety', *Trends Urology & Men Health*, 13: 30–33, https://doi.org/10.1002/tre.858.

Brenner LH, AT Brenner and D Horowitz (2009) 'Beyond Informed Consent: Educating the Patient', *Clinical Orthopaedics and Related Research*, 467(2): 348–351, https://doi.org/10.1007/s11999-008-0642-4.

Bright Data (2025) Data for AI 2025, https://brightdata.com/static/data_for_ai_report_2025.pdf?md5=740441-b83343c7.

Brown MT and JK Bussell (2011) 'Medication Adherence: WHO Cares?', *Mayo Clinic Proceedings*, 86(4): 304–314, https://doi.org/10.4065/mcp.2010.0575.

Brown RD (2021) 'Property Ownership and the Legal Personhood of Artificial Intelligence', *Information & Communications Technology Law*, 30(2): 208–234, https://doi.org/10.1080/13600834.2020.1861714.

Brożek B, M Furman, M Jakubiec and B Kucharzyk (2023) 'The Black Box Problem Revisited. Real And Imaginary Challenges for Automated Legal Decision Making', *Artif Intell Law*, 32: 427–440, https://doi.org/10.1007/s10506-023-09356-9.

Bryson J (2018) AI and Future Generations – Statement for UK Parliament APPG, https://joanna-bryson.blogspot.com/2018/07/ai-and-future-generations-statement-for.html.

Buchner B (2020) 'Informed Consent in Germany', in *Informed Consent and Health*, T Vansweevelt and N Glover-Thomas (Eds.) 216–234. Edward Elgar Publishing, Cheltenham.

Buiten M, A De Streel and M Peitz (2023) 'The Law and Economics of AI Liability', *Computer Law & Security Review*, 48, 105794, https://doi.org/10.1016/j.clsr.2023.105794.

Buiten MC (2024) 'Product Liability for Defective AI', *Eur J Law Econ*, 57: 239–273, https://doi.org/10.1007/s10657-024-09794-z.

Buonanno L (2019) Civil Liability in the Era of New Technology: The Influence of Blockchain, https://www.europeanlawinstitute.eu/fileadmin/user_upload/p_eli/YLA_Award/Submission_ELI_Young_Lawyers_Award_Luigi_Buonanno_ELI_2019.pdf.

Burrell J (2016) 'How the Machine 'Thinks': Understanding Opacity in Machine Learning Algorithms', *Big Data & Society*, 3(1), https://doi.org/10.1177/2053951715622512.

Bustamante CMV, K Alama-Maruta, C Ng and DDL Coppersmith (2022) 'Should Machines Be Allowed to 'Read Our Minds'? Uses and Regulation of Biometric Techniques that Attempt to Infer Mental States', *MIT Science Policy Review*, 3: 112–121, https://sciencepolicyreview.pubpub.org/pub/ogade56r.

Bygrave LA (2020) 'Article 22 Automated Individual Decision-Making, Including Profiling', in *The EU General Data Protection Regulation (GDPR): A Commentary*, C Kuner, LA Bygrave, C Docksey, L Drechsler (Eds.), 522–542. Oxford Academic, New York, https://doi.org/10.1093/oso/9780198826491.003.0055.

Canadian Agency for Drugs and Technologies in Health (2025) *2025 Watch List: Artificial Intelligence in Health Care*, Ottawa (ON), https://www.ncbi.nlm.nih.gov/books/NBK613808/?.

Campolo A, MR Sanfilippo, M Whittaker and K Crawford (2017) AI Now 2017 Report. AI Now Institute at New York University, https://ainowinstitute.org/AI_Now_2017_Report.pdf.

Carnevali D, L Zhong, E González-Almela, C Viana, M Rotkevich, A Wang, D Franco-Barranco, A Gonzalez-Marfil, MV Neguembor, A Castells-Garcia, I Arganda-Carreras and MP Cosma (2024) 'A Deep Learning Method that Identifies Cellular Heterogeneity Using Nanoscale Nuclear Features', *Nat Mach Intell*, 6: 1021–1033, https://doi.org/10.1038/s42256-024-00883-x.

Carter B (2020) 'Electronic Medical Records: A Prescription for Increased Medical Malpractice Liability?', *Vanderbilt Journal of Entertainment and Technology Law*, 13: 385–406.

Carter SM, W Rogers, KT Win, H Frazer, B Richards and N Houssami (2020) 'The Ethical, Legal and Social Implications of Using Artificial Intelligence Systems in Breast Cancer Care', *The Breast*, 49: 25–32, https://doi.org/10.1016/j.breast.2019.10.001.

Caruana R, Y Lou, J Gehrke, P Koch, M Sturm and N Elhadad (2015) 'Intelligible Models for Healthcare: Predicting Pneumonia Risk and Hospital 30-Day Readmission', *Proceedings of the 21th ACM SIGKDD International Conference on Knowledge Discovery and Data Mining*, 1721–1730, https://doi.org/10.1145/2783258.278861.

Casaburo D and L Gugliotta (2023) 'The EU AI Act Proposal(s): Context and Definition of AI', https://www.law.kuleuven.be/citip/blog/the-eu-ai-act-proposals-context-and-definition-of-ai/.

Castaneda C, K Nalley, C Mannion, P Bhattacharyya, P Blake, A Pecora, A Goy and KS Suh (2015) 'Clinical Decision Support Systems for Improving Diagnostic Accuracy and Achieving Precision Medicine', *Journal of Clinical Bioinformatics*, 5(1): 1–16.

Cauffman C (2018) 'Robo-Liability: The European Union in Search of the Best Way to Deal with Liability for Damage Caused by Artificial Intelligence', *Maastricht Journal of European and Comparative Law*, 25(5): 527–532, https://doi.org/10.1177/1023263X18812333.

Cè M, E Caloro, ME Pellegrino, M Basile, A Sorce, D Fazzini, G Oliva and M Cellina (2022) 'Artificial Intelligence in Breast Cancer Imaging: Risk Stratification, Lesion Detection and Classification, Treatment Planning and Prognosis – A Narrative Review', *Explor Target Antitumor Ther*, 3(6): 795–816, https://doi.org/10.37349/etat.2022.00113.

Cerrato P and J Halamka (2019) *The Transformative Power of Mobile Medicine: Leveraging Innovation, Seizing Opportunities and Overcoming Obstacles of Mhealth*. Academic Press, New York.

Chagal-Feferkorn KA (2020) 'How Can I Tell If My Algorithm Was Reasonable?', *Mich. Tech. L. Rev.*, 27: 213–241.

Challen R, J Denny, M Pitt, L Gompels, T Edwards and K Tsaneva-Atanasova (2019) 'Artificial Intelligence, Bias and Clinical Safety', *BMJ Qual Saf.*, 28(3): 231–237, https://doi.org/10.1136/bmjqs-2018-008370.

Chan B (2023) Black-Box Assisted Medical Decisions: AI Power vs. Ethical Physician Care', *Med Health Care and Philos*, 26: 285–292, https://doi.org/10.1007/s11019-023-10153-z.

Chan GKY (2021) 'Medical AI, Standard of Care in Negligence and Tort Law', in *AI, Data and Private Law: Translating Theory into Practice*, GKY Chan and M Yip (Eds.), 173–198, Oxford: Hart Publishing, https://doi.org/10.5040/9781509946860.ch-008.

Char DS, NH Shah and D Magnus (2018) 'Implementing Machine Learning in Health Care – Addressing Ethical Challenges', *N Engl J Med*, 378(11): 981–983, https://doi.org/10.1056/NEJMp1714229.

Char DS, MD Abramoff and C Feudtner (2020) 'Identifying Ethical Considerations for Machine Learning Healthcare Applications', *The American Journal of Bioethics*, 20(11): 7–17, https://doi.org/10.1080/15265161.2020.1819469.

Chauhan SP and D Chauhan (2007) 'Emotional Intelligence: Does It Influence Decision Making and Role Efficacy?', *Indian Journal of Industrial Relations*, 43(2): 217–38.

Chen J, A Beam, S Saria and EA Mendonça (2019) 'Potential Trade-Offs and Unintended Consequences of AI', in *Artificial Intelligence in Health Care: The Hope, the Hype, the Promise, the Peril*, MS Matheny, T Israni, M Ahmed and D Whicher (Eds.), 89–118. NAM Special Publication. Washington, DC.

Chen M (2020) 'A Tale of Two Deficits: Causality and Care in Medical AI', *Philos. Technol.* 33: 245–267, https://doi.org/10.1007/s13347-019-00359-6.

Cheong BC (2024) 'Transparency and Accountability in AI Systems: Safeguarding Wellbeing in the Age of Algorithmic Decision-Making', *Frontiers in Human Dynamics*, 6, 1421273, https://doi.org/10.3389/fhumd.2024.1421273.

Chinen MA (2016) 'The Co-Evolution of Autonomous Machines and Legal Responsibility', *Va J Law Technol.*, 20(2): 338–393.

Chinta SV, Z Wang, A Palikhe, X Zhang, A Kashif, MA Smith, J Liu, and W Zhang (2025) 'AI-Driven Healthcare: Fairness in AI Healthcare: A Survey', *PLOS Digital Health*, 4(5): e0000864.

Chipidza FE, RS Wallwork and TA Stern (2015) 'Impact of the Doctor-Patient Relationship', *The Primary Care Companion for CNS Disorders*, 17(5), https://doi.org/10.4088/PCC.15f01840.

Chopard B and O Musy (2023) 'Market for Artificial Intelligence In Health Care And Compensation for Medical Errors', *International Review of Law & Economics*, 75, https://doi.org/10.1016/j.irle.2023.106153.

Chung J (2017) 'What Should We Do About Artificial Intelligence in Health Care?', *Health L.J.*, 22: 37–43.

Chung J and A Zink (2018) 'Hey Watson Can Sue you for Malpractice Examining the Liability of Artificial Intelligence in Medicine', *Asia Pacific Journal of Health Law & Ethics*, 11(2): 51–80.

Cirillo D, S Catuara-Solarz, C Morey, E Guney, L Subirats, S Mellino, A Gigante, A Valencia, MJ Rementeria, AS Chadha and N Mavridis (2020) 'Sex and Gender Differences and Biases in Artificial Intelligence for Biomedicine and Healthcare', *npj Digit. Med.*, 3, 81, https://doi.org/10.1038/s41746-020-0288-5.

Clarke DM (2014) 'Causation and Liability in Tort Law', *Jurisprudence*, 5(2): 217–243.

Coeckelbergh M (2014) 'Good Healthcare Is in the 'How': The Quality of Care, the Role of Machines, and the Need for New Skills', in *Machine Medical Ethics*, SP van Rysewyk and M Pontier (Eds.) 33–47. Springer, Cham.

Cohen IG (2020) 'Informed Consent and Medical Artificial Intelligence: What to Tell the Patient?', *Geo. LJ*, 108, 1425–1469.

Collingridge D (1980) *The Social Control of Technology*. Frances Pinter, London.

Colombo S, L Rampino and F Zambrelli (2021) 'The Adaptive Affective Loop: How AI Agents Can Generate Empathetic Systemic Experiences', in *Advances in Information and Communication: Proceedings of the 2021 Future of Information and Communication Conference (FICC)*, Volume 1, K Arai (ed.), 547–559. Springer International Publishing, Cham.

Combs V (2019) 'South African Clinics Use Artificial Intelligence to Expand HIV Treatment', TechRepublic, https://www.techrepublic.com/article/south-african-clinics-use-artificialintelligence-to-expand-hiv-treatment/.

Comer AR (2025) 'Which Values Should Guide Evidence-Based Practice?', *AMA Journal of Ethics*, 27(1): E21–26, https://doi.org/10.1001/amajethics.2025.21.

Conner-Simons A (2017) 'Using Artificial Intelligence to Improve Early Breast Cancer Detection, *MIT News*, https://news.mit.edu/2017/artificial-intelligence-early-breast-cancer-detection-1017.

Contaldo MT, G Pasceri, G Vignati, L Bracchi, S Triggiani and G Carrafiello (2024) 'AI in Radiology: Navigating Medical Responsibility', *Diagnostics*, 14(14), 1506, https://doi.org/10.3390/diagnostics14141506.

Coppola F, L Faggioni, M Gabelloni, F De Vietro, V Mendola, A Cattabriga, MA Cocozza, G Vara, A Piccinino, S Lo Monaco, LV Pastore, M Mottola, S Malavasi, A Bevilacqua, E Neri and R Golfieri (2021) 'Human, All Too Human? An All-Around Appraisal of the "Artificial Intelligence Revolution" in Medical Imaging', *Frontiers in Psychology*, 12, 710982.

Coulter A (1997) 'Partnerships with Patients: The Pros and Cons of Shared Clinical Decision-Making', *Journal of Health Services Research & Policy*, 2(2): 112–21, https://doi.org/10.1177/135581969700200209.

Council of Europe (2023) 'Guide to Health Literacy Contributing to Trust Building And Equitable Access to Healthcare', https://rm.coe.int/inf-2022-17-guide-health-literacy/1680a9cb75.

Couture V, M Roy, I Ganache, M Goetghebeur and IG Cohen (2021) 'What Makes Artificial Intelligence Exceptional in Health Technology Assessment?', *Frontiers in Artificial Intelligence*, 4, 736697, https://doi.org/10.3389/frai.2021.736697.

Crawford K and R Calo (2016) 'There is a Blind Spot in AI Research', *Nature*, 538: 311–313, https://doi.org/10.1038/538311a.

Crew B (2020) 'Worth the Cost? A Closer Look at the Da Vinci Robot's Impact on Prostate Cancer Surgery', *Nature*, 580(7804): S5–S5, https://doi.org/10.1038/d41586-020-01037-w.

Crigger E, K Reinbold, C Hanson, A Kao, K Blake and M Irons (2022) 'Trustworthy Augmented Intelligence in Health Care', *J Med Syst*, 46, 12, https://doi.org/10.1007/s10916-021-01790-z.

Crook B, M Schlüter and T Speith (2023) 'Revisiting the Performance-Explainability Trade-Off in Explainable Artificial Intelligence (XAI)', *2023 IEEE 31st International Requirements Engineering Conference Workshops (REW)*, Hannover, Germany, 316–324, https://doi.org/10.1109/REW57809.2023.00060.

Curca GC (2010) 'Concepts of Deontological Responsibility and Medical Malpractice', *Revista Romana de Bioetica*, 8(1): 161–169.

Currie G, E Rohren and KE Hawk (2024) 'The Role of Artificial Intelligence in Supporting Person-Centred Care', in *Person-Centred Care in Radiology*, G Currie, E Rohren and KE Hawk (Eds.), 343–362. CRC Press, Boca Raton, https://doi.org/10.1201/9781003310143.

Cusenza GG (2024) 'Litigating Governmental Use of AI', *Ohio State Technology Law Journal, Forthcoming*, 2181: 2–46.

Dalton-Brown S (2020) 'The Ethics of Medical AI and the Physician-Patient Relationship', *Camb Q Healthc Ethics*, 29(1): 115–121, https://doi.org/10.1017/S0963180119000847.

Danish Medical Agency (2023) 'CE Marking', https://laegemiddelstyrelsen.dk/en/devices/ce-marking/.

Davahli MR, W Karwowski, K Fiok, T Wan and HR Parsaei (2021) 'Controlling Safety of Artificial Intelligence-Based Systems in Healthcare', *Symmetry*, 13(1), 102, https://doi.org/10.3390/sym13010102.

Davenport T and R Kalakota (2019) 'The Potential for Artificial Intelligence in Healthcare', *Future Healthc J.*, 6(2): 94–98, https://doi.org/10.7861/futurehosp.6-2-94.

Davis JL, R Armstrong, A Groggel, S Doolan, J Sheedy, TP Love and D Herath (2023) 'Role-taking and Robotic Form: An Exploratory Study of Social Connection in Human-Robot Interaction', *International Journal of Human-Computer Studies*, 178, 103094, https://doi.org/10.1016/j.ijhcs.2023.103094.

De Bruyne J (2019) 'A Conceptual and Comparative Analysis of the Obligations of Third-Party Certifiers', *Ohio Northern University Law Review*, 44(2): 203–261.

De Cruz P (2021) Comparative Healthcare Law. Routledge-Cavendish, London.

De Miguel Beriain I (2020) 'Should we Have A Right to Refuse Diagnostics and Treatment Planning by Artificial Intelligence?', *Med Health Care and Philos*, 23: 247–252, https://doi.org/10.1007/s11019-020-09939-2.

De Sio FS and JV Den Hoven (2018) 'Meaningful Human Control Over Autonomous Systems: A Philosophical Account', *Frontiers in Robotics and AI*, 5, 15, https://doi.org/10.3389/frobt.2018.00015.

Dekker J, J Karchoud, AM Braamse, H Buiting, IR Konings, ME Van Linde, CS Schuurhuizen, MA Sprangers, AT Beekman and HM Verheul (2020) 'Clinical Management of Emotions in Patients with Cancer: Introducing the Approach

"Emotional Support and Case Finding"', *Translational Behavioral Medicine*, 10(6): 1399–1405, https://doi.org/10.1093/tbm/ibaa115.

Del Riego A (2021) 'Deconstructing Fallacies in Products Liability Law to Provide a Remedy for Economic Loss', *American Business Law Journal*, 58(2): 387–447, https://doi.org/10.1111/ablj.12185.

Deloitte (2024) Providing Insurance Coverage for Artificial Intelligence May Be a Blue Ocean Opportunity, 29 May 2024, https://www2.deloitte.com/us/en/insights/industry/financial-services/financial-services-industry-predictions/2024/risk-insurance-for-ai.html.

Delpierre C and T Lefèvre (2023) 'Precision and Personalized Medicine: What their Current Definition Says and Silences About the Model of Health they Promote. Implication for the Development of Personalized Health', *Frontiers in Sociology*, 8, https://doi.org/10.3389/fsoc.2023.1112159.

Demetzou K (2019) 'Data Protection Impact Assessment: A tool for Accountability and the Unclarified Concept of "High Risk" in the General Data Protection Regulation', *Computer Law & Security Review*, 35(6), 105342, https://doi.org/10.1016/j.clsr.2019.105342.

Denecke K and CR Baudoin (2022) 'A Review of Artificial Intelligence and Robotics in Transformed Health Ecosystems', *Frontiers in Medicine*, 9, 795957, https://doi.org/10.3389/fmed.2022.795957.

Deniz S, M Akbolat, M Çimen and O Ünal (2021) 'The Mediating Role of Shared Decision-Making in the Effect of the Physician–Doctor Relationship on Compliance with Treatment', *Journal of Patient Experience*, 8, https://doi.org/10.1177/23743735211018066.

Dhillon G, G Harpreet, V Buddhavarapu, A Virmani, S Surani and R Kashyap (2024) 'Humanity at the Crossroads: Ethical Implications of AI in Medicine,' The Hospitalist, https://www.the-hospitalist.org/hospitalist/article/36988/hospital-medicine/humanity-at-the-crossroads-ethical-implications-of-ai-in-medicine/.

Díaz-Rodríguez N, J Del Ser, M Coeckelbergh, M López de Prado, E Herrera-Viedma and F Herrera (2023) 'Connecting the Dots in Trustworthy Artificial Intelligence: From AI Principles, Ethics, and Key Requirements to Responsible AI Systems and Regulation', *Information Fusion*, 99, 101896, https://doi.org/10.1016/j.inffus.2023.101896.

Douglas DM, J Lacey and D Howard (2023) 'Ethical Risks of AI-Designed Products: Bespoke Surgical Tools as a Case Study', *AI Ethics*, 3: 1117–1133, https://doi.org/10.1007/s43681-022-00219-8.

Doximity (2022) Physician Learning Preferences, https://press.doximity.com/reports/physician-learning-report-2022.pdf.

Drabu H, A McAdams, J Bonser and C Carr (2021) 'The EU Proposal for Regulation of Artificial Intelligence: Meaningful Steps Toward Grasping the Medico-Legal Nettle?', https://www.dacbeachcroft.com/en/gb/articles/2021/april/the-eu-pro

posal-for-regulation-of-artificial-intelligence-meaningful-steps-toward-grasping-the-medico-legal-nettle/.

Driever EM, AM Stiggelbout and PL Brand (2020) 'Shared Decision Making: Physicians' Preferred Role, Usual Role and their Perception of its Key Components', *Patient Education and Counseling*, 103(1): 77–82, https://doi.org/10.1016/j.pec.2019.08.004.

Drossman DA, L Chang, JK Deutsch, AC Ford, A Halpert, K Kroenke, S Nurko, J Ruddy, J Snyder and A Sperber (2021) 'A Review of the Evidence and Recommendations on Communication Skills and the Patient–Provider Relationship: A Rome Foundation Working Team Report', *Gastroenterology*, 161(5): 1670–1688.e7, https://doi.org/10.1053/j.gastro.2021.07.037.

Dueno T (2020) 'Racist Robots and the Lack of Legal Remedies in the Use of Artificial Intelligence in Healthcare', *Connecticut Insurance Law Journal*, 27(1): 337–358.

Duffourc M and S Gerke (2023) 'Decoding US Tort Liability in Healthcare's Black-Box AI Era: Lessons from the European Union', *Stan. Tech. L. Rev.*, 27: 1–70.

Dugdale DC, R Epstein and SZ Pantilat (1999) 'Time and the Physician-Doctor Relationship', *J Gen Intern Med.*, 14(1): S34–40, https://doi.org/10.1046/j.1525-1497.1999.00263.x.

Durán JM and N Formanek (2018) 'Grounds for Trust: Essential Epistemic Opacity and computational Reliabilism', *Minds Mach*, 28(4): 645–666.

Durán JM (2021) 'Dissecting Scientific Explanation in AI (sXAI): A Case for Medicine and Healthcare', *Artificial Intelligence*, 297, 103498, https://doi.org/10.1016/j.artint.2021.103498.

Durán JM and KR Jongsma (2021) 'Who is Afraid of Black Box Algorithms? On the Epistemological and Ethical Basis of Trust in Medical AI', *Journal of Medical Ethics*, 47: 329–335.

Dyschkant A (2015) 'Legal Personhood: How We Are Getting It Wrong', *U Illinois L Ver*, 2075–2110.

Earl J (2019) 'Innovative Practice, Clinical Research, and the Ethical Advancement of Medicine', *The American Journal of Bioethics: AJOB*, 19(6), 7, https://doi.org/10.1080/15265161.2019.1602175.

Editorial (2024) 'Empathic AI Can't Get under the Skin', *Nat Mach Intell*, 6, 495, https://doi.org/10.1038/s42256-024-00850-6.

Edwards L and M Veale (2017) 'Slave to the Algorithm? Why a "Right to an Explanation" Is Probably not the Remedy you Are Looking For', *Duke Law and Technology Review*, 16(1): 1–65.

Ehteshami BB, M Veta, PJ Van Diest, B Van Ginneken, N Karssemeijer, G Litjens, J Van der Laak, and the CAMELYON16 Consortium (2017) 'Diagnostic Assessment of Deep Learning Algorithms for Detection of Lymph Node Metastases in Women with Breast Cancer', *JAMA*, 318(22): 2199–2210, https://doi.org/10.1001/jama.2017.14585.

El Mestari SZ, G Lenzini and H Demirci (2024) 'Preserving Data Privacy in Machine Learning Systems', *Computers & Security*, 137, 103605, https://doi.org/10.1016/j.cose.2023.103605.

Eldakak A, A Alremeithi, E Dahiyat, H Mohamed, and MI Abdulrahim Abdulla (2024) 'Civil Liability for the Actions of Autonomous AI in Healthcare: An Invitation to Further Contemplation', *Humanities and Social Sciences Communications*, 11(1): 1–8, https://doi.org/10.1057/s41599-024-02806-y.

Elish MC (2019) 'Moral Crumple Zones: Cautionary Tales in Human-Robot Interaction', Engaging Science, *Technology and Society*, 5: 40–60.

Encord (2023), 'What is Ensemble Learning?', Encord Blog, November 24, 2023, https://encord.com/blog/what-is-ensemble-learning/.

Enqvist L (2023) 'Human Oversight' in the EU Artificial Intelligence Act: What, When and By Whom?', *Law, Innovation and Technology*, 15(2): 508–535, https://doi.org/10.1080/17579961.2023.2245683.

Epstein RJ (2023) 'To Find Fault Is Easy, to Find No-Fault Is Fair', *Future Healthc J*, 10(1): 85–89, https://doi.org/10.7861/fhj.2022-0049.

Epstein RM (2006) 'Making Communication Research Matter: What Do Patients Notice, What Do Patients Want, and What Do Patients Need?', *Patient Education and Counseling*, 60(3): 272–278.

Epstein RM and RL Street (2011) 'The Values and Value of Patient-Centered Care', *The Annals of Family Medicine*, 9(2): 100–103.

Esmaeilzadeh P (2024) 'Challenges and Strategies for Wide-Scale Artificial Intelligence (AI) Deployment in Healthcare Practices: A Perspective for Healthcare Organizations', *Artificial Intelligence in Medicine*, 151, 102861, https://doi.org/10.1016/j.artmed.2024.102861.

Esmaeilzadeh P, T Mirzaei and S Dharanikota (2021) 'Patients' Perceptions Toward Human–Artificial Intelligence Interaction in Health Care: Experimental Study', *Journal of Medical Internet Research*, 23(11), https://doi.org/10.2196/25856.

Esteva A, B Kuprel, RA Novoa, J Ko, SM Swetter, HM Blau and S Thrun (2017) 'Dermatologist-level Classification of Skin Cancer with Deep Neural Networks', *Nature*, 542: 115–118.

European Commission (2019) Defining Value In Value-Based Healthcare – Report of the Expert Panel on Effective Ways of Investing in Health (EXPH), Luxembourg: Publications Office of the European Union, https://health.ec.europa.eu/document/download/634e0a0c-4bff-4050-ad5f-b80381f36f33_en.

European Commission (2020) Report on the Safety and Liability Implications of Artificial Intelligence, the Internet of Things and Robotics, COM/2020/64 final, 19/02/2020, https://eur-lex.europa.eu/legalcontent/EN/TXT/PDF/?uri=CELEX:52020DC0064&rid=4.

European Commission (2022) The Digital Europe Programme, 7 June 2022, https://digital-strategy.ec.europa.eu/en/activities/digital-programme.

European Commission – Directorate-General for Justice and Consumers (2019) Liability for Artificial Intelligence and Other Emerging Digital Technologies, Publications Office, https://data.europa.eu/doi/10.2838/573689.

European Economic and Social Committee (2013) *European Economic and Social Committee: The Future of the European Union – Key Messages from Civil Society*. Luxembourg: Publications Office of the European Union, https://www.eesc.europa.eu/sites/default/files/resources/docs/qe-01-13-678-en-c.pdf.

European Health Parliament (2016) Digital Skills for Health Professionals Committee on Digital Skills for Health Professionals, https://www.healthparliament.eu/wp-content/uploads/2017/09/Digital-skills-for-health-professionals.pdf.

European Parliament (2017) Report with Recommendations to the Commission on Civil Law Rules on Robotics, 27.1.2017 – (2015/2103(INL)), https://www.europarl.europa.eu/doceo/document/A-8-2017-0005_EN.html.

European Parliament (2020) Artificial Intelligence in a Digital Age, European Parliament Resolution of 3 May 2022 on Artificial Intelligence in a Digital Age (2020/2266(INI)), https://www.europarl.europa.eu/doceo/document/TA-9-2022-0140_EN.html.

European Parliament (2022) Artificial Intelligence in Healthcare – Applications, Risks, and Ethical and Societal Impacts, https://www.europarl.europa.eu/RegData/etudes/STUD/2022/729512/EPRS_STU(2022)729512_EN.pdf.

European Parliament – Expert Group on Liability and New Technologies (2019) 'Liability For Artificial Intelligence And Other Emerging Digital Technologies, European Union, https://www.europarl.europa.eu/meetdocs/2014_2019/plmrep/COMMITTEES/JURI/DV/2020/01-09/AI-report_EN.pdf.

European Society for Emergency Medicine (2019) 'Recommendations on Informed Consent for European Emergency Departments for Adults and Children', https://eusem.org/images/Patient_Informed_Consent_in_the_ED-FINAL.pdf.

European Society of Radiology (2019) What the Radiologist Should Know about Artificial Intelligence – An ESR White Paper. Insights Imaging, 10, 44, https://doi.org/10.1186/s13244-019-0738-2.

Evans BJ and F Pasquale (2022) 'Product Liability Suits for FDA-Regulated AI/ML Software', in *The Future of Medical Device Regulation: Innovation and Protection*, IG Cohen, T Minssen, WN Price, C Robertson and C Shachar (eds), 22–35. Cambridge University Press, Cambridge.

Evans H and D Snead (2024) 'Why do Errors Arise in Artificial Intelligence Diagnostic Tools in Histopathology and How Can We Minimize Them?', *Histopathology*, 84(2): 279–287, https://doi.org/10.1111/his.15071.

Expert Group on Liability and New Technologies – New Technologies Formation (2019) 'Report on Liability for Artificial Intelligence and Other Emerging Digital Technologies', https://www.europarl.europa.eu/meetdocs/2014_2019/plmrep/COMMITTEES/JURI/DV/2020/01-09/AI-report_EN.pdf.

Fahim S (2024) *Ethico-Legal Aspect of AI-driven Driverless Cars: Comparing Autonomous Vehicle Regulations in Germany, California, and India*. Springer Nature Singapore, Singapore.

Farah L, JM Murris, I Borget, A Guilloux, NM Martelli and SIM Katsahian (2023) 'Assessment of Performance, Interpretability, and Explainability in Artificial Intelligence-Based Health Technologies: What Healthcare Stakeholders Need to Know', *Mayo Clin Proc Digital Health*, 1(2): 120–138, https://doi.org/10.1016/j.mcpdig.2023.02.004.

Farahmand S, O Shabestari, M Pakrah, H Hossein-Nejad, M Arbab and S Bagheri-Hariri (2017) 'Artificial Intelligence-Based Triage for Patients with Acute Abdominal Pain in Emergency Department; A Diagnostic Accuracy Study', *Adv J EmergMed*, 1: e5, https://doi.org/10.22114/AJEM.v1i1.11.

Farayola OA, OL Olorunfemi and PO Shoetan (2024) 'Data Privacy and Security in It: A Review of Techniques and Challenges', *Computer Science & IT Research Journal*, 5(3): 606–615.

Farhud DD and S Zokaei (2021) 'Ethical Issues of Artificial Intelligence in Medicine and Healthcare', *Iranian Journal of Public Health*, 50(11), https://doi.org/10.18502/ijph.v50i11.7600.

Farrell AM, S Devaney and A Dar (2010) 'No-Fault Compensation Schemes for Medical Injury: A Review', https://ssrn.com/abstract=2221836 or http://dx.doi.org/10.2139/ssrn.2221836.

Faure M and S Li (2022) 'Artificial Intelligence and (Compulsory) Insurance', *Journal of European Tort Law*, 13(1): 1–24.

Federation of State Medical Boards (2024) Navigating the Responsible and Ethical Incorporation of Artificial Intelligence into Clinical Practice, Adopted by FSMB House of Delegates, https://www.fsmb.org/siteassets/advocacy/policies/incorporation-of-ai-into-practice.pdf.

Feldman R, E Aldana and K Stein (2019) 'Artificial Intelligence in the Health care Space: How We Can Trust What We Cannot Know', *Stan. L.& Pol'y Rev.*, 30: 399–419.

Ferreira RM (2023) 'New Evidence-Based Practice: Artificial Intelligence as a Barrier Breaker', *World Journal of Methodology*, 13(5): 384–389, https://doi.org/10.5662/wjm.v13.i5.384.

Ferretti A, M Schneider and A Blasimme (2018) 'Machine Learning in Medicine: Opening the New Data Protection Black Box', *Eur Data Protect Law Rev (EDPL)*, 4(3): 320–332.

Ferrucci D, A Levas, S Bagchi, D Gondek and ET Mueller (2013) 'Watson: Beyond Jeopardy!', *Artificial Intelligence*, 199–200, https://doi.org/10.1016/j.artint.2012.06.009.

Finck M and F Pallas (2020) 'They Who Must Not Be Identified- Distinguishing Personal from Non-Personal Data under the GDPR', *International Data Privacy Law*, 10(1): 11–36.

Fisk MJ (2024) 'Computing, Data, and the Role of General Practitioners and General Practice in England', *Intelligent Medicine*, https://doi.org/10.1016/j.imed.2024.04.001.

Fitzpatrick PJ (2023) 'Improving Health Literacy Using the Power of Digital Communications to Achieve Better Health Outcomes for Patients and Practitioners', *Frontiers in Digital Health*, 5, https://doi.org/10.3389/fdgth.2023.1264780.

Fogel AL and JC Kvedar (2018) 'Artificial Intelligence Powers Digital Medicine', *NPJ Digital Medicine*, 1(1): 5.

Formosa P and M Ryan (2021) 'Making Moral Machines: Why We Need Artificial Moral Agents', *AI & Soc*, 36: 839–885, https://doi.org/10.1007/s00146-020-01089-6.

Fosch-Villaronga E and T Mahler (2021) 'Cybersecurity, Safety and Robots: Strengthening the Link Between Cybersecurity and Safety in the Context of Care Robots', *Computer Law & Security Review*, 41: 105528, https://doi.org/10.1016/j.clsr.2021.105528.

Foss-Solbrekk K (2023) 'Searchlights Across the Black Box: Trade Secrecy Versus Access to Information', *Computer Law & Security Review*, 50, 105811, https://doi.org/10.1016/j.clsr.2023.105811.

Fraser AG, E Biasin, B Bijnens, N Bruining, EG Caiani, K Cobbaert, RH Davies, SH Gilbert, L Hovestadt, E Kamenjasevic, Z Kwade, G McGauran, G O'Connor, Be Vasey and FE Rademakers (2023) 'Artificial Intelligence in Medical Device Software and High-Risk Medical Devices – A Review of Definitions, Expert Recommendations and Regulatory Initiatives', *Expert Review of Medical Devices*, 20(6): 467–491, https://doi.org/10.1080/17434440.2023.2184685.

Freeman K, J Geppert, C Stinton, D Todkill, S Johnson, A Clarke and S Taylor-Phillips (2021) 'Use of Artificial Intelligence for Image Analysis in Breast Cancer Screening Programmes: Systematic Review of Test Accuracy', *BMJ*, 1;374: n1872, https://doi.org/10.1136/bmj.n1872.

French-Mowat E and J Burnett 'How Are Medical Devices Regulated in the European Union?', *Journal of the Royal Society Medicine*, 105(1): S22–S28.

Friedman LM, C Furberg, DL DeMets, DM Reboussin and CB Granger (2015) *Fundamentals of Clinical Trials*. Springer, Cham.

Friedrich AB, J Mason and JR Malone (2022) 'Rethinking Explainability: Toward a Postphenomenology of Black-Box Artificial Intelligence in Medicine', *Ethics Inf Technol*, 24 8, https://doi.org/10.1007/s10676-022-09631-4.

Froomkin AM, I Kerr and J Pineau (2019) 'When AIs Outperform Doctors: Confronting the Challenges of a Tort-Induced Over-Reliance on Machine Learning', *Arizona Law Review*, 61: 33–100.

Funer F and U Wiesing (2024) 'Physician's Autonomy in the Face of AI Support: Walking The Ethical Tightrope', *Frontiers in Medicine*, 11, 1324963, https://doi.org/10.3389/fmed.2024.1324963.

Gadepally R and P Lakhani (2022) Overfitting and Generalization of AI in Medical Imaging, Data Science Institute – American College of Radiology.

Galbois-Lehalle D (2021) AI and Civil Liability: Welcomed but Perfectible Recommendations of the European Parliament, Chair Legal and Regulatory Implications of Artificial Intelligence, MIAI Grenoble Alpes, https://ai-regulation.com/wp-content/uploads/2021/01/AI-and-Civil-Liability-Welcomed-but-Perfectible-Recommendations-of-the-European-Parliament-DGL.pdf.

Galeon D (2016) 'IBM's Watson AI Recommends Same Treatment as Doctors in 99% of Cancer Cases', *Futurism*, futurism.com/ibms-watson-ai-recommends-same-treatment-as-doctors-in-99-of-cancer-cases.

Galletta M, MF Piazza, SL Meloni, E Chessa, I Piras and JE Arnetz (2022) 'Patient Involvement in Shared Decision-Making: Do Patients Rate Physicians and Nurses Differently?', *International Journal of Environmental Research and Public Health*, 19(21), https://doi.org/10.3390/ijerph192114229.

Garber S (2013) Economic Effects of Product Liability and Other Litigation Involving the Safety and Effectiveness of Pharmaceuticals. Rand Corporation, https://www.rand.org/content/dam/rand/pubs/monographs/MG1200/MG1259/RAND_MG1259.sum.pdf.

Garcia-Micó TG (2023), Quem Controla os Controladores? Uma Teoria Sobre a Responsabilidade Civil Dos Organismos Notificados (Who Controls the Controllers? A Theory on the Civil Liability of Notified Bodies) (October 19, 2023). JusGov Research Paper No. 2024-02, 2024, Available at SSRN: https://ssrn.com/abstract=4856249 or http://dx.doi.org/10.2139/ssrn.4856249.

Gaylord CL (1972) 'Fault, No Fault or Strict Liability?', *American Bar Association Journal*, 58(6): 589–593.

Gen M and L Lin (2023) 'Genetic Algorithms and their Applications', in *Springer Handbook of Engineering Statistics*, H Pham (Ed.), 635–674. Springer London, London.

Georgieva L and C Kuner (2020) 'Article 9 Processing of Special Categories of Personal Data', in *The EU General Data Protection Regulation: A Commentary*, C Kuner, L Bygrave, C Docksey, and L Drechsler, 365–384. Oxford University Press, Oxford.

Gerke S, B Babic, T Evgeniou and IG Cohen (2020) 'The Need for a System View to Regulate Artificial Intelligence/Machine Learning-Based Software as Medical Device', *NPJ Digit. Med.*, 3, 53, https://doi.org/10.1038/s41746-020-0262-2.

Gerke S, T Minssen, and IG Cohen (2020) 'Ethical and Legal Challenges of Artificial Intelligence-Driven Healthcare', *Artificial Intelligence in Healthcare*, 295–336, https://doi.org/10.1016/B978-0-12-818438-7.00012-5.

Ghassemi M, L Oakden-Rayner and AL Beam (2021) 'The False Hope of Current Approaches to Explainable Artificial Intelligence in Health Care', *The Lancet Digital Health*, 3(11): e745–e750.

Gianfrancesco MA, S Tamang, J Yazdany and G Schmajuk (2018) 'Potential Biases in Machine Learning Algorithms Using Electronic Health Record Data', *JAMA Intern Med.*, 178(11): 1544–1547, https://doi.org/10.1001/jamainternmed.2018.3763.

Gichoya JW, I Banerjee, AR Bhimireddy, JL Burns, LA Celi, LC Chen, Ramon Correa, N Dullerud, M Ghassemi, SC Huangj, PC Kuo, MP Lungren, LJ Palmer, BJ Price, S Purkayastha, AT Pyrros, L Oakden-Rayner, C Okechukwu, L Seyyed-Kalantari, H Trivedi, R Wang, Z Zaimanb and H Zhang (2022) 'AI Recognition of Patient Race in Medical Imaging: A Modelling Study', *The Lancet Digital Health*, 4(6): e406–e414.

Giliker P (2010) *Vicarious Liability in Tort: A Comparative Perspective*. Cambridge University Press, Cambridge.

Gill SS, M Xu, C Ottaviani, P Patros, R Bahsoon, A Shaghaghi, M Golec, V Stankovski, H Wu, A Abraham, M Singh, H Mehta, SK Ghosh, T Baker, AK Parlikad, H Lutfiyya, SS Kanhere, R Sakellariou, S Dustdar, O Rana and S Uhlig (2022) 'AI for Next Generation Computing: Emerging Trends and Future Directions', *Internet of Things*, 19: 100514.

Gillner S (2024) 'We're Implementing AI Now, so Why not Ask Us What to Do? – How AI Providers Perceive and Navigate the Spread of Diagnostic AI in Complex Healthcare Systems', *Social Science & Medicine*, 340, 116442, https://doi.org/10.1016/j.socscimed.2023.116442.

Gillon R (2015) 'Defending the Four Principles Approach as a Good Basis for Good Medical Practice and Therefore for Good Medical Ethics', *J Med Ethics*, 41: 111–116.

Giuffrida I and T Treece (2020) 'Keeping AI under Observation: Anticipated Impacts on Physicians' Standard of Care', *Tulane Journal of Technology and Intellectual Property* 22: 111–123.

Goddard K, A Roudsari and JC Wyatt (2012) 'Automation Bias: A Systematic Review of Frequency, Effect Mediators, and Mitigators', *Journal of the American Medical Informatics Association-JAMIA*, 19(1): 121–127, https://doi.org/10.1136/amiajnl-2011-000089.

Goh E, R Gallo, J Hom, E Strong, Y Weng, H Kerman, JA Cool, Z Kanjee, AS Parsons, N Ahuja, E Horvitz, D Yang, A Milstein, APJ Olson, A Rodman and JH Chen (2024) 'Large Language Model Influence on Diagnostic Reasoning: A Randomized Clinical Trial', *JAMA Network Open*, 7(10), e2440969, https://doi.org/10.1001/jamanetworkopen.2024.40969.

Goldberg JCP, AJ Sebok, BC Zipursky and M Kendrick (2021) *Tort Law: Responsibilities and Redress*. Aspen Publishing, Aspen.

Golden G, C Popescu, S Israel, K Perlman, C Armstrong, R Fratila, M Tanguay-Sela and D Benrimoh (2024) 'Applying Artificial Intelligence to Clinical Decision Support in Mental Health: What Have We Learned?', *Health Policy and Technology*, 100844, https://doi.org/10.1016/j.hlpt.2024.100844.

Gómez-González E and E Gómez (2020) *Artificial Intelligence in Medicine and Healthcare: Applications, Availability and Societal Impact*, Publications Office of the European Union, Luxembourg, https://doi.org/10.2760/047666.

Goodman KW (2007) 'Ethical and Legal Issues in Decision Support', in *Clinical Decision Support Systems: Theory and Practice*, 2nd ed., ES Berner (ed.), 126–39. Springer, Cham.

Gordijn B and H Ten Have (2023) 'What's Wrong with Medical Black Box AI?', *Med Health Care and Philos*, 26: 283–284, https://doi.org/10.1007/s11019-023-10168-6.

Gornet M and W Maxwell (2024) 'The European Approach to Regulating AI Through Technical Standards', *Internet Policy Review*, 13(3), https://doi.org/10.14763/2024.3.1784.

Green B (2022) 'The Flaws of Policies Requiring Human Oversight of Government Algorithms', *Computer Law & Security Review*, 45, 105681, https://doi.org/10.1016/j.clsr.2022.105681.

Greenberg MD (2009) 'Medical Malpractice and New Devices: Defining an Elusive Standard of Care', *Health Matrix Clevel*, 19(2): 423–445.

Griffin F (2021) 'Artificial Intelligence and Liability in Health Care', *Health Matrix*, 31: 65–106.

Grote T and P Berens (2020) 'On the Ethics of Algorithmic Decisionmaking in Healthcare', *Journal of Medical Ethics*, 46: 205–211.

Group of the European Parliament (2024) Protect Consumers from Flawed and Broken Products, https://www.eppgroup.eu/newsroom/protect-consumers-from-flawed-and-broken-products.

Grozdanovski Y (2024) 'The Explanations One Needs for the Explanations One Gives. The Necessity of Explainable AI (XAI) for Causal Explanations of AI-related Harm – Deconstructing the "Refuge of Ignorance" in the EU's AI Liability Regulation', JUST-AI Jean Monnet Research Papers.

Gualdi F and A Cordella (2021) 'Artificial Intelligence And Decision-Making: The Question of Accountability', *Proceedings of the 54th Hawaii International Conference on System Sciences*, 2297–2306, https://eprints.lse.ac.uk/110995/1/0226.pdf.

Guerra A, F Parisi and D Pi (2022) 'Liability for Robots I: Legal Challenges', *Journal of Institutional Economics*, 18(3): 331–343, https://doi.org/10.1017/S1744137421000825.

Guerrero Quiñones JL (2024) 'Using Artificial Intelligence to Enhance Patient Autonomy in Healthcare Decision-Making', *AI & Soc*, https://doi.org/10.1007/s00146-024-01956-6.

Guingrich RE and MS Graziano (2024) 'Ascribing Consciousness to Artificial Intelligence: Human-AI Interaction and its Carry-Over Effects on Human-Human Interaction', *Frontiers in Psychology*, 15, 1322781, https://doi.org/10.3389/fpsyg.2024.1322781.

Gulley A and A Hilliard (2024) Lost in Transl(A)t(I)on: Differing Definitions of AI [Updated], https://www.holisticai.com/blog/ai-definition-comparison.

Gunasekara SR, HNTK Kaldera and MB Dissanayake (2021) 'A Systematic Approach for MRI Brain Tumor Localization and Segmentation Using Deep Learning and Active Contouring', *Journal of Healthcare Engineering*, 2021(1), 6695108, https://doi.org/10.1155/2021/6695108.

Gupta S, S Kamboj and S Bag (2021) 'Role of Risks in the Development of Responsible Artificial Intelligence in the Digital Healthcare Domain', *Information Systems Frontiers*, 25: 2257–2274.

Gursoy D, OH Chi, L Lu and R Nunkoo (2019) 'Consumers Acceptance of Artificially Intelligent (AI) Device Use in Service Delivery', *International Journal of Information Management*, 49: 157–169, https://doi.org/10.1016/j.ijinfomgt.2019.03.008.

Gvozdenovic M (2022) 'Standard of Care in Medical Malpractice: Deference, Daubert, or Different Direction', *Journal of Law & Medicine*, 29(4): 1220–1235.

Habli I, T Lawton and Z Porter (2020) Artificial Intelligence in Health Care: Accountability and Safety', *Bull World Health Organ*, 1;98(4): 251–256, https://doi.org/10.2471/BLT.19.237487.

Hacker P (2022) The European AI Liability Directives – Critique of a Half-Hearted Approach and Lessons for the Future, CoRR abs/2211.13960.

Hacker P, R Krestel, S Grundmann and F Naumann (2020) 'Explainable AI Under Contract and Tort Law: Legal Incentives and Technical Challenges', *Artificial Intelligence and Law*, 28: 415–439.

Hagen T and R Narozniak (2019) 'IBM Seeks Niche for Watson for Oncology', *Oncology Live*®, 20(20).

Hall JW, PY Shuck and JR Canlas (2020) 'Medical Device Design and Control in the Hospital', in *Clinical Engineering Handbook*, 2nd edition, E Iadanza (ed.), 436–445. Academic Press, Cambrdige MA, https://doi.org/10.1016/B978-0-12-813467-2.00067-5.

Hallows R, L Glazier, MS Katz, M Aznar and M Williams (2022) 'Safe and Ethical Artificial Intelligence in Radiotherapy and Lessons Learned from the Aviation Industry', *Clinical Oncology*, 34, 99e101, https://doi.org/10.1016/j.clon.2021.11.019.

Han R, JN Acosta, Z Shakeri, JPA Ioannidis, EJ Topol and P Rajpurkar (2024) 'Randomised Controlled Trials Evaluating Artificial Intelligence in Clinical Practice: A Scoping Review', *The Lancet – Digital Health*, 6(5): e367–e373, https://doi.org/10.1016/S2589-7500(24)00047-5.

Haraldseid-Driftland C, S Billett, V Guise V Guise, L Schibevaag, J Alsvik, B Fagerdal, H Lyng and S Wiig (2022) 'The Role of Collaborative Learning in Resilience in Healthcare – A Thematic Qualitative Meta-Synthesis of Resilience Narratives', *BMC Health Serv Res*, 22, 1091, https://doi.org/10.1186/s12913-022-08451-y.

Hardavella G, A Aamli-Gaagnat, A Frille, N Saad, A Niculescu and P Powell (2017) 'Top Tips to Deal with Challenging Situations: Doctor–Patient Interactions', *Breathe*, 13(2): 129–135, https://doi.org/10.1183/20734735.006616.

Hardt M and MH Chin (2020) 'It is Time for Bioethicists to Enter the Arena of Machine Learning Ethics', *American Journal of Bioethics*, 20(11): 18–20, https://doi.org/10.1080/15265161.2020.1820115.

Harned Z, M Lungren and P Rajpurkar (2019) 'Machine Vision, Medical AI, and Malpractice', *Harvard Journal of Law & Technology Digest*, https://jolt.law.harvard.edu/digest/machine-vision-medical-ai-and-malpractice.

Harrold LR, TS Field and JH Gurwitz (1999) 'Knowledge, Patterns of Care, and Outcomes of Care for Generalists and Specialists', *J Gen Intern Med*, 14: 499–511, https://doi.org/10.1046/j.1525-1497.1999.08168.x.

Harry A (2023) 'Revolutionizing Healthcare: The Transformative Role of Artificial Intelligence in the Health Sector', *Bullet: Jurnal Multidisiplin Ilmu*, 2(2): 326–335.

Harvey HB and V Gowda (2022) 'Clinical Applications of AI in MSK Imaging: A Liability perspective', *Skeletal Radiol*, 51: 235–238, https://doi.org/10.1007/s00256-021-03782-z.

Harwood T, J Maltby and EB Mukaetova-Ladinska (2019) 'Role of Artificial Intelligence (AI) Art in Care of Ageing Society: Focus on Dementia', *OBM Geriatrics*, 3(3): 062, https://doi.org/10.21926/obm.geriatr.1903062.

Hatherley JJ (2020) 'Limits of Trust in Medical AI', Journal of Medical Ethics, 46: 478–481.

Hauglid MK and T Mahler (2023) 'Doctor Chatbot: The EU's Regulatory Prescription for Generative Medical AI', *Oslo Law Review*, 1: 1–23.

Health Action International (2023) 'The EU AI Liability Directive and Product Liability Directive', https://haiweb.org/wp-content/uploads/2023/06/AI-Liability_Background.pdf.

Health Action International (2023) 'The EU Medical Devices Regulation and the EU AI Act: A Short Comparison', https://haiweb.org/wp-content/uploads/2023/03/MDR-AIAct_OnePager_FINAL.pdf.

Heaven D (2018) 'Your Next Doctor's Appointment Might Be With an AI', *MIT Tech. Rev.* https://www.technologyreview.com/s/612267/your-next-doctors-appointment might-be-with-an-ai.

Hendrycks D (2025) Introduction to AI Safety, Ethics, and Society. Routledge, (1st ed.). CRC Press/Taylor & Francis, Boca Raton, FL, https://doi.org/10.1201/9781003530336.

Hengstler M, E Enkel and S Duelli (2016) 'Applied Artificial Intelligence and Trust – The Case of Autonomous Vehicles and Medical Assistance Devices', *Technol Forecast Soc Chang*, 105: 105–120.

Herington J, MD McCradden, K Creel, R Boellaard, EC Jones, AK Jha, A Rahmim, PJH Scott, JJ Sunderland, RL Wahl, S Zuehlsdorff and B Saboury (2023) 'Ethical Considerations for Artificial Intelligence in Medical Imaging: Deployment and Governance', *J Nucl Med.*, 64(10): 1509–1515, https://doi.org/10.2967/jnumed.123.266110.

Heubel F (2000) 'Patients or Customers: Ethical Limits of Market Economy in Health Care', *The Journal of Medicine and Philosophy*, 25(2): 240–253.

High-Level Expert Group on AI – AI HLEG (2019) Ethics Guidelines for Trustworthy AI, https://www.aepd.es/sites/default/files/2019-12/ai-ethics-guidelines.pdf.

Hill JC (2005) 'The Learned Intermediary Doctrine and Beyond', *Def. Counsel J.*, 72: 362–379.

Hill, RK (2016) 'What an Algorithm Is', *Phil & Technol*, 29(1): 35–59.

Hinteregger M (2023) 'Art 4:102 Principles of European Tort Law: An Objective or Subjective Standard of Fault – Does the Difference Really Matter?', *Journal of European Tort Law*, 14(1): 61–72, https://doi.org/10.1515/jetl-2023-0005.

Hintze M (2020) 'Automated Individual Decisions to Disclose Personal Data: Why GDPR Article 22 Should Not Apply', https://ssrn.com/abstract=3630026 or http://dx.doi.org/10.2139/ssrn.3630026.

Hirmiz R (2023) 'Against the Substitutive Approach to AI In Healthcare', *AI Ethics*, 4: 1507–1518, https://doi.org/10.1007/s43681-023-00347-9.

Hodge SD (2022) 'The Medical and Legal Implications of Artificial Intelligence in Health Care – An Area of Unsettled Law', *Rich. J.L. & Tech*, 28: 405–468.

Hoffman BL, EL Rosenthal, JB Colditz, R McGarry and BA Primack (2018) 'Use of Twitter to Assess Viewer Reactions to the Medical Drama, Code Black', *Journal of Health Communication*, 23(3): 244–253.

Hoffman S and A Podgurski (2009) 'E-Health Hazards: Provider Liability & Electronic Health Record Systems', *Berkeley Tech. L.J.*, 24: 1523–1582.

Hoffman S and A Podgurski (2013) 'The Use and Misuse of Biomedical Data: Is Bigger Really Better?', *Am. J.L. & Med.*, 29: 497–538, http://doi.org/10.1177/009885881303900401.

Holm S, C Stanton and B Bartlett (2021) 'A New Argument for No-Fault Compensation in Health Care: The Introduction of Artificial Intelligence Systems', *Health Care Anal*, 29(3): 171–188, https://doi.org/10.1007/s10728-021-00430-4.

Holzinger A, G Langs, H Denk, K Zatloukal and H Müller (2019) 'Causability and Explainability of Artificial Intelligence in Medicine', *Wiley Interdiscip Rev Data Min Knowl* Discov, 9(4): e1312, https://doi.org/10.1002/widm.1312.

Holzinger A, A Carrington and H Müller (2020) 'Measuring the Quality of Explanations: The System Causability Scale (SCS)', *KI – Künstliche Intelligenz*, 34(2): 193–198, https://doi.org/10.1007/s13218-020-00636-z.

Holzinger A, Zatloukal K and H Müller (2025) 'Is Human Oversight to AI Systems Still Possible?', *New Biotechnology*, 85: 59–62, https://doi.org/10.1016/j.nbt.2024.12.003.

Howard A and J Borenstein (2018) 'The Ugly Truth about Ourselves and our Robot Creations: The Problem of Bias and Social Inequity', *Sci Eng Eth.*, 24(5): 1521–136, https://doi.org/10.1007/s11948-017-9975-2.

Howard A, T McWilliams, G Hannant, N Saghir, P Giannoudis and J Scott (2019) 'Could No-Fault Compensation for Medical Errors Improve Care and Reduce Costs?', *Br J Hosp Med (Lond)*, 2;80(7): 387–390, https://doi.org/10.12968/hmed.2019.80.7.387.

Huang J, MT Wittbrodt, CN Teague, E Karl, G Galal, M Thompson, A Chapa, ML Chiu, B Herynk, R Linchangco, A Serhal, JA Heller, SF Abboud and M Etemadi (2025) 'Efficiency and Quality of Generative AI – Assisted Radiograph Reporting', *JAMA Netw Open*, 8(6): e2513921, https://doi.org/10.1001/jamanetworkopen.2025.13921.

Huang KA, HK Choudhary and PC Kuo (2024) 'Artificial Intelligent Agent Architecture and Clinical Decision-Making in the Healthcare Sector', *Cureus*, 16(7): e64115, https://doi.org/10.7759/cureus.64115.

Huang X, H Wang, C She, J Feng, X Liu, X Hu, L Chen and Y Tao (2022) 'Artificial Intelligence Promotes the Diagnosis and Screening of Diabetic Retinopathy', *Front Endocrinol (Lausanne)*, 13:946915, https://doi.org/10.3389/fendo.2022.946915.

Hulsen T (2023) 'Explainable Artificial Intelligence (XAI): Concepts and Challenges in Healthcare', *AI*, 4(3): 652–666.

Hundt A, A Murali, P Hubli, R Liu, N Gopalan, M Gombolay and GD Hager (2021) '"Good Robot! Now Watch This!": Repurposing Reinforcement Learning for Task-to-Task Transfer', *5th Annual Conference on Robot Learning*, 164: 1564–1574.

Husgen J (2014) 'Product Liability Suits Involving Drug or Device Manufacturers and Physicians: The Learned Intermediary Doctrine and the Physician's Duty to Warn', *Missouri Medicine*, 111(6): 478–481.

IBM (n.d.) IBM Watson for Oncology: What Watson for Oncology Can Do for your Organization, https://www.ibm.com/products/clinical-decision-supportoncology.

Inglada Galiana L, L Corral Gudino and P Miramontes González (2024) 'Ethics and Artificial Intelligence', *Revista ClíNica Española* (English Edition), 224(3): 178–186, https://doi.org/10.1016/j.rceng.2024.02.003.

Inoue K, C Sasaki and M Nakamura (2015) 'Communication Robots for Elderly People and Their Families to Support Their Daily Lives – Case Study of Two Families Living with the Communicaton Robot', *Stud Health Technol Inform*, 217: 980–983.

Institute of Medicine (US) Committee on Quality of Health Care in America (2000) *To Err is Human: Building a Safer Health System*, LT Kohn, JM Corrigan and MS Donaldson (Eds.). National Academies Press (US), Washington (DC).

Iserson KV (2024) 'Informed Consent for Artificial Intelligence in Emergency Medicine: A Practical Guide', *The American Journal of Emergency Medicine*, 76: 225–230, https://doi.org/10.1016/j.ajem.2023.11.022.

Issa H, J Jaber and H Lakkis (2024) 'Navigating AI Unpredictability: Exploring Technostress in AI-Powered Healthcare Systems', *Technological Forecasting and Social Change*, 202, 123311, https://doi.org/10.1016/j.techfore.2024.123311.

Italian Committee for Bioethics (2020) Artificial Intelligence and Medicine: Some Ethical Aspects, http://bioetica.governo.it/en/opinions/joint-opinions-icbicbbsl/artificial-intelligence-and-medicine-some-ethical-aspects/.

Jackson BW (2019) 'Artificial Intelligence and the Fog of Innovation: A Deep-Dive on Governance', *Santa Clara High Tech. L.J.*, 35(4): 35–63.

Jackson C (2013) 'Reasonable Persons, Reasonable Circumstances', *San Diego Law Review*, 50(3): 651–706.

Jakubik J, M Vössing, N Kühl, J Walk and G Satzger (2024) 'Data-Centric Artificial Intelligence', *Bus Inf Syst Eng*, https://doi.org/10.1007/s12599-024-00857-8.

James JT (2013) 'A New, Evidence-based Estimate of Patient Harms Associated with Hospital Care', *Journal of Patient Safety*, 9(3): 122–128, https://doi.org/10.1097/PTS.0b013e3182948a69.

Jassar S, SJ Adams, A Zarzeczny and BE Burbridge (2022) 'The Future of Artificial Intelligence in Medicine: Medical-Legal Considerations for Health Leaders', *Healthcare Management Forum*, 35(3): 185–189, https://doi.org/10.1177/08404704221082069.

Javaid M, A Haleem, R Pratap Singh, R Suman and S Rab (2022) 'Significance of Machine Learning in Healthcare: Features, Pillars and Applications', *International Journal of Intelligent Networks*, 3: 58–73, https://doi.org/10.1016/j.ijin.2022.05.002.

Jesso ST, A Kelliher, H Sanghavi, T Martin and SH Parker (2022) 'Inclusion of Clinicians in the Development and Evaluation of Clinical Artificial Intelligence Tools: A Systematic Literature Review', *Frontiers in Psychology*, 13, https://doi.org/10.3389/fpsyg.2022.830345.

Jiang ZZ (2024) 'Self-Disclosure to AI: The Paradox of Trust and Vulnerability in Human-Machine Interactions' ArXiv, https://arxiv.org/abs/2412.20564.

Jiang S, Z Wu, X Zhang, Y Ji, J Xu, P Liu, Y Liu, J Zheng, L Zhao and J Chen (2024) 'How Does Patient-Centered Communication Influence Patient Trust?: The Roles of Patient Participation and Patient Preference', *Patient Education and Counseling*, 122, 108161, https://doi.org/10.1016/j.pec.2024.108161.

Jobin A, M Ienca and E Vayena (2019) 'The Global Landscape of AI Ethics Guidelines', *Nature Machine Intelligence*, 1(9): 389–399, https://doi.org/10.1038/s42256-019-0088-2.

Jogan M, S Kurada, S Vasisht, V Singh and DA Hashimoto (2024) 'Quality over Quantity? The Role of Data Quality and Uncertainty for AI in Surgery', *Global Surg Educ*, 3, 79, https://doi.org/10.1007/s44186-024-00276-z.

Johnson DG (2015) 'Technology with No Human Responsibility?', *J. Bus. Ethics*, 127: 707–715.

Johnson SJL (2019) 'AI, Machine Learning, and Ethics in Health Care', *Journal of Legal Medicine*, 39(4): 427–441, https://doi.org/10.1080/01947648.2019.1690604.

Jongsma KR and M Sand (2022) 'Agree to Disagree: The Symmetry of Burden of Proof in Human-AI Collaboration', *Journal of Medical Ethics*, 48: 230–231.

Jorgensen S (1974–1975), 'Liability and Fault', *Tulane Law Review*, 49(2): 329–345.

Jorstad KT (2020) 'Intersection of Artificial Intelligence and Medicine: Tort Liability in the Technological Age', *J Med Artif Intell*, 3:17, https://doi.org/10.21037/jmai-20-57.

Juravle G, A Boudouraki, M Terziyska and C Rezlescu (2020) 'Trust in Artificial Intelligence for Medical Diagnoses', *Prog Brain Res*, 253: 263–282, https://doi.org/10.1016/bs.pbr.2020.06.006.

Kachalia A (2004) 'Studdert DM. Professional Liability Issues in Graduate Medical Education', *JAMA*, 292(9): 1051–1056, https://doi.org/10.1001/jama.292.9.1051.

Kamensky S (2020) 'Artificial Intelligence and Technology in Health Care: Overview and Possible Legal Implications', *Depaul J. Health Care L.*, 21(3), https://via.library.depaul.edu/jhcl/vol21/iss3/3.

Karthik A (2023) 'Reimagining the Data Subject in GDPR', *GRACE: Global Review of AI Community Ethics*, 1(1)

Karwaki TE (2020) 'Establishing a Patient-Pharmacist Relationship: Clarifying Duties and Improving Patient', *Baylor L. Rev.*, 72: 507–563.

Keating GC (2022) *Reasonableness and Risk: Right and Responsibility in the Law of Torts*. Oxford Academic, Oxford, https://doi.org/10.1093/oso/9780190867942.003.0007.

Kelly SM (2017) 'A Robotic Crib Rocked My Baby to Sleep for Months', CNN Business, https://money.cnn.com/2017/08/10/technology/gadgets/snooreview/index.html.

Kempt H and SK Nagel (2022) 'Responsibility, Second Opinions and Peer-Disagreement: Ethical and Epistemological Challenges of Using AI in Clinical Diagnostic Contexts', *Journal of Medical Ethics*, 48: 222–229.

Kempt H, JC Heilinger and SK Nagel (2023) '"I'm Afraid I Can't Let You Do that, Doctor": Meaningful Disagreements with AI In Medical Contexts', *AI & Soc*, 38: 1407–1414, https://doi.org/10.1007/s00146-022-01418-x.

Kendir C, R Fujisawa, OB Fernandes, K De Bienassis and N Klazinga (2023) 'Patient Engagement for Patient Safety: The Why, What, and How of Patient Engagement for Improving Patient Safety', OECD Health Working Papers, No. 159, OECD Publishing, Paris, https://doi.org/10.1787/5fa8df20-en.

Keown J (1998) 'Reining in the Bolam Test', *The Cambridge Law Journal*, 57(2): 235–273.

Kerasidou A (2020) 'Artificial Intelligence and the Ongoing Need for Empathy, Compassion and Trust in Healthcare', *Bulletin of the World Health Organization*, 98(4): 245–250, https://doi.org/10.2471/BLT.19.237198.

Kerasidou A (2021) 'Ethics of Artificial Intelligence in Global Health: Explainability, Algorithmic Bias and Trust', *Journal of Oral Biology and Craniofacial Research*, 11: 612–614, https://doi.org/10.1016/j.jobcr.2021.09.004.

Khan U and A Alamäki (2022) 'Trustworthy Artificial Intelligence in Healthcare', *eSignals PRO*, http://urn.fi/URN:NBN:fi-fe2022111665723.

Khanal S, H Zhang and A Taeihagh (2025) 'Why and How Is the Power of Big Tech Increasing in the Policy Process? The Case of Generative AI', *Policy and Society*, 44(1): 52–69, https://doi.org/10.1093/polsoc/puae012.

Khanna NN, MA Maindarkar, V Viswanathan, JF Fernandes, S Paul, M Bhagawati, P Ahluwalia, Z Ruzsa, A Sharma, R Kolluri, IM Singh, JR Laird, M Fatemi, A Alizad, L Saba, V Agarwal, A Sharma, JS Teji, M Al-Maini, … JS Suri (2022) 'Economics of Artificial Intelligence in Healthcare: Diagnosis vs. Treatment', *Healthcare*, 10(12), https://doi.org/10.3390/healthcare10122493.

Khosla V (2017) 'Here's why One Tech Investor Thinks Some Doctors Will Be Obsolete in Five Years', CNBC, www.cnbc.com/2017/04/07/vinod-khosla-radiologists-obsolete-five-years.html.

Kiener M (2021) 'Artificial Intelligence in Medicine and the Disclosure of Risks', *AI & Soc*, 36: 705–713, https://doi.org/10.1007/s00146-020-01085-w.

Kim D (2023) 'Artificial Intelligence Should Not Become a "Black Hole" for Human Agency in Tort Law', *The Tort Law Review*, 29(2): 52–168.

Kim HS (2020) 'Decision-Making in Artificial Intelligence: Is It Always Correct?', *Journal of Korean Medical* Science, 35(1): 1–3.

King OC and M Mertens (2023) 'Self-Fulfilling Prophecy in Practical and Automated Prediction', *Ethic Theory Moral Prac*, 26: 127–152, https://doi.org/10.1007/s10677-022-10359-9.

Kiseleva A, D Kotzinos and P De Hert (2022) 'Transparency of AI in Healthcare as a Multilayered System of Accountabilities: Between Legal Requirements and Technical Limitations', *Front Artif Intell*, 30;5, 879603, https://doi.org/10.3389/frai.2022.879603.

Kitsios F, M Kamariotou, AI Syngelakis and MA Talias (2023) 'Recent Advances of Artificial Intelligence in Healthcare: A Systematic Literature Review', *Applied Sciences*, 13(13), 7479, https://doi.org/10.3390/app13137479.

Kloss D (1984) 'The Duty of Care: Medical Negligence', *British Medical Journal*, 289(6436): 66–68, https://doi.org/10.1136/bmj.289.6436.66.

Knoedler L, A Vogt, M Alfertshofer, JM Camacho, D Najafali, A Kehrer, L Prantl, J Iske, J Dean, S Hoefer, C Knoedler and S Knoedler (2024) 'The Law Code of ChatGTP and Artificial Intelligence – How To Shield Plastic Surgeons and Reconstructive Surgeons Against Justitia's Sword', *Frontiers in Surgery*, 11, 1390684, https://doi.org/10.3389/fsurg.2024.1390684.

Koch BA, JS Borghetti, P Machnikowski, P Pichonnaz, T Rodríguez de las Heras Ballell, C Twigg-Flesner and C Wendehorst (2022) 'Response of the European Law Institute to the Public Consultation on Civil Liability – Adapting Liability Rules to the Digital Age and Artificial Intelligence', *Journal of European Tort Law*, 13(1): 25–63, https://doi.org/10.1515/jetl-2022-0002.

Kokoulina O (2023) 'Transparency of Algorithmic Decision-Making: Limits Posed by IPRS And Trade Secrets', in *The Exploitation of Intellectual Property Rights*, J Schovsbo (Ed.), 28–56. Edward Elgar Publishing, Cheltenham.

Komorowski M, LA Celi, O Badawi, AC Gordon and A Aldo Faisal (2018) 'The Artificial Intelligence Doctor Learns Optimal Treatment Strategies for Sepsis in Intensive Care', *Nature Medicine*, 24: 1716–1720, https://doi.org/10.1038/s41591-018-0213-5.

Koplin JJ, C Gyngell, J Savulescu and DF Vears (2022) 'Moving from 'Fully' to 'Appropriately' Informed Consent in Genomics: The PROMICE Framework', *Bioethics*, 36(6), 655–665. https://doi.org/10.1111/bioe.13027.

Korteling JH, GC Van de Boer-Visschedijk, RA Blankendaal, RC Boonekamp and AR Eikelboom (2021) 'Human Versus Artificial Intelligence', *Frontiers in Artificial Intelligence*, 4, 622364, https://doi.org/10.3389/frai.2021.622364.

Kostenko OM, KI Bieliakov, OO Tykhomyrov and IV Aristova (2024) '"Legal Personality" of Artificial Intelligence: Methodological Problems of Scientific Reasoning by Ukrainian and EU experts', *AI & Society*, 39(4): 1683–1693.

Kotschy W (2020) 'Article 6 – Lawfulness of Processing', in *The EU General Data Protection Regulation: A Commentary*, C Kuner, L Bygrave, C Docksey and L Drechsler, 321–344. Oxford University Press, Oxford.

Krishnan G, S Singh, M Pathania, S Gosavi, S Abhishek, A Parchani and M Dhar (2023) Artificial Intelligence in Clinical Medicine: Catalyzing a Sustainable Global Healthcare Paradigm', *Frontiers in Artificial Intelligence*, 6, https://doi.org/10.3389/frai.2023.1227091.

Kudina O (2021) 'Regulating AI in Health Care: The Challenges of Informed User Engagement', *Hastings Cent Rep*, 51(5): 6–7, https://doi.org/10.1002/hast.1263.

Kühne S, J Jacobsen, N Legewie and J Dollmann (2025) 'Attitudes Toward AI Usage in Patient Health Care: Evidence from a Population Survey Vignette Experiment', *Journal of Medical Internet Research*, 27, e70179, https://doi.org/10.2196/70179.

Kumar Y, A Koul, R Singla and MF Ijaz (2023) 'Artificial Intelligence in Disease Diagnosis: A Systematic Literature Review, Synthesizing Framework and Future Research Agenda', *J Ambient Intell Humaniz Comput*, 14(7): 8459–8486, https://doi.org/10.1007/s12652-021-03612-z.

Kumar S, AA Abdelhamid and Z Tarek (2023) 'Visualising the Unseen: Exploring GRAD-CAM for Interpreting Convolutional Image Classifiers', *Journal of Artificial Intelligence and Metaheuristics (JAIM)*, 4(1): 34–42.

Kung TH, M Cheatham, A Medenilla, C Sillos, L De Leon, C Elepaño, M Madriaga, R Aggabao, G Diaz-Candido, J Maningo and V Tseng (2023) 'Performance of ChatGPT on USMLE: Potential for AI-Assisted Medical Education Using Large Language Models', *PLOS Digit Health*, 2(2): e0000198, https://doi.org/10.1371/journal.pdig.0000198.

Kuo RYL, CJ Harrison, BE Jones, L Geoghegan and D Furniss (2021) 'Perspectives: A Surgeon's Guide to Machine Learning', *International Journal of Surgery*, 94: 106133, https://doi.org/10.1016/j.ijsu.2021.106133.

Kurki VAJ (2023) *Legal Personhood*. Cambridge University Press, Cambridge.

La Diega GN and LCT Bezerra (2024) 'Can there Be Responsible AI Without AI Liability? Incentivizing Generative AI Safety Through Ex-Post Tort Liability Under the EU AI Liability Directive', *International Journal of Law and Information Technology*, 32(1): eaae021, https://doi.org/10.1093/ijlit/eaae021.

La Puma J, DL Schiedermayer, S Toulmin, SH Miles and JA McAtee (1988) 'The Standard of Care: A Case Report and Ethical Analysis', *Ann Intern Med*, 108(1): 121–124, https://doi.org/10.7326/0003-4819-108-1-121.

LaCroix T and AS Luccioni (2025) 'Metaethical Perspectives on "Benchmarking" AI Ethics', *AI Ethics*, 5, 4029–4047, https://doi.org/10.1007/s43681-025-00703-x.

Lagioia F and G Contissa (2020) 'The Strange Case of Dr. Watson: Liability Implications of AI Evidence-Based Decision Support Systems in Health Care', *Eur. J. Legal Stud*, 12: 245–289.

Lai A (2021) 'Artificial Intelligence, LLC: Corporate Personhood as Tort Reform', *Michigan State Law Review*, 2: 597–654.

Lambert SI, M Madi, S Sopka, A Lenes, H Stange, C Buszello and A Stephan (2023) 'An Integrative Review on the Acceptance of Artificial Intelligence Among Healthcare Professionals in Hospitals', *Npj Digital Medicine*, 6(1): 1–14, https://doi.org/10.1038/s41746-023-00852-5.

Langer A, MP Marshall and S Levy-Tzedek (2023) 'Ethical Considerations In Child-Robot Interactions', *Neuroscience & Biobehavioral Reviews*, 151: 105230, https://doi.org/10.1016/j.neubiorev.2023.105230.

Langer M and RN Landers (2021) 'The Future of Artificial Intelligence at Work: A Review on Effects of Decision Automation and Augmentation on Workers Targeted by Algorithms and Third-Party Observers', *Computers in Human Behavior*, 123: 106878, https://doi.org/10.1016/j.chb.2021.106878.

Langlois S (2020) 'Collective Competence: Moving from Individual to Collaborative Expertise', *Perspectives on Medical Education*, 9(2): 71–73, https://doi.org/10.1007/s40037-020-00575-3.

Langman S, N Capicotto, Y Maddahi and K Zareinia (2021) 'Roboethics Principles and Policies in Europe and North America', *SN Appl. Sci.*, 3, 857, https://doi.org/10.1007/s42452-021-04853-5.

LaRosa E and D Danks (2018) 'Impacts on Trust of Healthcare AI', in *Proceedings of the 2018 AAAI/ACM Conference on AI, Ethics, and Society* (*AIES '18*), 210–215. Association for Computing Machinery, New York, NY, USA, https://doi.org/10.1145/3278721.3278771.

Laurie G, S Harmon and G Porter (2016) *Mason and McCall Smith's Law and Medical Ethics*, 10th. Oxford University Press.

Lauritsen SM, ME Kalør, EL Kongsgaard, KM Lauritsen, MJ Jørgensen, J Lange and B Thiesson (2020) 'Early Detection of Sepsis Utilizing Deep Learning on Electronic Health Record Event Sequences', *Artificial Intelligence in Medicine*, 104, 101820, https://doi.org/10.1016/j.artmed.2020.101820.

Laux J (2023) 'Institutionalised Distrust and Human Oversight of Artificial Intelligence: Towards a Democratic Design of AI Governance Under the European Union AI Act', *AI & Soc*, 39: 2853–2866, https://doi.org/10.1007/s00146-023-01777-z.

Lawton G (2012) Healthcare Has Mixed Feelings about AI, IEEE Intelligent Systems http://www.computer.org/intelligent.

Lawton T, P Morgan, Z Porter, S Hickey, A Cunningham, N Hughes, I Iacovides, Y Jia, V Sharma and I Habli (2024) 'Clinicians Risk Becoming 'Liability Sinks' for

Artificial Intelligence', *Future Healthcare Journal*, 11(1), https://doi.org/10.1016/j.fhj.2024.100007.

Lazcoz G and P De Hert (2023) 'Humans in the GDPR and AIA Governance of Automated and Algorithmic Systems. Essential Pre-Requisites Against Abdicating Responsibilities', *Computer Law & Security Review*, 50, 105833, https://doi.org/10.1016/j.clsr.2023.105833.

Lebovitz S, N Levina and H Lifshitz-Assa (2021) 'Is AI Ground Truth Really True? The Dangers of Training and Evaluating AI Tools Based on Experts' Know-What', *MIS Quarterly*, 45 (3b), 1501–1525, https://doi.org/10.25300/MISQ/2021/16564.

Ledley FD, SS McCoy, G Vaughan and EG Cleary (2020) 'Profitability of Large Pharmaceutical Companies Compared with Other Large Public Companies', *JAMA*, 323(9): 834–843, https://doi.org/10.1001/jama.2020.0442.

Lee CH and HJ Yoon (2017) 'Medical Big Data: Promise and Challenges', *Kidney Res Clin Pract*, 36(1): 3–11, https://doi.org/10.23876/j.krcp.2017.36.1.3.

Lee D and SN Yoon (2021) 'Application of Artificial Intelligence-Based Technologies in the Healthcare Industry: Opportunities and Challenges', *International Journal of Environmental Research and Public Health*, 18(1), 271, https://doi.org/10.3390/ijerph18010271.

Leslie D (2019) 'Understanding Artificial Intelligence Ethics and Safety: A Guide for the Responsible Design and Implementation of AI Systems in the Public Sector', The Alan Turing Institute, https://doi.org/10.5281/zenodo.3240529.

Levine M, R Philpot, SJ Nightingale and A Kordoni (2024) 'Visual Digital Data, Ethical Challenges, and Psychological Science', *American Psychologist*, 79(1): 109–122, https://doi.org/10.1037/amp0001192.

Li J, Y Mao and J Zhang (2022) 'Maintenance and Quality Control of Medical Equipment Based on Information Fusion Technology', *Computational Intelligence and Neuroscience*, 2022(1), 9333328, https://doi.org/10.1155/2022/9333328.

Li S and B Schütte (2024) 'The Proposed EU Artificial Intelligence Liability Directive: Does/Will Its Content Reflect Its Ambition?', *Technology and Regulation*, 2024: 143–151, https://doi.org/10.26116/techreg.2024.014.

Li YF and DM Liang (2019) 'Safe Semi-Supervised Learning: A Brief Introduction', *Frontiers of Computer Science*, 13: 669–676.

Lior A (2020) 'The AI Accident Network: Artificial Intelligence Liability Meets Network Theory', *Tul. L. Rev.*, 95: 1103–1166.

Lior A (2021) 'Insuring AI: The Role of Insurance in Artificial Intelligence Regulation', *Harvard Journal of Law & Technology*, 35(2): 468–530.

Liu S, AP Wright, BL Patterson, JP Wanderer, RW Turer, SD Nelson, AB McCoy, DF Sittig and A Wright (2023) 'Using AI-Generated Suggestions from ChatGPT to Optimize Clinical Decision Support', *Journal of the American Medical Informatics Association*, 30(7): 1237–1245.

London AJ (2019) 'Artificial Intelligence and Black-Box Medical Decisions: Accuracy versus Explainability', *Hastings Cent Rep.*, 49(1): 15–21, https://doi.org/10.1002/hast.973.

Longoni C, A Bonezzi and CK Morewedge (2019) 'Resistance to Medical Artificial Intelligence', *Journal of Consumer Research*, 46(4): 629–650, https://doi.org/10.1093/jcr/ucz013.

Lorenzini G, LA Ossa, DM Shaw and BS Elger (2023) 'Artificial Intelligence and the Doctor–Patient Relationship Expanding the Paradigm of Shared Decision Making', *Bioethics*, 37(5): 424–429, https://doi.org/10.1111/bioe.13158.

Lu H, Y Li, M Chen, H Kim and S Serikawa (2018) Brain Intelligence: Go Beyond Artificial Intelligence', *Mobile Netw Appl*, 23: 368–375, https://doi.org/10.1007/s11036-017-0932-8.

Lu Q and PJ Schulz (2023) 'How Physicians Tackle Internet-Misinformed Patients: Going Beyond Traditional Patient-Centered Communication – A Study Protocol', *Advances in Medical Education and Practice*, 14: 983–988, https://doi.org/10.2147/AMEP.S425434.

Ludvigsen K, S Nagaraja and A Daly (2022) 'When Is Software a Medical Device? Understanding and Determining the "Intention" and Requirements for Software as a Medical Device in European Union Law', *European Journal of Risk Regulation*, 13(1): 78–93, https://doi.org/ 10.1017/err.2021.45.

Lühnen J, I Mühlhauser and A Steckelberg (2018) 'The Quality of Informed Consent Forms – A Systematic Review and Critical Analysis', *Deutsches ÄRzteblatt International*, 115(22): 377–383, https://doi.org/10.3238/arztebl.2018.0377.

Lupton M (2018) 'Some Ethical and Legal Consequences of the Application of Artificial Intelligence in the Field of Medicine', *Trends Med Volume*, 18(4): 2–7, https://doi.org/ 10.15761/TiM.1000147.

Luxton DD (2014) 'Recommendations for the Ethical Use and Design of Artificial Intelligent Care Providers', *Artificial Intelligence in Medicine*, 62: 1–10.

Lyons JB, K Sycara, M Lewis and A Capiola (2021) 'Human-Autonomy Teaming: Definitions, Debates, and Directions', *Front. Psychol.*, 12, 589585, https://doi.org/10.3389/fpsyg.2021.589585.

Lysaght T, HY Lim, V Xafis and KY Ngiam (2019) 'AI-Assisted Decision-making in Healthcare: The Application of an Ethics Framework for Big Data in Health and Research', *Asian Bioethics Review*, 11(3): 299–314, https://doi.org/10.1007/s41649-019-00096-0.

Madan R, N Das, R Patley, N Nagpal, Y Malik and SB Math (2024) 'Consequences of Medical Negligence and Litigations on Health Care Providers – A Narrative Review', *Indian Journal of Psychiatry*, 66(4): 317–325.

Madani A, R Arnaout, M Mofrad and R Arnaout (2018) 'Fast and Accurate View Classification of Echocardiograms Using Deep Learning', *Npj Digital Medicine*, 1(1): 1–8, https://doi.org/10.1038/s41746-017-0013-1.

Maddox TM, JS Rumsfeld JS and PRO Payne (2019) 'Questions for Artificial Intelligence in Health Care', *JAMA*, 321(1): 31–32, https://doi.org/10.1001/jama.2018.18932.

Madiega T (2023) Artificial Intelligence Liability Directive, Briefing EU Legislation in Progress, EPRS – European Parliamentary Research Service, https://www.europarl.europa.eu/RegData/etudes/BRIE/2023/739342/EPRS_BRI(2023)739342_EN.pdf.

Maestre JM (2016) 'Healthcare Professional Competencies in Legal Doctrine', *Colombian Journal of Anesthesiology*, 44(1): 36–39.

Magrani E (2019) 'New Perspectives on Ethics and the Laws of Artificial Intelligence', *Internet Policy Ver*, 8(3): 1–19.

Mahmood U, H Chan, K Drukker, RK Samala, Q Chen, D Vergara, H Greenspan, N Petrick, B Sahiner, Z Huo, RM Summers, KH Cha, G Tourassi, TM Deserno, KT Grizzard, JJ Näppi, H Yoshida, D Regge, R Mazurchuk, … L Hadjiiski (2024) 'Artificial Intelligence in Medicine: Mitigating Risks and Maximizing Benefits Via Quality Assurance, Quality Control, and Acceptance Testing', *BJR|Artificial Intelligence*, 1(1), https://doi.org/10.1093/bjrai/ubae003.

Mahmud H, AN Islam, SI Ahmed and K Smolander (2022) 'What Influences Algorithmic Decision-Making? A Systematic Literature Review on Algorithm Aversion', *Technological Forecasting and Social Change*, 175, 121390, https://doi.org/10.1016/j.techfore.2021.121390.

Mai JE (2016) 'Big Data Privacy: The Datafication of Personal Information', *The Information Society*, 32(3): 192–199.

Maliha G, S Gerke, IG Cohen and RB Parikh (2021) 'Artificial Intelligence and Liability in Medicine: Balancing Safety and Innovation', *Milbank Q.*, 99(3): 629–647, https://doi.org/10.1111/1468-0009.12504.

Mann S, CT Berdahl, L Baker and F Girosi (2022) 'Artificial Intelligence Applications Used in the Clinical Response to COVID-19: A Scoping Review', *PLOS Digit Health*, 1(10), e0000132, https://doi.org/10.1371/journal.pdig.0000132.

Mann FK (1945) 'The Socialization of Risks', *The Review of Politics*, 7(1): 43–57.

Mantelero A (2024) 'The Fundamental Rights Impact Assessment (FRIA) in the AI Act: Roots, Legal Obligations and Key Elements for a Model Template', *Computer Law & Security Review*, 54, 106020, https://doi.org/10.1016/j.clsr.2024.106020.

Marchant GE and LM Tournas (2019) 'AI Health Care Liability: From Research Trials to Court Trials', *J. Health & Life Sci. L.*, 12(2): 23–42.

Marchisio E (2020) 'Medical Civil Liability Without Deterrence: Preliminary Remarks for Future Research', *J Civil Law Stud*, 13(1): 87–118.

Marchisio E (2021) 'In Support of "No-Fault" Civil Liability Rules for Artificial Intelligence', *SN Soc Sci.*, 1(2): 54, https://doi.org/10.1007/s43545-020-00043-z.

Marcos M, M Balser, A Ten Teije, F Van Harmelen and C Duelli (2003) 'Experiences in the Formalisation and Verification of Medical Protocols', in *Artificial Intelligence in Medicine: 9th Conference on Artificial Intelligence*, AIME 2003, Protaras, Cyprus, October 18–22, Proceedings 9, 132–141. Springer, Berlin – Heidelberg.

Marotta A (2022) 'When AI Is Wrong: Addressing Liability Challenges in Women's Healthcare', *Journal of Computer Information Systems*, 62, 6: 1310–1319, https://doi.org/10.1080/08874417.2022.2089773.

Martin-Lapoirie D (2022) 'Teamwork in Health Care and Medical Malpractice Liability: An Experimental Investigation', *Eur J Law Econ*, 54: 251–282, https://doi.org/10.1007/s10657-021-09708-3.

Martin LR, SL Williams, KB Haskard and MR DiMatteo (2005) 'The Challenge of Patient Adherence', *Therapeutics and Clinical Risk Management*, 1(3): 189–199.

Martinho A, A Poulsen, M Kroesen and C Chorus (2021) 'Perspectives about Artificial Moral Agents', *AI Ethics*, 1: 477–490, https://doi.org/10.1007/s43681-021-00055-2.

Marwaha JS, AB Landman, GA Brat, T Dunn and WJ Gordon (2022) 'Deploying Digital Health Tools Within Large, Complex Health Systems: Key Considerations for Adoption and Implementation', *npj Digit. Med.*, 5, 13, https://doi.org/10.1038/s41746-022-00557-1.

Matsuzaki T (2018) Ethical Issues of Artificial Intelligence in Medicine', *California Western Law Review*, 55(1): 255–274.

Matthias A (2004) 'The Responsibility Gap: Ascribing Responsibility for the Actions of Learning Automata', *Ethics Inf Technol*, 6: 175–183, https://doi.org/10.1007/s10676-004-3422-1.

Matulionyte R, P Nolan, F Magrabi and A Beheshti (2022) 'Should AI-Enabled Medical Devices Be Explainable?', *International Journal of Law and Information Technology*, 30(2): 151–180.

Mazzini G (2019) 'A System of Governance for Artificial Intelligence Through the Lens of Emerging Intersections Between AI and EU Law', Digital Revolution – New Challenges for Law, https://papers.ssrn.com/sol3/papers.cfm?abstract_id=3369266.

McAndrew S (2019) 'Internal Morality of Medicine and Physician Autonomy', *Journal of Medical Ethics*, 45(3): 198–203, https://doi.org/10.1136/medethics-2018-105069.

McCarthy J, ML Minsky, N Rochester and CE Shannon (2006) 'A Proposal for the Dartmouth Summer Research Project on Artificial Intelligence, August 31, 1955', *AI Magazine*, 27(4), 12, https://doi.org/10.1609/aimag.v27i4.1904.

McCradden MD, S Joshi, M Mazwi and JA Anderson (2020) 'Ethical Limitations of Algorithmic Fairness Solutions in Health Care Machine Learning', *Lancet Digit Health*, 2(5): e221–e223, https://doi.org/10.1016/S2589-7500(20)30065-0.

McDougall RJ (2019) 'Computer Knows Best? The Need for Value-Flexibility in Medical AI', *Journal of Medical Ethics*, 45: 156–160.

McGrail S (2020) 'Diagnostic Robotics, Mayo Clinic Bring Triage Platform to Patients', Hit Infrastructure, https://hitinfrastructure.com/news/diagnostic-robotics-mayo-clinic-bring-triage-platform-to-patients.

McKeown A, M Mourby, P Harrison, S Walker, M Sheehan and I Singh (2021) 'Ethical Issues in Consent for the Reuse of Data in Health Data Platforms', *Science and Engineering Ethics*, 27(1): 1–21.

McKinney SM, M Sieniek, V Godbole, J Godwin, N Antropova, H Ashrafian, T Back, M Chesus, GS Corrado, A Darzi, M Etemadi, FJ Gilbert, D Hassabis, S Jansen, A Karthikesalingam, CJ Kelly, D King, JR Ledsam, D Melnick … S Shetty (2020) 'International Evaluation of an AI System for Breast Cancer Screening', *Nature*, 577(7788): 89–94, https://doi.org/10.1038/s41586-019-1799-6.

McLean S (2010) *Autonomy, Consent and the Law.* Routledge-Cavendish, London.

McStay A (2024) *Automating Empathy: Decoding Technologies that Gauge Intimate Life.* Oxford University Press, Oxford.

Medical Device Coordination (2021) Group Document MDCG 2021–24 – Guidance on Classification of Medical Devices, https://health.ec.europa.eu/system/files/2021-10/mdcg_2021-24_en_0.pdf.

Mehta N and MV Devarakonda (2019) 'Machine Learning, Natural Language Programming, and Electronic Health Records: The Next Step in the Artificial Intelligence Journey?', *Journal of Allergy and Clinical Immunology*, 141(6): 2019–2021.

Mehta D (2023) 'The Role of Artificial Intelligence in Healthcare and Medical Negligence', *Liverpool Law Rev*, 45(1): 125–142, https://doi.org/10.1007/s10991-023-09340-y.

Melchiori FA (2020) 'Las Teorías De La Causalidad en el Daño: Equivalencia de las Condiciones, Causalidad Adecuada e Imputación Objetiva en la Doctrina del Tribunal Supremo', *JM Bosch*, https://doi.org/10.2307/j.ctv14t46hg.

Mello MM (2001) 'Of Swords and Shields: The Role of Clinical Practice Guidelines in Medical Malpractice Litigation', *University of Pennsylvania Law Review*, 149(3): 645–671.

Mello MM and N Guha (2024) 'Understanding Liability Risk from Using Health Care Artificial Intelligence Tools', *New England Journal of Medicine*, 390(3): 271–278.

Mello MM, DM Studdert, AB Kachalia and TA Brennan (2006) '"Health Courts" and Accountability for Patient Safety', *The Milbank Quarterly*, 84(3): 459–492.

Mendel R, E Traut-Mattausch, E Jonas, S Leucht, JM Kane, K Maino, W Kissling and J Hamann (2011) 'Confirmation Bias: Why Psychiatrists Stick To Wrong Preliminary Diagnoses', *Psychol Med.*, 41(12): 2651–2659, https://doi.org/10.1017/S0033291711000808.

Mengoni M, S Ceccacci and A Generosi (2024) 'Emotion Recognition and Affective Computing', in *Interaction Techniques and Technologies in Human-Computer Interaction*, C Stephanidis and G Salvendy (Eds.), 325–354. CRC Press.

Mennella C, U Maniscalco, G De Pietro and M Esposito (2024) 'Ethical and Regulatory Challenges of AI Technologies in Healthcare: A Narrative Review', *Heliyon*, 10(4), e26297, https://doi.org/10.1016/j.heliyon.2024.e26297.

Merat N, B Seppelt, T Louw, J Engström, JD Lee, E Johansson, Charles A. Green, S Katazaki, C Monk, M Itoh, D McGehee, T Sunda, K Unoura, T Victor, A Schieben and A Keinath (2019) 'The "Out-of-the-Loop" Concept in Automated Driving: Proposed Definition, Measures and Implications', *Cognition, Technology & Work*, 21: 87–98.

Merritt SM, WJ Bryant, A Staley, M McKenna, A Leone and L Shirase (2019) 'Automation-Induced Complacency Potential: Development and Validation of a New Scale', *Frontiers in Psychology*, 10, 417917, https://doi.org/10.3389/fpsyg.2019.00225.

Meskó B (2023) 'The Impact of Multimodal Large Language Models on Health Care's Future', *Journal of Medical Internet Research*, 25, https://doi.org/10.2196/52865.

Meskó B, G Hetényi and Z Győrffy (2018) 'Will Artificial Intelligence Solve the Human Resource Crisis in Healthcare?', *BMC Health Serv Res*, 18: 545.

Meskó B and EJ Topol (2023) 'The Imperative for Regulatory Oversight of Large Language Models (or Generative AI) in Healthcare', *npj Digit Med*, 6: 120, https://doi.org/10.1038/s41746-023-00873-0.

Methnani L, A Aler Tubella, V Dignum and A Theodorou (2021) 'Let Me Take Over: Variable Autonomy for Meaningful Human Control', *Front. Artif. Intell.*, 4, 737072, https://doi.org/10.3389/frai.2021.737072.

Millar J and IR Kerr (2016) 'Delegation, Relinquishment, and Responsibility: The Prospect of Expert Robots', in *Robot Law*, R Calo, AM Froomkin and I Kerr (Eds.), 102–128. Edward Elgar Edward Elgar Publishing, Cheltenham.

Miller AP (2018) 'Want Less-Biased Decisions? Use Algorithms', *Harvard Business Review*, https://hbr.org/2018/07/want-less-biased-decisions-usealgorithms.

Miller KW (2010) 'It's Not Nice to Fool Humans', *IT Professional*, 12(1): 51–52, https://doi.org/10.1109/MITP.2010.32.

Mills S and HS Sætra (2024) 'The Autonomous Choice Architect', *AI & Society*, 39(2): 583–595.

Milne-Ives M, C De Cock, E Lim, MH Shehadeh, N De Pennington, G Mole, E Normando and E Meinert (2020) 'The Effectiveness of Artificial Intelligence Conversational Agents in Health Care: Systematic Review', *J Med Internet Res.*, 22(10): e20346, https://doi.org/10.2196/20346.

Minssen T, S Gerke, M Aboy, N Price and G Cohen (2020) 'Regulatory Responses to Medical Machine Learning', *Journal of Law and the Biosciences*, 7(1): lsaa002, https://doi.org/10.1093/jlb/lsaa002.

Miotto R, L Li, BA Kidd and JT Dudley (2016) 'Deep Patient: An Unsupervised Representation to Predict the Future of Patients from the Electronic Health Records', *Sci Rep.*, 17, 6: 26094, https://doi.org/10.1038/srep26094.

Mittelman M, S Markham and M Taylor (2018) 'Patient Commentary: Stop Hyping Artificial Intelligence-Patients Will Always Need Human Doctors', *BMJ 363*, k4669.

Mittelstadt B (2021) The Impact of Artificial Intelligence on the Doctor-Patient Relationship, Council of Europe, https://www.coe.int/en/web/bioethics/-/new-report-on-the-impact-of-artificial-intelligence-on-the-doctor-patient-relationship.

Mittelstadt B, P Allo, M Taddeo, S Wachter and L Floridi (2016) 'The Ethics of Algorithms: Mapping the Debate', *Big Data & Society*, 3(2), https://doi.org/10.1177/2053951716679679.

Mittermaier M, M Raza and JC Kvedar (2023) 'Collaborative Strategies for Deploying AI-Based Physician Decision Support Systems: Challenges and Deployment Approaches', *Npj Digital Medicine*, 6(1): 1–2, https://doi.org/10.1038/s41746-023-00889-6.

Moffett P and G Moore (2011) 'The Standard of Care: Legal History and Definitions: The Bad and Good News', *West J Emerg Med*, 12(1): 109–112.

Montagnani ML, M Najjar and A Davola (2024) 'The EU Regulatory Approach(Es) to AI Liability, and its Application to the Financial Services Market', *Computer Law & Security Review*, 53, 105984, https://doi.org/10.1016/j.clsr.2024.105984.

Montagnani ML and M Cavallo (2021) 'Liability and Emerging Digital Technologies: An EU Perspective', *Notre Dame Journal of International & Comparative Law*, 11(2): 208–231.

Montavon G, W Samek and KR Müller (2017) 'Methods for Interpreting and Understanding Deep Neural Networks', *Digital Signal Processing*, 73: 1–15, https://doi.org/10.1016/j.dsp.2017.10.011.

Montemayor C, J Halpern and A Fairweather (2022) 'In Principle Obstacles for Empathic AI: Why We Can't Replace Human Empathy in Healthcare', *AI & Soc*, 37(4): 1353–1359, https://doi.org/10.1007/s00146-021-01230-z.

Montori VM, MM Ruissen, IG Hargraves, JP Brito and M Kunneman (2023) 'Shared Decision-Making as a Method of Care', *BMJ Evidence-Based Medicine*, 28(4): 213–217, https://doi.org/10.1136/bmjebm-2022-112068.

Mora-Cantallops M, S Sánchez-Alonso, E García-Barriocanal and MA Sicilia (2021) 'Traceability for Trustworthy AI: A Review of Models and Tool', *Big Data Cogn. Comput.*, 5(2), 20.

Morley J, CCV Machado, C Burr, J Cowls, I Joshi, M Taddeo and L Floridi (2020) 'The Ethics of AI in Health Care: A Mapping Review', *Soc Sci Med.*, 260: 113172, https://doi.org/10.1016/j.socscimed.2020.113172.

Mosqueira-Rey E, E Hernández-Pereira, D Alonso-Ríos, J Bobes-Bascarán and A Fernández-Leal (2023) 'Human-in-the-Loop Machine Learning: A State of the Art', *Artif Intell Rev*, 56: 3005–3054, https://doi.org/10.1007/s10462-022-10246-w.

Moudatsou M, A Stavropoulou, A Philalithis and S Koukouli (2020) 'The Role of Empathy in Health and Social Care Professionals', *Healthcare*, 8(1), https://doi.org/10.3390/healthcare8010026.

Moulton B and JS King (2010) 'Aligning Ethics with Medical Decision-Making: The Quest for Informed Patient Choice', *J Law Med Ethics*, 38(1): 85–97, https://doi.org/10.1111/j.1748-720X.2010.00469.x.

Mourby M, KÓ Cathaoir and CB Collin (2021) 'Transparency of Machine-Learning in Healthcare: The GDPR & European Health Law', *Computer Law & Security Review*, 43: 105611.

Mulheron R (2020) *Principles of Tort Law* (2nd ed). Cambridge University Press, Cambridge.

Muraille E (2019) 'Ethical Control of Innovation in a Globalized and Liberal World: Is Good Science Still Science?' *Endeavour*, 43(4), 100709, https://doi.org/10.1016/j.endeavour.2020.100709.

Murray B (2012) 'Informed Consent: What Must a Physician Disclose to a Patient?', *Virtual Mentor*, 14(7): 563–566, https://doi.org/10.1001/virtualmentor.2012.14.7.hlaw1-1207.

Naeini FB, AM AlAli, R Al-Husari, A Rigi, MK Al-Sharman, D Makris and Y Zweiri (2019) 'A Novel Dynamic-Vision-Based Approach for Tactile Sensing Applications', *IEEE Transactions on Instrumentation and Measurement*, 69(5): 1881–1893.

Nagarathinam A, A Chellasamy, N Elangovan and S Rengasamy (2023) 'Modeling the Intention to Use AI Healthcare Chabot's in the Indian Context', in *Proceedings of the 2nd International Conference on Emerging Technologies and Intelligent Systems*, MA Al-Sharafi, M Al-Emran, MN Al-Kabi and K Shaalan (Eds.). ICETIS 2022. Lecture Notes in Networks and Systems, vol 573. Springer, Cham. https://doi.org/10.1007/978-3-031-20429-6_59.

Nagwani NK and JS Suri (2023) 'An Artificial Intelligence Framework on Software Bug Triaging, Technological Evolution, and Future Challenges: A Review', *International Journal of Information Management Data Insights*, 3(1): 100153.

Naik N, BMZ Hameed, DK Shetty, D Swain, M Shah, R Paul, K Aggarwal, S Ibrahim, V Patil, K Smriti, S Shetty, BP Rai, P Chlosta and BK Somani (2022) 'Legal and Ethical Consideration in Artificial Intelligence in Healthcare: Who Takes Responsibility?', *Front. Surg.*, 9:862322, https://doi.org/10.3389/fsurg.2022.862322.

Naik BS and A Basu (2022) 'The Doctrine of "Reasonable Doctor" in Medical Negligence: Need to Be More Reasonable', *Journal of Forensic and Legal Medicine*, 92, 102451, https://doi.org/10.1016/j.jflm.2022.102451.

Naikar N, A Brady, G Moy and HW Kwok (2023) 'Designing Human-AI Systems for Complex Settings: Ideas from Distributed, Joint, and Self-Organising Perspectives of Sociotechnical Systems and Cognitive Work Analysis', *Ergonomics*, 66(11): 1669–1694, https://doi.org/10.1080/00140139.2023.2281898.

Najjar R (2023) 'Redefining Radiology: A Review of Artificial Intelligence Integration in Medical Imaging', *Diagnostics*, 13(17): 2760, https://doi.org/10.3390/diagnostics13172760.

Nehez J (2020) 'AI Can Never Think: The Uniqueness of Human Thought', *Studia Gilsoniana*, 9(3): 467–492.

Neufeld E and S Finnestad (2020) 'In Defense of the Turing Test', *AI & Soc*, 35: 819–827, https://doi.org/10.1007/s00146-020-00946-8.

Newdick C (2012) *Who Should We Treat? Rights, Rationing, and Resources in the* NHS, 2nd ed. Oxford, Oxford Academic, Oxford, https://doi.org/10.1093/acprof:oso/9780199264186.003.0006.

Nguyen A, J Yosinski and J Clune (2015) 'Deep Neural Networks are Easily Fooled: High Confidence Predictions for Unrecognizable Images', *IEEE Comp. Vision & Pattern Recognition*, 427–436.

Nguyen TBA (2019) 'No-Fault Versus Strict Liability Compensation Systems in Medical Malpractice Law in Vietnam in Comparision with Belgium, France, and England', *Asian Journal of Law and Economics*, 10(1): 20180004, https://doi.org/10.1515/ajle-2018-0004.

Nickel PJ (2022) 'Trust in Medical Artificial Intelligence: A Discretionary Account', *Ethics Inf Technol*, 24(7), https://doi.org/10.1007/s10676-022-09630-5.

Nikolinakos N (2023) *EU Policy and Legal Framework for Artificial Intelligence, Robotics and Related Technologies – The AI Act*. Springer International Publishing, Cham.

Nishant R, D Schneckenberg and M Ravishankar (2024) 'The Formal Rationality of Artificial Intelligence-Based Algorithms and the Problem of Bias', *Journal of Information Technology*, 39(1): 19–40, https://doi.org/10.1177/02683962231176842.

Novelli C, F Casolari, A Rotolo, M Taddeo and L Floridi (2024) 'AI Risk Assessment: A Scenario-Based, Proportional Methodology for the AI Act', *DISO*, 3, 13, https://doi.org/10.1007/s44206-024-00095-1.

Novelli C, G Bongiovanni and G Sartor (2022) 'A Conceptual Framework for Legal Personality and its Application to AI', *Jurisprudence*, 13(2): 194–219, https://doi.org/10.1080/20403313.2021.2010936.

Novelli C, M Taddeo and L Floridi (2023) 'Accountability in Artificial Intelligence: What It Is and How It Works', *AI & Soc*, https://doi.org/10.1007/s00146-023-01635-y.

Nuffield Council on Bioethics (2018) Artificial Intelligence (AI) in Healthcare and Research, https://www.nuffieldbioethics.org/wp-content/uploads/Artificial-Intelligence-AI-in-healthcare-and-research.pdf.

Nys H (2005) Report on Medical Liability in Council of Europe Member States, Council of Europe, https://rm.coe.int/1680700281.

O'Brien D (2023) 'Digital Love: Love Through the Screen/Of the Screen', in *Love and the Politics of Intimacy: Bodies, Boundaries, Liberation*, S Dikova, W McMahon and J Savage (Eds.), 111–126. Bloomsbury, London.

O'Connor O and TP McVeigh (2025) 'Increasing Use of Artificial Intelligence in Genomic Medicine for Cancer Care – The Promise and Potential Pitfalls', *BJC Rep* 3, 20, https://doi.org/10.1038/s44276-025-00135-4.

O'Sullivan S, N Nevejans, C Allen, A Blyth, S Leonard, U Pagallo, K Holzinger, A Holzinger, MI Sajid and H Ashrafian (2019) 'Legal, Regulatory, and Ethical Frameworks for Development of Standards in Artificial Intelligence (AI) and Autonomous Robotic Surgery', *The International Journal of Medical Robotics and Computer Assisted Surgery*, 15(1), e1968, https://doi.org/10.1002/rcs.1968.

Obermeyer Z, B Powers, C Vogeli and S Mullainathan (2019) 'Dissecting Racial Bias in an Algorithm Used to Manage the Health of Populations', *Science*, 366: 447–53, https://doi.org/10.1126/science.aax2342.

OECD (2020) Trustworthy AI in Health, https://www.oecd.org/health/trustworthy-artificial-intelligence-in-health.pdf.

Oikonomou EK and R Khera (2024) 'Designing Medical Artificial Intelligence Systems for Global Use: Focus on Interoperability, Scalability, and Accessibility', *Hellenic Journal of Cardiology*, 81, 9–17, https://doi.org/10.1016/j.hjc.2024.07.003.

Osasona F, OO Amoo, A Atadoga, TO Abrahams, OA Farayola and BS Ayinla (2024) 'Reviewing the Ethical Implications of AI in Decision Making Processes', *International Journal of Management & Entrepreneurship Research*, 6(2): 322–335.

Ossa LA, G Starke, G Lorenzini, JE Vogt, DM Shaw and BS Elger (2022) 'Re-Focusing Explainability in Medicine', *Digital Health,* https://doi.org/10.1177/20552076221074488.

Oussous A, FZ Benjelloun, AA Lahcen and S Belfkih (2018) 'Big Data Technologies: A Survey', *Journal of King Saud University-Computer and Information Sciences*, 30(4): 431–448.

Oyekunle D, U Matthew, D Preston and D Boohene (2024) 'Trust beyond Technology Algorithms: A Theoretical Exploration of Consumer Trust and Behavior in Technological Consumption and AI Projects', *Journal of Computer and Communications*, 12: 72–102, https://doi.org/10.4236/jcc.2024.126006.

Pagallo U (2017) 'From Automation to Autonomous Systems: A Legal Phenomenology with Problems of Accountability), in *IJCAI International Joint Conference on Artificial Intelligence, International Joint Conferences on Artificial Intelligence*, 17–23, https://www.ijcai.org/Proceedings/2017/0003.pdf.

Pal M, S Parija, G Panda, K Dhama and RK Mohapatra (2022) 'Risk Prediction of Cardiovascular Disease Using Machine Learning Classifiers', *Open Med* (*Wars*), 17(1): 1100–1113, https://doi.org/10.1515/med-2022-0508.

Pallikarakis N and Z Bliznakov (2016) 'Hospital Technology Management', *Clinical Engineering*, 47–61, https://doi.org/10.1016/B978-0-12-803767-6.00005-2.

Pallocci M, M Treglia, P Passalacqua, R Tittarelli, C Zanovello, L De Luca, V Caparrelli, V De Luna, AM Cisterna, G Quintavalle and LT Marsella (2023) 'Informed Consent: Legal Obligation or Cornerstone of the Care Relationship?', *International Journal of Environmental Research and Public Health*, 20(3): 2118, https://doi.org/10.3390/ijerph20032118.

Palmieri S and T Goffin (2023) 'A Blanket That Leaves the Feet Cold: Exploring the AI Act Safety Framework for Medical AI', *European Journal of Health Law*, 30(4): 406–427, https://doi.org/10.1163/15718093-bja10104.

Palmieri S, P Walraet and T Goffin (2021) 'Inevitable Influences: AI-Based Medical Devices at the Intersection of Medical Devices Regulation and the Proposal for AI Regulation', *European Journal of Health Law*, 28(4): 341–358.

Panch T, H Mattie and LA Celi (2019) 'The "Inconvenient Truth" About AI in Healthcare', *npj Digit. Med.* 2, 77, https://doi.org/10.1038/s41746-019-0155-4.

Pande V (2018) 'Artificial Intelligence's "Black Box" Is Nothing to Fear', N.Y. Times, https://www.nytimes.com/2018/01/25/opinion/artificial-intelligence-black-box.html.

Panel for the Future of Science and Technology (2022) Artificial Intelligence in Healthcare – Applications, Risks, and Ethical and Societal Impacts, https://www.europarl.europa.eu/thinktank/en/document/EPRS_STU(2022)729512.

Panteli D, H Legido-Quigley, C Reichebner, G Ollenschläger, C Schäfer and R Busse (2019) 'Clinical Practice Guidelines as a Quality Strategy', in *Improving Healthcare Quality in Europe: Characteristics, Effectiveness and Implementation of Different Strategies*, R Busse.

N Klazinga, D Panteli and W Quentin (Eds.) 233–263. European Observatory on Health Systems and Policies, Copenhagen, Denmark.

Paranjape K, M Schinkel, R Nannan Panday, J Car and P Nanayakkara (2019) 'Introducing Artificial Intelligence Training in Medical Education', *JMIR Med Educ*, 5(2): e16048, https://doi.org/10.2196/16048.

Parikh RB, S Teeple and AS Navathe (2019) 'Addressing Bias in Artificial Intelligence in Health Care', *JAMA*, 322: 2377–2378, https://doi.org/10.1001/jama.2019.18058.

Park HJ (2024) 'Patient Perspectives on Informed Consent for Medical AI: A Web-Based Experiment', *Digital Health*, 10, https://doi.org/10.1177/20552076241247938.

Parker, LJ (1999) 'Managing and Maintaining a Safe Environment in the Hospital Setting', *British Journal of Nursing*, 8(16): 1053–1066.

Pasquale F (2020) 'When Medical Robots Fail: Malpractice Principles for an Era of Automation', https://www.brookings.edu/techstream/when-medical-robots-fail-malpractice-principles-for-an-era-of-automation/.

Pasquale F (2022) 'The Price of Autonomy: Liability Standards for Complementary and Substitutive Medical Robotics and Artificial Intelligence', *Ius et Praxis*, 28(1): 3–19, https://doi.org/10.4067/S0718-00122022000100003.

Paterick TJ, GV Carson, MC Allen and TE Paterick (2008) 'Medical Informed Consent: General Considerations for Physicians', *Mayo Clinic Proceedings*, 83(3): 313–319, https://doi.org/10.4065/83.3.313.

Paterick TE (2019) 'The Standard of Care: Medical Errors/Violations and the Law', *The Journal of Medical Practice Management: MPM*, 35(1): 15–22.

Paterick ZR, NJ Patel, E Ngo, K Chandrasekaran, AJ Tajik and TE Paterick (2018) 'Medical Liability in the Electronic Medical Records Era', *Baylor U. Med. Ctr. Proc.*, 31: 558–561.

Patterson D (2015) 'Consciousness and Moral Responsibility', *Journal of Law and the Biosciences*, 2(3): 771–774, https://doi.org/10.1093/jlb/lsv037.

Paul M, L Maglaras, MA Ferrag and I Almomani (2023) 'Digitization of Healthcare Sector: A Study on Privacy and Security Concerns', *ICT Express*, 9(4): 571–588, https://doi.org/10.1016/j.icte.2023.02.007.

Pearce FJ, S Cruz Rivera, X Liu, E Manna, A Denniston and M Calvert (2023) 'The Role of Patient-Reported Outcome Measures in Trials of Artificial Intelligence Health Technologies: Analysis of Clinicaltrials. Gov Records (1997–2022)', *The Lancet Digital Health*, 5(3): e160.

Pearson SD and LH Raeke (2000) 'Patients' Trust in Physicians: Many Theories, Few Measures, and Little Data', *Journal of General Internal Medicine*, 15(7): 509–513, https://doi.org/10.1046/j.1525-1497.2000.11002.x.

Pellegatta S (2019) 'Autonomous Driving and Civil Liability: The Italian Perspective', *Rivista di Diritto dell'Economia, dei Trasporti e dell'Ambiente*, 135–161.

Pereira AGD (2013) 'Medical Liability: Comparing "Civil Law" and "Common Law"', in *Legal and Forensic Medicine*, R Beran (Ed.), 591–404. Springer, Berlin, Heidelberg, https://doi.org/10.1007/978-3-642-32338-6_61.

Pereira AGD (2015) *Direitos dos Pacientes e Responsabilidade Médica*. Coimbra Editora, Coimbra.

Pearce FJ, S Cruz Rivera, X Liu, E Manna, A Denniston and M Calvert (2023) 'The Role of Patient-Reported Outcome Measures in Trials of Artificial Intelligence Health Technologies: Analysis of Clinicaltrials. Gov Records (1997–2022)', *The Lancet Digital Health*, 5(3), e160.

Phillips L and M Scheffmann-Petersen (2020) 'Minding the Gap Between the Policy and Practice of Patient-Centeredness: Cocreating a Model for Tensional Dialogue in the "Active Patient Support" Program', *Qualitative Health Research*, 30(9): 1419–1430.

Pieper M and R Gleasure (2025) 'How AI Helps to Compile Human Intelligence: An Empirical Study of Emerging Augmented Intelligence for Medical Image Scanning', *Information Systems Journal*, https://doi.org/10.1111/isj.12585.

Piletskene E (2019) 'Consciousness and Unconsciousness of Artificial Intelligence', *Future Human Image*, 11: 66–71, https://doi.org/10.29202/fhi/11/7.

Pinevich Y, KJ Clark, AM Harrison, BW Pickering and V Herasevich (2021) 'Interaction Time with Electronic Health Records: A Systematic Review', *Appl Clin Inform*, (4): 788–799, https://doi.org/10.1055/s-0041-1733909.

Plaza C (2020) 'Cutting out the Middleman: Empirically Testing the Continued Applicability of the Learned Intermediary Rule in the Age of Direct-to-Consumer Advertising of Prescription Pharmaceuticals', *Quinnipiac Health LJ*, 24: 393–420.

Ploug T and S Holm (2020) 'The Right to Refuse Diagnostics and Treatment Planning by Artificial Intelligence', *Med Health Care and Philos*, 23: 107–114, https://doi.org/10.1007/s11019-019-09912-8.

Polat C (2023a) 'A Critical Appraisal of the Proposal for an AI Liability Directive: Is It Fit for Its Own Purpose? Is the EU's AI Liability Directive Really Promoting the Roll-out of Trustworthy AI? This Paper Provides Insights into the Proposal's Limitations', *Zeitschrift für das Privatrecht der Europäischen Union*, 20(3): 114–121.

Polat C (2023b) 'Great Unclear Expectations: A Study on the Defectiveness of Autonomous Systems', *Journal of Consumer Policy*, 43(1): 145–175.

Polito C and L Pupillo (2024) 'Artificial Intelligence and Cybersecurity', *Intereconomics*, 59(1): 10–13.

Poon AIF and JJY Sung (2021) 'Opening the Black Box of AI-Medicine', *J Gastroenterol Hepatoly*, 36(3): 581–584, https://doi.org/10.1111/jgh.15384.

Popper AF (2023) 'In Defense of Deterrence', *Albany Law Review*, 75(1): 181–204.

Potočnik J, S Foley and E Thomas (2023) 'Current and Potential Applications of Artificial Intelligence in Medical Imaging Practice: A Narrative Review', *Journal of Medical Imaging and Radiation Sciences*, 54(2): 376–385, https://doi.org/10.1016/j.jmir.2023.03.033.

Powell A (2025) 'Machine Healing: How AI Is Transforming Medicine and Healthcare', *Harvard Gazette*, news.harvard.edu/gazette/story/2025/03/how-ai-is-transforming-medicine-healthcare/.

Powell J (2019) 'Trust Me, I'm a Chatbot: How Artificial Intelligence in Health Care Fails the Turing Test', *J Med Internet Res*, 21(10): e16222, https://doi.org/10.2196/16222.

Prabhu SP (2019) 'Ethical Challenges of Machine Learning and Deep Learning Algorithms', *Lancet Oncol.*, 20(5): 621–622, https://doi.org/10.1016/S1470-2045%2819%2930230-X.

Prabhudesai S, L Yang, S Asthana, X Huan, QV Liao and N Banovic (2023) 'Understanding Uncertainty: How Lay Decision-Makers Perceive and Interpret Uncertainty in Human-AI Decision Making', *Proceedings of the 28th International Conference on Intelligent User Interfaces*, 379–396, https://doi.org/10.1145/3581641.3584033.

Prakash S, JN Balaji, A Joshi and KM Surapaneni (2022) 'Ethical Conundrums in the Application of Artificial Intelligence (AI) in Healthcare – A Scoping Review of Reviews', *J. Pers. Med.*, 12(11): 1914, https://doi.org/10.3390/jpm12111914.

Price II WN (2015) 'Black-Box Medicine', *Harv. J.L. & Tech*, 28(2): 419–467.

Price II WN (2017) 'Regulating Black-Box Medicine', *Mich. L. Rev*, 116(3): 421–474, https://doi.org/10.36644/mlr.116.3.

Price II WN (2018) 'Medical Malpractice and Black-Box Medicine', in *Big Data, Health Law, and Bioethics*, IG Cohen, HF Lynch, E Vayena and U Gasser (Eds.), 295–306. Cambridge University Press, New York, https://doi.org/10.1017/9781108147972.027.

Price II WN (2019) 'Medical AI and Contextual Bias', *Harvard Journal of Law & Technology*, 33(1): 65–116.

Price II WN and IG Cohen (2024) 'Locating Liability for Medical AI', *DePaul Law Review*, 73(2): 339–368.

Price II WN, S Gerke and IG Cohen (2019) 'Potential Liability for Physicians Using Artificial Intelligence', *JAMA*, 12;322(18): 1765–1766, https://doi.org/10.1001/jama.2019.15064.

Price II WN, S Gerke and IG Cohen (2024) 'Liability for Use of Artificial Intelligence in Medicine', in *Research Handbook on Health, AI and the Law*, B Solaiman and IG Cohen (Eds.), 150–166. Edward Elgar Publishing, Cheltenham.

Procter R, P Tolmie and M Rouncefield (2023) 'Holding AI to Account: Challenges for the Delivery of Trustworthy AI in healthcare', *ACM Transactions on Computer-Human Interaction*, 30(2): 1–34.

Prosser WL (1936) 'The Procedural Effect of Res Ipsa Loquitur', *Minnesota Law Review*, 2527: 241–271.

Pruski M (2024) 'AI-Enhanced Healthcare: Not a new Paradigm for Informed Consent', *Bioethical Inquiry*, 21, 475–489, https://doi.org/10.1007/s11673-023-10320-0.

Pu L (2021) 'Fairness of the Distribution of Public Medical and Health Resources', *Frontiers in Public Health*, 9, https://doi.org/10.3389/fpubh.2021.768728.

Purtova N (2018) 'The Law of Everything. Broad Concept of Personal Data and Future of EU Data Protection Law', *Law, Innovation and Technology*, 10(1): 40–81, https://doi.org/10.1080/17579961.2018.1452176.

Quinn TP, S Jacobs, M Senadeera, V Le and S Coghlan (2022) 'The Three Ghosts of Medical AI: Can the Black-Box Present Deliver?', *Artif Intell Med*, 124: 102158, https://doi.org/10.1016/j.artmed.2021.102158.

Quinn TP, M Senadeera, S Jacobs, S Coghlan and V Le (2021) 'Trust and Medical AI: The Challenges We Face and the Expertise Needed to Overcome Them', *Journal of the American Medical Informatics Association: JAMIA*, 28(4): 890–894, https://doi.org/10.1093/jamia/ocaa268.

Quinn TO, S Jacobs, M Senadeera, V Le and S Coghlan (2022) 'The Three Ghosts of Medical AI: Can the Black-Box Present Deliver?', *Artificial Intelligence in Medicine*, 124: 102158.

Rafiei A, M Ghiasi Rad, A Sikora and R Kamaleswaran (2024) 'Improving Mixed-Integer Temporal Modeling by Generating Synthetic Data Using Conditional Generative Adversarial Networks: A Case Study of Fluid Overload Prediction in the Intensive Care Unit', *Computers in Biology and Medicine*, 168, 107749, https://doi.org/10.1016/j.compbiomed.2023.107749.

Rahman MA, E Victoros, J Ernest, R Davis, Y Shanjana and MR Islam (2024) 'Impact of Artificial Intelligence (AI) Technology in Healthcare Sector: A

Critical Evaluation of Both Sides of the Coin', *Clinical Pathology*, 17, https://doi.org/10.1177/2632010X241226887.

Raji ID, A Smart, RN White, M Mitchell, T Gebru, B Hutchinson, J Smith-Loud, D Theron, and P Barnes (2020) 'Closing the AI Accountability Gap: Defining an End-To-End Framework for Internal Algorithmic Auditing', in *Proceedings of the 2020 Conference on Fairness, Accountability, and Transparency* (*FAT* * '20), Association for Computing Machinery. New York, NY, USA, 33–44, https://doi.org/10.1145/3351095.3372873.

Rajpurkar P, E Chen, O Banerjee and EJ Topol (2022) 'AI in Health and Medicine', *Nat Med*, 28: 31–38, https://doi.org/10.1038/s41591-021-01614-0.

Ramsey L (2023) 'AI Algorithm Could Soon Be Used to Rapidly Diagnose Heart Attacks With Better Accuracy' News Medical, https://www.news-medical.net/news/20230511/AI-algorithm-could-soon-be-used-to-rapidly-diagnose-heart-attacks-with-better-accuracy.aspx.

Randolph J (2020) 'Why Healthcare Leaders Must Take Risks and Not Have Regrets', Healthcare Innovation, https://www.hcinnovationgroup.com/clinical-it/digital-health-innovation/article/21158902/why-healthcare-leaders-must-take-risks-and-not-have-regrets.

Raposo VL (2012) 'To Act or Not to Act, that Is the Question: Informed Consent in a Criminal Perspective', *European Journal of Health Law*, 19(4): 379–390.

Raposo VL (2013) *Do Ato Médico ao Problema Jurídico*. Almedina, Coimbra.

Raposo VL (2014) 'Prescrição e Uso Off-Label de Medicamentos e de Dispositivos Médicos', *Lex Medicinae Revista Portuguesa de Direito da Saúde*, 11(21–22): 35–75.

Raposo VL (2015) 'Electronic Health Records: Is It A Risk Worth Taking in Healthcare Delivery?', *GMS Health Technol Assess*, 10;11, https://doi.org/10.3205/hta000123.

Raposo VL (2016a) 'Telemedicine: The Legal Framework (or the Lack of It) in Europe', *GMS Health Technol Assess*, 16;12, https://doi.org/10.3205/hta000126.

Raposo VL (2016b) 'The Unbearable Lightness of Culpability: The Compensation for Damages in the Practice of Medicine', *Saúde E Sociedade*, 25(1): 57–69, https://doi.org/10.1590/S0104-12902016144195.

Raposo VL (2017a) 'Is it Possible to Achieve a Common Standard of Care for Cyber-medicine within the EU?', *Medicine and Law*, 36(2): 29–42.

Raposo VL (2017b) 'What Is Off-Label Prescription and How Is It (Non)Regulated in China', *Asia Pacific Journal of Health Law and Ethics*, 1–21.

Raposo VL (2018) *Danos Causados por Medicamentos – Enquadramento Jurídico à Luz do Ordenamento Europeu*. Almedina, Coimbra.

Raposo VL (2019) 'Defensive Medicine and the Imposition of a More Demanding Standard of Care', *J Leg Med*, 39(4): 401–416, https://doi.org/10.1080/01947648.2019.1677273.

Raposo VL (2024a) 'Da IA como Pessoa Jurídica (Fundamentos e Consequências): O Verdadeiro Jogo da Imitação', in *Estudos em Homenagem ao Prof. Doutor António*

Pinto Monteiro, F. Cassiano dos Santos et al. (Eds.), Vol. I, 1049–1072. Instituto Juridico, Coimbra.

Raposo VL (2024b) 'How is 'Unexplainable' and Non-transparent AI Affecting Healthcare Delivery?', *Oslo Law Review*, 11(1): 1–12, https://doi.org/10.18261/olr.11.1.6.

Raposo VL (2025) 'The Fifty Shades of Black: About Black Box AI and Explainability in Healthcare', *Medical Law Review*, 33(1), https://doi.org/10.1093/medlaw/fwaf005.

Rasheed K, A Qayyum, M Ghaly, A Al-Fuqaha, A Razi and J Qadir (2022) 'Explainable, Trustworthy, and Ethical Machine Learning for Healthcare: A Survey', *Computers in Biology and Medicine*, 149: 106043.

Raskulla S (2023) 'Hybrid Theory of Corporate Legal Personhood and its Application to Artificial Intelligence', *SN Soc Sci*, 3, 78, https://doi.org/10.1007/s43545-023-00667-x.

Ratti E and M Graves (2022) 'Explainable Machine Learning Practices: Opening Another Black Box for Reliable Medical AI', *AI Ethics*, 2: 801–814, https://doi.org/10.1007/s43681-022-00141-z.

Raub M (2018) 'Bots, Bias and Big Data: Artificial Intelligence, Algorithmic Bias and Disparate Impact Liability in Hiring Practices', *Arkansas Law Review*, 71(2): 529–570.

Rawas S (2024) 'AI: The Future of Humanity', *Discov Artif Intell*, 4, 25, https://doi.org/10.1007/s44163-024-00118-3.

Read-Coop SCE (2023) 'What is Ground Truth?', https://readcoop.eu/what-is-ground-truth/.

Reardon S (2019) 'Rise of Robot Radiologists', *Nature*, 576(7787): S54-S58, https://doi.org/10.1038/d41586-019-03847-z.

Reddy S, W Rogers, VP Makinen, E Coiera, P Brown, M Wenzel, E Weicken, S Ansari, P Mathur, A Casey and B Kelly (2021) 'Evaluation Framework to Guide Implementation of AI Systems into Healthcare Settings', *BMJ Health Care Inform.*, 28(1): e100444, https://doi.org/10.1136/bmjhci-2021-100444.

Reddy K, P Gharde, H Tayade, M Patil, LS Reddy and D Surya (2023) 'Advancements in Robotic Surgery: A Comprehensive Overview of Current Utilizations and Upcoming Frontiers', *Cureus*, 15(12), https://doi.org/10.7759/cureus.50415.

Reddy S, S Allan, S Coghlan and P Cooper (2020) 'A Governance Model for the Application of AI in Health Care', *Journal of the American Medical Informatics Association*, 27(3): 491–497, https://doi.org/10.1093/jamia/ocz192.

Rego ML (2025) 'Bidding Farewell to Good Family Fathers: Advancing the Role of Private Law in the Promotion of Just and Inclusive Societies', *Int J Semiot Law*, https://doi.org/10.1007/s11196-025-10296-7.

Renda A, J Arroyo, Ra Fanni, M Laurer, A Sipiczki, T Yeung, G Maridis, M Fernandes, G Endrodi, S Milio, V Devenyi, S Georgiev, G De Pierrefeu (2021) Study to Support an Impact Assessment of the Regulatory Requirements for Artificial Intelligence in Europe (Final Report), European Commission, Publications Office of the European

Union, https://op.europa.eu/en/publication-detail/-/publication/55538b70-a638-11eb-9585-01aa75ed71a1.

Resneck J (2023) 'Physicians, Not Judges, Should Direct Patient Care', American Medical Association, https://www.ama-assn.org/about/leadership/physicians-not-judges-should-direct-patient-care.

Ribeiro J (2016) 'AlphaGo's Unusual Moves Prove its AI Prowess, Experts Say', *Computer World*, https://www.computerworld.com/article/3043667/alphagos-unusual-moves-prove-its-ai-prowess-experts-say.html.

Ribeiro MT, S Singh and C Guestrin (2016) 'Why Should I Trust You?: Explaining the Predictions of Any Classifier' in Proceedings of the 22nd ACM SIGKDD International Conference on Knowledge Discovery and Data Mining (2016), https://www.kdd.org/kdd2016/papers/files/rfp0573-ribeiroA.pdf.

Roberts JL, LR Fowler and PS Appelbaum (2022) 'Unlimited Medical Liability?', *Emory Law Journal Online*, 72: 42.

Rodziewicz TL, B Houseman and JE Hipskind (2023) 'Medical Error Reduction and Prevention', StatPearls. StatPearls Publishing, Treasure Island (FL), https://www.ncbi.nlm.nih.gov/books/NBK499956/.

Rolls-Royce (2020) The Aletheia Framework, https://www.rollsroyce.com/w/media/Files/R/RollsRoyce/documents/stand-alone-pages/aletheia-frameworkworksheet-v1.pdf 2020.

Rosenblatt F (1958) 'The Perceptron: A Probabilistic Model for Information Storage and Organization in the Brain', *Psychol. Rev.*, 65(6): 386–408, https://doi.org/10.1037/h0042519.

Ross C and I Swetlitz (2017) 'IBM Pitched Watson as a Revolution in Cancer Care. It is Nowhere Close', STAT News, https://www.statnews.com/2017/09/05/watson-ibm-cancer/.

Ross C and I Swetlitz (2018) 'IBM's Watson Supercomputer Recommended 'Unsafe And Incorrect' Cancer Treatments, Internal Documents Show', STAT News, https://www.statnews.com/wp-content/uploads/2018/09/IBMs-Watson-recommended-unsafe-and-incorrect-cancer-treatments-STAT.pdf.

Rossi F (2016) Artificial Intelligence: Potential Benefits and Ethical Considerations. European Parliament, https://www.europarl.europa.eu/RegData/etudes/BRIE/2016/571380/IPOL_BRI(2016)571380_EN.pdf.

Roter D (2000) 'The Enduring and Evolving Nature of the Patient-Physician Relationship', *Patient Educ Couns January*, 39(1): 5–15, https://doi.org/10.1016/s0738-3991(99)00086-5.

Rothlisberger PN (2024) 'AI-Powered Patient-Centered Care: A Call to Action for Innovation', *Journal of Healthcare Management*, 69(4): 255–266.

Rowland SP, JE Fitzgerald, M Lungren, E Lee, Z Harned and AH McGregor (2022) 'Digital Health Technology-Specific Risks for Medical Malpractice Liability', *NPJ Digital Medicine*, 5(1): 157, https://doi.org/10.1038/s41746-022-00698-3.

Rubin J and KM Prager (2018) 'Guide to Considering Nonpsychiatric Medical Intervention Over Objection for the Patient Without Decisional Capacity', *Mayo Clinic Proceedings*, 93(7): 826–829, https://doi.org/10.1016/j.mayocp.2018.04.005.

Ruschemeier H (2023) 'AI as a Challenge for Legal Regulation – The Scope of Application of the Artificial Intelligence Act Proposal', *Era Forum*, 23(3): 361–376.

Safi S, T Thiessen and KJ Schmailzl (2018) 'Acceptance and Resistance of New Digital Technologies in Medicine: Qualitative Study', *JMIR Research Protocols*, 7(12), https://doi.org/10.2196/11072.

Sakr S (2020) *Big Data 2.0 Processing Systems*. Springer International Publishing.

Samoili S, ML Cobo, E Gómez, G De Prato, F Martínez-Plumed and B Delipetrev (2020) 'AI Watch. Defining Artificial Intelligence. Towards an Operational Definition and Taxonomy of Artificial Intelligence', https://eprints.ugd.edu.mk/28047/1/3.%20jrc118163_ai_watch._defining_artificial_intelligence_1.pdf.

Santoni de Sio F and G Mecacci (2021) 'Four Responsibility Gaps with Artificial Intelligence: Why they Matter and How to Address them', *Philos. Technol*, 34: 1057–1084, https://doi.org/10.1007/s13347-021-00450-x.

Sarela AI (2024) 'Consent for Medical Treatment: What is "Reasonable"?', *Health Care Anal*, 32(1): 47–62, https://doi.org/10.1007/s10728-023-00466-8.

Sarker IH (2021) 'Machine Learning: Algorithms, Real-World Applications and Research Directions', *Sn Comput. Sci.*, 2, 160, https://doi.org/10.1007/s42979-021-00592-x.

Sasseville M, S Ouellet, C Rhéaume, M Sahlia, V Couture, P Després, JS Paquette, D Darmon, F Bergeron and MP Gagnon (2025) 'Bias Mitigation in Primary Health Care Artificial Intelligence Models: Scoping Review', *J Med Internet Res*, 27: e60269, https://doi.org/10.2196/60269.

Sauerbrei A, A Kerasidou, F Lucivero and N Hallowell (2023) 'The Impact of Artificial Intelligence on the Person-Centred, Doctor-Patient Relationship: Some Problems and Solutions', *BMC Medical Informatics and Decision Making*, 23, 73, https://doi.org/10.1186/s12911-023-02162-y.

Savage N (2022) 'Breaking into the Black Box of Artificial Intelligence', *Nature*, https://doi.org/10.1038/d41586-022-00858-1.

Savage N (2023) 'Synthetic Data Could Be Better than Real Data', *Nature*, https://doi.org/10.1038/d41586-023-01445-8.

Sayburn A (2017) 'Will the Machines Take Over Surgery?', *The Bulletin of the Royal College of Surgeons of England*, 99(3): 88–90, https://doi.org/10.1308/rcsbull.2017.87.

Scherer MU (2016) 'Regulating Artificial Intelligence Systems: Risks, Challenges, Competencies, and Strategies', *Harvard Journal of Law & Technology*, 29(2): 359–360.

Schiff D and J Borenstein (2019) 'How Should Clinicians Communicate with Patients About the Roles of Artificially Intelligent Team Members?', *AMA J Ethics*, 21(2): E138–145, https://doi.org/10.1001/amajethics.2019.138.

Schoenfeld EM, S Mader, C Houghton, R Wenger, MA Probst, DA Schoenfeld, PK Lindenauer and KM Mazor (2019) 'The Effect of Shared Decisionmaking on Patients' Likelihood of Filing a Complaint or Lawsuit: A Simulation Study', *Ann Emerg Med.*, 74(1): 126–136, https://doi.org/10.1016/j.annemergmed.2018.11.017.

Schönberger D (2019) 'Artificial Intelligence in Healthcare: A Critical Analysis of the Legal and Ethical Implications', *International Journal of Law and Information Technology*, 27(2): 171–203, https://doi.org/10.1093/ijlit/eaz004.

Schröttner J and C Baumgartner (2023) 'The Notified Body: The Conformity Assessment Body for Medical Devices in Europe', in *Medical Devices and In Vitro Diagnostics: Requirements in Europe*, C Baumgartner, J Harer, J Schrttner (Eds.), 39–61. Springer International Publishing, Cham.

Schütte B and L Majewski (2022) 'Private Liability for AI-Related Harm: Towards More Predictable Rules for the Single Market', *Market and Competition Law Review*, vi(1): 123–159.

Schwartz R, A Vassilev, K Greene, L Perine, A Burt and P Hall (2022) *Towards a Standard for Identifying and Managing Bias in Artificial Intelligence.* US Department of Commerce, National Institute of Standards and Technology, NIST Special Publication 1270, https://doi.org/10.6028/NIST.SP.1270.

Schwartz WF (1989) 'Objective and Subjective Standards of Negligence: Defining the Reasonable Person to Induce Optimal Care and Optimal Populations of Injurers and Victims', *Georgetown Law Journal*, 78(2): 241–280.

Schweikart SJ (2021) 'Who Will Be Liable for Medical Malpractice in the Future? How the Use of Artificial Intelligence in Medicine Will Shape Medical Tort Law', *Minnesota Journal of Law, Science and Technology*, 22(2): 1–22.

Seavey WA (1919–1920) 'The Rationale of Agency', *Yale Law Journal*, 29(8): 859–895.

See B (2021) 'Paging Doctor Robot: Medical Artificial Intelligence, Tort Liability, and Why Personhood May Be the Answer', *Brook. L. Rev.*, 87: 417–443.

Selbst AD and J Powles (2017) 'Meaningful Information and the Right to Explanation', *International Data Privacy Law*, 7(4): 233–242, https://doi.org/10.1093/idpl/ipx022.

Selbst AD (2020) 'Negligence and AI's Human Users', *Boston University Law Review*, 100: 1315–1376.

Sendak MP, MC Elish, M Gao, JD Futoma, W Ratliff, M Nichols, AD Bedoya, S Balu and C O'Brien (2019) 'The Human Body Is a Black Box: Supporting Clinical Decision-Making with Deep Learning', *Proceedings of the 2020 Conference on Fairness, Accountability, and Transparency*, 99–109.

Senger JM and P O'Leary (2018) 'Big Data and Human Medical Judgment Regulating Next-Generation Clinical Decision Support', in *Big Data, Health Law, and Bioethics*,

IG Cohen, HF Lynch, E Vayena and U Gasser (Eds.), 283–294. Cambridge University Press, Cambridge.

Sezgin E (2023) 'Artificial Intelligence in Healthcare: Complementing, Not Replacing, Doctors and Healthcare Providers', *Digital Health*, 9, https://doi.org/10.1177/20552076231186520.

Sgaier SK, V Huang and G Charles (2020) 'The Case for Causal AI', *Stanford Social Innovation Review*, 18(3): 50–55, https://doi.org/10.48558/KT81-SN73.

Shachar C and S Gerke (2023) 'Prevention of Bias and Discrimination in Clinical Practice Algorithms', *JAMA*, 329(4): 283–284.

Shah AB (2020) 'Artificial Intelligence in Healthcare: Can Regulation Catch up with Innovation?', *RAIL: The Journal of Robotics, Artificial Intelligence & Law*, 3(1): 41–46.

Shakya AK, G Pillai and S Chakrabarty (2023) 'Reinforcement Learning Algorithms: A Brief Survey', *Expert Systems with Applications*, 231, 120495, https://doi.org/10.1016/j.eswa.2023.120495.

Shalak M, MA Shariff, V Doddapaneni and N Suleman (2022) 'The Truth, the Whole Truth, and Nothing but the Truth: Therapeutic Privilege', *J Postgrad Med.*, 68(3): 152–155, https://doi.org/10.4103/jpgm.jpgm_1146_21.

Sharma S and V Kumar (2022) 'Application of Genetic Algorithms in Healthcare: A Review', in *Next Generation Healthcare Informatics. Studies in Computational Intelligence*, BK Tripathy, P Lingras, AK Kar and CL Chowdhary (Eds.). Springer, Singapore, https://doi.org/10.1007/978-981-19-2416-3_5.

Shaughnessy A (2022) 'How Does Acetaminophen Work?', Tufts Now, https://now.tufts.edu/2022/09/14/how-does-acetaminophen-work.

Shavell S (2018) 'The Mistaken Restriction of Strict Liability to Uncommon Activities', *Journal of Legal Analysis*, 10: 1–45, https://doi.org/10.1093/jla/lay004.

Shaw J (2019) 'Artificial Intelligence and Ethics', *Harvard Magazine*, 30: 1–11.

Sheeran N, L Jones, R Pines, B Jin, A Pamoso, J Eigeland and M Benedetti (2022) 'How Culture Influences Patient Preferences for Patient-Centered Care with Their Doctors', *Journal of Communication in Healthcare*, 16(2): 186–196, https://doi.org/10.1080/17538068.2022.2095098.

Shelton D (2015) 'Nature as a Legal Person', *VertigO – La Revue Électronique En Sciences De L'environnement*, 22, https://doi.org/10.4000/vertigo.16188.

Shenoy A, G Shenoy and G Shenoy (2021) 'Respondeat Superior in Medicine and Public Health Practice: The Question Is – Who Is Accountable for Whom?', *Ethics, Medicine and Public Health*, 17, 100634, https://doi.org/10.1016/j.jemep.2021.100634.

Shepherd SM, C Willis-Esqueda, Y Paradies, D Sivasubramaniam, J Sherwood and T Brockie (2018) 'Racial and Cultural Minority Experiences and Perceptions of Health Care Provision in A Mid-Western Region', *Int J Equity Health*, 17, 33, https://doi.org/10.1186/s12939-018-0744-x.

Shoabjareh AH, M Ghasri, T Roberts, A Lapworth, N Dobos and C Boshuijzen-van Burken (2024) 'The Role of Trust and Distrust in Technology Usage: An In-Depth Investigation of Traffic Information Apps Usage for Mandatory and Non-Mandatory Trips', *Travel Behaviour and Society*, 37, 100816, https://doi.org/10.1016/j.tbs.2024.100816.

Shortliffe EH and MJ Sepúlveda (2018) 'Clinical Decision Support in the Era of Artificial Intelligence', *JAMA*, 320:2199–200.

Shuren Y (2023) 'Towards Trustworthy and Understandable AI: Unraveling Explainability Strategies on Simplifying Algorithms, Appropriate Information Disclosure, and High-level Collaboration', in *Proceedings of the 26th International Academic Mindtrek Conference (Mindtrek '23)*, 133–143. Association for Computing Machinery, New York, NY, USA, https://doi.org/10.1145/3616961.3616965.

Siau K and W Wang (2018) 'Building Trust in Artificial Intelligence, Machine Learning, and Robotics', *Cutter Business Technology Journal*, 31(2): 47–53.

Silver K (2024) 'When Should the Master Answer? Respondeat Superior and the Criminal Law', *Criminal Law and Philosophy*, 18: 89–108, https://doi.org/10.1007/s11572-023-09659-7.

Simon DA and AS Kesselheim (2024) 'Physician and Device Manufacturer Tort Liability for Remote Patient Monitoring Devices', in *Digital Health Care Outside of Traditional Clinical Settings: Ethical, Legal, and Regulatory Challenges and Opportunities*, IG Cohen, DB Kramer, J Adler-Milstein, C Shachar (Eds.), 109–122. Cambridge University Press, Cambridge.

Singh S and YK Loke (2012) 'Drug Safety Assessment in Clinical Trials: Methodological Challenges and Opportunities', *Trials*, 13, 138, https://doi.org/10.1186/1745-6215-13-138.

Singhal K, S Azizi, T Tu and V Natarajan (2023) 'Large Language Models Encode Clinical Knowledge', *Nature*, 620: 172–180, https://doi.org/10.1038/s41586-023-06291-2.

Singhal S and N Patel (2025) 'What to Expect in US Healthcare in 2025 and Beyond', Mckinsey and Company, https://www.mckinsey.com/industries/healthcare/our-insights/what-to-expect-in-us-healthcare-in-2025-and-beyond.

Sinsky C, L Colligan, L Li, M Prgomet, S Reynolds, L Goeders, J Westbrook, M Tutty and G Blike (2016) 'Allocation of Physician Time in Ambulatory Practice: A Time and Motion Study in 4 Specialties', *Ann Intern Med.*,165(11): 753–760, https://doi.org/10.7326/M16-0961.

Slack D, S Hilgard, E Jia, S Singh and H Lakkaraju (2020) 'Fooling LIME and SHAP: Adversarial Attacks on Post Hoc Explanation Methods', in *Proceedings of the AAAI/ACM Conference on AI, Ethics, and Society*, 180–186. Association for Computing Machinery, New York, NY, USA.

Smith H and K Fotheringham (2020) 'Artificial Intelligence in Clinical Decision-Making: Rethinking Liability', *Medical Law International*, 20(2): 131–154, https://doi.org/10.1177/0968533220945766.

Smith H (2021) 'Clinical AI: Opacity, Accountability, Responsibility and Liability', *AI & Soc*, 36: 535–545, https://doi.org/10.1007/s00146-020-01019-6.

Smith PC, A Stepan, V Valdmanis and P Verheyen (1997) 'Principal-Agent Problems in Health Care Systems: An International Perspective', *Health Policy*, 41(1): 37–60, https://doi.org/10.1016/S0168-8510(97)00012-2.

Smith TM (2024) 'Health Care AI Is No Longer Theoretical. It's Totally Practical', AMA, https://www.ama-assn.org/practice-management/digital/health-care-ai-no-longer-theoretical-it-s-totally-practical.

Soffar H (2025) 'RIBA Robotic Bear Nurse – Features, Advantages, Disadvantages', Online Sciences, https://www.online-sciences.com/robotics/riba-robotic-bear-nurse-features-advantages-disadvantages/.

Sokol DK (2013) '"First Do No Harm" Revisited', *BMJ*, 347: f6426, https://doi.org/10.1136/bmj.f6426.

Solaiman B and A Malik (2024) 'Regulating Algorithmic Care in the European Union: Evolving Doctor-Patient Models Through the Artificial Intelligence Act (AI-Act) and the Liability Directives', *Medical Law Review*, fwae033, https://doi.org/10.1093/medlaw/fwae033.

Solaiman SM (2017) 'Legal Personality of Robots, Corporations, Idols and Chimpanzees: A Quest for Legitimacy', *Artif Intell Law*, 25: 155–179, https://doi.org/10.1007/s10506-016-9192-3.

Solum L (1992) 'Legal Personhood for Artificial Intelligence', *North Carolina L Rev*, 70(4): 1231–1287.

Song X, B Xu and Z Zhao (2022) 'Can People Experience Romantic Love for Artificial Intelligence? An Empirical Study of Intelligent Assistants', *Information & Management*, 59(2), 103595, https://doi.org/10.1016/j.im.2022.103595.

Sparrow R and L Sparrow (2006) 'In the Hands of Machines? The Future of Aged Care', *Minds & Machines*, 16: 141–161, https://doi.org/10.1007/s11023-006-9030-6.

Spiro H (2009) 'Commentary: The Practice of Empathy', *Acad. Med.*, 84: 1177–1179, https://doi.org/10.1097/ACM.0b013e3181b18934.

Starke C, J Baleis, B Keller and F Marcinkowski (2022) 'Fairness Perceptions of Algorithmic Decision-Making: A Systematic Review of the Empirical Literature', *Big Data & Society*, 9(2), https://doi.org/10.1177/20539517221115189.

Stern AD, A Goldfarb, T Minssen and WN Price II (2022) 'AI Insurance: How Liability Insurance Can Drive the Responsible Adoption of Artificial Intelligence in Health Care', *New England Journal of Medicine Catalyst*, 4(4), https://doi.org/10.1056/CAT.21.0242.

Stöger K, D Schneeberger and A Holzinger (2021) 'Medical Artificial Intelligence: The European Legal Perspective', *Commun. ACM*, 64, 11: 34–36, https://doi.org/10.1145/3458652.

Stoumpos AI, F Kitsios and MA Talias (2023) 'Digital Transformation in Healthcare: Technology Acceptance and Its Applications', *International Journal of Environmental Research and Public Health*, 20(4), https://doi.org/10.3390/ijerph20043407.

Strickland E (2019) 'How IBM Watson Overpromised and Underdelivered on at Health Care', IEEE SPECTRUM, https://spectrum.ieee.org/how-ibm-watson-over promised-and-underdelivered-on-ai-health-care.

Strohm L, C Hehakaya, ER Ranschaert, WPC Boon and EHM Moors (2020) 'Implementation of Artificial Intelligence (AI) Applications in Radiology: Hindering and Facilitating Factors', *European Radiology*, 30: 5525–5532.

Stuart GR, D Cope, J Ip, M Lotosh, A Shah, J Weng and R Tang (2021) '"Garbage in, Garbage Out" Revisited: What Do Machine Learning Application Papers Report About Human-Labeled Training Data?', *Quantitative Science Studies*, 2(3): 795–827, https://doi.org/10.1162/qss_a_00144.

Sukel K (2017) 'With a Little Help from AI Friends', *New Scientist*, 235(3134): 36–39.

Sullivan HR and SJ Schweikart (2019) 'Are Current Tort Liability Doctrines Adequate for Addressing Injury Caused by AI?', *Ama J. Ethics*, 21: E160–E166.

Sulmasy LS and TA Bledsoe for the ACP Ethics, Professionalism and Human Rights Committee (2019) 'American College of Physicians Ethics Manual: Seventh Edition', *Ann Intern Med.*, 170: S1–S32, https://doi.org/10.7326/M18-2160.

Swan M (2013) 'The Quantified Self: Fundamental Disruption in Big Data Science and Biological Discovery', *Big Data*, 1(2): 85–99.

Tanaka M, S Matsumura and S Bito (2023) 'Roles and Competencies of Doctors in Artificial Intelligence Implementation: Qualitative Analysis Through Physician Interviews', *JMIR Form Res*, 7: e46020, https://doi.org/10.2196/46020.

Tang X (2019) 'The Role of Artificial Intelligence in Medical Imaging Research', *BJR Open*, 2(1): 20190031, https://doi.org/10.1259/bjro.20190031.

Taulli T (2019) *Artificial Intelligence Basics: A Non-Technical Introduction.* Appres, Monrovia, CA.

Taye MM (2023) 'Understanding of Machine Learning with Deep Learning: Architectures, Workflow, Applications and Future Directions', *Computers*, 12(5): 91, https://doi.org/10.3390/computers12050091.

Taylor I (2024) 'Is Explainable AI Responsible AI?', *AI & Soc*, 40, 1695–1704, https://doi.org/10.1007/s00146-024-01939-7.

Tech Desck (2019) 'Watch: This Talking Robot Helps Senior Citizens Fight Loneliness and Isolation', Republicworld.com.

TerKonda SP and EM Fish (2023) 'Artificial Intelligence Viewed Through the Lens of State Regulation', *Intelligence-Based Medicine*, 7, 100088, https://doi.org/10.1016/j.ibmed.2023.100088.

Terry N (2019) 'Of Regulating Healthcare AI and Robots', *Yale Journal of Law and Technology*, 21(Special Issue): 133–190.

Terry NP and LF Wiley (2016) 'Liability for Mobile Health and Wearable Technologies', *Annals Health L.*, 25: 62–97.

The European Consumer Organisation (2023) Proposal for an AI Liability Directive – BEUC position paper, Ref: BEUC-X-2023-050-02/05/2023, https://www.beuc.eu/sites/default/files/publications/BEUC-X-2023-050_Proposal_for_an_AI_Liability_Directive.pdf.

Thelisson E and H Verma (2024) 'Conformity Assessment Under the EU AI Act General Approach', *AI Ethics*, 4: 113–12, https://doi.org/10.1007/s43681-023-00402-5.

Thom DH, RL Kravitz, RA Bell, E Krupat and R Azari (2002) 'Patient Trust in the Physician: Relationship to Patient Requests', *Family Practice*, 19(5): 476–483, https://doi.org/10.1093/fampra/19.5.476.

Thomas S (2017) 'Artificial Intelligence and Medical Liability (Part II), BiLL HEALTH, http://blog.petrieflom.law.harvard.edu/2017/02/10/artificial-intelligence-and-medical-liability-part-ii/.

Thompson RB (1994) 'Unpacking Limited Liability: Direct and Vicarious Liability of Corporate Participants for Torts of the Enterprise', *Vanderbilt Law Review*, 47(1): 1–42.

Thornton RG (2003) 'The Learned Intermediary Doctrine and its Effects on Prescribing Physicians', *Proc* (*Bayl Univ Med Cent*), 16(3): 359–361, https://doi.org/10.1080/08998280.2003.11927929.

Timmermans S (2005) 'From Autonomy to Accountability: The Role of Clinical Practice Guidelines in Professional Power', *Perspectives in Biology and Medicine*, 48(4): 490–501, https://doi.org/10.1353/pbm.2005.0096.

Topol EJ (2019) *Deep Medicine: How Artificial Intelligence Can Make Healthcare Human Again*. Basic Books, Hachette UK.

Topol EJ (2019) 'High-Performance Medicine: The Convergence of Human and Artificial Intelligence', *Nat Med*, 25: 44–56, https://doi.org/10.1038/s41591-018-0300-7.

Torrent-Sellens J, AI Jiménez-Zarco and F Saigí-Rubió (2021) 'Do People Trust in Robot-Assisted Surgery? Evidence from Europe', *International Journal of Environmental Research and Public Health*, 18(23), https://doi.org/10.3390/ijerph182312519.

Tozzo P, F Angiola, A Gabbin, C Politi and L Caenazzo (2021) 'The Difficult Role of Artificial Intelligence in Medical Liability: To Err Is Not Only Human', *Clin Ter*, 172(6): 527–528, https://doi.org/10.7417/CT.2021.2372.

Trofimiuk E (2024) 'Benefits of Conversational AI in Healthcare: 10 Use Case' Eliftech, https://www.eliftech.com/insights/conversational-ai-in-healthcare/.

Truog RD (2012) 'Patients and Doctors: Evolution of a Relationship', *N Engl J Med*, 366: 581–585, https://doi.org/10.1056/NEJMp1110848.

Tschider CA (2024) 'Humans Outside the Loop', *Yale Journal of Law & Technology*, 26(2), 324–429.

Tsopra R, X Fernandez, C Luchinat and A Burgun (2021) 'A Framework for Validating AI in Precision Medicine: Considerations from the European Itfoc Consortium', *BMC Medical Informatics and Decision Making*, 21, https://doi.org/10.1186/s12911-021-01634-3.

Tunkel AR, BJ Hartman, SL Kaplan, BA Kaufman, KL Roos, WM Scheld and RJ Whitley (2004) 'Practice Guidelines for the Management of Bacterial Meningitis', *Clinical Infectious Diseases*, 39(9): 1267–1284, https://doi.org/10.1086/425368.

Turing A (1936) 'On Computable Numbers, with an Application to the Entscheidungsproblem', *Proc Lond Math Soc 2nd Ser*, 42: 230–265.

Turing AM (1950) 'I.-Computing Machinery and Intelligence', *Mind*, LIX(236): 433–460, https://doi.org/10.1093/mind/LIX.236.433.

U.S. Congress, Office of Technology Assessment (1994) Defensive Medicine and Medical Malpractice, OTA-H-602.

U.S. Food and Drug Administration (2021) Zimmer Biomet Recalls ROSA One 3.1 Brain Application Due to Error in Software, https://www.accessdata.fda.gov/scripts/cdrh/cfdocs/cfres/res.cfm?id=189497.

UK Parliament (2020) 'AI and Healthcare' Postnote, Number 637, https://researchbriefings.files.parliament.uk/documents/POST-PN-0637/POST-PN-0637.pdf.

Ulfbeck V (2020) 'Product Liability in a World of Automated Products and Digital Distribution Channels', *Católica Law Review*, 4(2): 191–200, https://doi.org/10.34632/catolicalawreview.2020.9325.

Umoh EE (2024) 'Reliability of AI Algorithms in Safety Application', *International Journal of Engineering and Advanced Technology Studies*, 12(2): 74–85.

University of California (2025) 'AI Could Help Improve Early Detection of Interval Breast Cancers', https://www.sciencedaily.com/releases/2025/05/250505121428.htm.

Usemahu SA and JD Panjaitan (2024) 'Analyst Types of Malpractice in Health Law', *International Journal of Social Research*, 2(3): 130–140.

Vale SB and G Zanfir-Fortuna (2022) 'Automated Decision-Making Under the GDPR: Practical Cases from Courts and Data Protection Authorities', *Future of Privacy Forum*, https://fpf.org/wp-content/uploads/2022/05/FPF-ADM-Report-R2-singles.pdf.

Van der Stappen A and M Funk (2021) 'Towards Guidelines for Designing Human- In-The-Loop Machine Training Interfaces', in *26th International Conference on Intelligent User Interfaces*, 514–519, https://doi.org/10.1145/3397481.3450668.

Van Kolfschooten H and J Van Oirschot (2024) 'The EU Artificial Intelligence Act (2024): Implications for Healthcare', *Health Policy*, 149: 105152.

Van Maris A, N Zook, P Caleb-Solly, M Studley, A Winfield and S Dogramadzi (2020) 'Designing Ethical Social Robots-A Longitudinal Field Study with Older Adults', *Front Robot AI*, 7(1), https://doi.org/10.3389/frobt.2020.00001.

Van Norman GA (2016) 'Drugs, Devices, and the FDA: Part 1: An Overview of Approval Processes for Drugs', *JACC: Basic to Translational Science*, 1(3): 170–179, https://doi.org/10.1016/j.jacbts.2016.03.002.

Van Rooij B and M Brownlee (2021) 'Does Tort Deter? Inconclusive Empirical Evidence about the Effect of Liability in Preventing Harmful Behaviour', in *The Cambridge Handbook of Compliance*, B Van Rooij and DD Sokol (Eds.), 311–325. Cambridge University Press, Cambridge.

Vanderpool D (2021) 'The Standard of Care', *Innovations in Clinical Neuroscience*, 18(7–9), 50–51, https://www.ncbi.nlm.nih.gov/pmc/articles/PMC8667701/.

Vannuccini S and E Prytkova (2023) 'Artificial Intelligence's New Clothes? A System Technology Perspective', *Journal of Information Technology*, 39(2): 317–338, https://doi.org/10.1177/02683962231197824.

Varelius J (2006) 'The Value of Autonomy in Medical Ethics', *Medicine, Health Care, and Philosophy*, 9(3): 377–388, https://doi.org/10.1007/s11019-006-9000-z.

Varkey B (2021) 'Principles of Clinical Ethics and their Application to Practice', *Medical Principles and Practice*, 30(1): 17–28, https://doi.org/10.1159/000509119.

Varnosfaderani SM and M Forouzanfar (2024) 'The Role of AI in Hospitals and Clinics: Transforming Healthcare in the 21st Century', *Bioengineering*, 11(4), https://doi.org/10.3390/bioengineering11040337.

Vaswani PA, TF Tropea and N Dahodwala (2020) 'Overcoming Barriers to Parkinson Disease Trial Participation: Increasing Diversity and Novel Designs for Recruitment and Retention', *Neurotherapeutics*, 17(4): 1724–1735, https://doi.org/10.1007/s13311-020-00960-0.

Vayena E, A Blasimme and IG Cohen (2018) 'Machine Learning in Medicine: Addressing Ethical Challenges', *PLoS Med*, 15(11): e1002689, https://doi.org/10.1371/journal.pmed.1002689.

Verdicchio M and A Perin (2022) 'When Doctors and AI Interact: on Human Responsibility for Artificial Risks', *Philos. Technol.*, 35, 11, https://doi.org/10.1007/s13347-022-00506-6.

Veruggio G (2006) 'The EURON Roboethics Roadmap', *2006 6th IEEE-RAS International Conference on Humanoid Robots*, 612–617, https://doi.or/10.1109/ICHR.2006.321337.

Viljanen M (2017) 'A Cyborg Turn in Law?', *German Law Journal*, 18(5): 1277–1308, https://doi.org/10.1017/S2071832200022331.

Villasenor J and V Foggo (2020) 'Artificial Intelligence, Due Process and Criminal Sentencing', *Michigan State Law Review*, 2020(2): 295–354.

Visweswaran S, AJ King and GF Cooper (2022) 'Integration of AI for Clinical Decision Support', in *Intelligent Systems in Medicine and Health. Cognitive Informatics in Biomedicine and Healthcare*, TA Cohen, VL Patel and EH Shortliffe (Eds.). Springer, Cham, https://doi.org/10.1007/978-3-031-09108-7_10.

Vladeck DC (2014) 'Machines Without Principles: Liability Rules and Artificial Intelligence', *Wash. L. Rev.*, 89: 117–150.

Volovici V, NL Syn, A Ercole, JJ Zhao and N Liu (2022) 'Steps to Avoid Overuse and Misuse of Machine Learning in Clinical Research', *Nat Med*, 28: 1996–1999, https://doi.org/10.1038/s41591-022-01961-6.

Vyas DA, LG Eisenstein and DS Jones (2020) 'Hidden in Plain Sight – Reconsidering the Use of Race Correction in Clinical Algorithms', *New England Journal of Medicine*, 383(9): 874–882.

Wachter S, B Mittelstadt and L Floridi (2017a) 'Transparent, Explainable, and Accountable AI for Robotics', *Science Robotics*, 2(6): eaan6080.

Wachter S, B Mittelstadt and L Floridi (2017b) 'Why a Right to Explanation of Automated Decision-Making Does Not Exist in the General Data Protection Regulation', *International Data Privacy Law*, 7(2), 76–99, https://doi.org/10.1093/idpl/ipx005.

Wadden JJ (2021) 'What Kind of Artificial Intelligence Should We Want for Use in Healthcare Decision-Making Applications', *Canadian Journal of Bioethics / Revue Canadienne de Bioethique*, 4(1): 94–100.

Wade B, J Abraham, M Coder and Chief Policy Officer (2023) 'Guidance to Industry. Classification of Digital Health Technologies', Health Advances – Digital Therapeutic Alliance https://dtxalliance.org/wp-content/uploads/2023/06/Guidance-to-Industry-Classification-of-Digital-Health-Technologies-2023Jun05.pdf.

Wagner JB (2019) 'Artificial Intelligence in Medical Imaging', *Radiologic Technology*, 90(5): 489–501.

Walker S (2014) 'Human Rights Impact Assessments: Emerging Practice and Challenges', in *Economic, Social, and Cultural Rights in International Law: Contemporary Issues and Challenges*, E Riedel, G Giacca and C Golay (Eds.). Oxford Academic, Oxford, https://doi.org/10.1093/acprof:oso/9780199685974.003.0014.

Wallis KA (2017) 'No-Fault, No Difference: No-Fault Compensation for Medical Injury and Healthcare Ethics and Practice', *The British Journal of General Practice*, 67(654): 38–39, https://doi.org/10.3399/bjgp17X688777.

Walter Y (2024) 'Managing the Race to the Moon: Global Policy and Governance in Artificial Intelligence Regulation – A Contemporary Overview and an Analysis of Socioeconomic Consequences', *Discov Artif Intell*, 4, 14, https://doi.org/10.1007/s44163-024-00109-4.

Wang B, J Zhou, Y Li and F Chen (2023) 'Impact of Fidelity and Robustness of Machine Learning Explanations on User Trust', in *Australasian Joint Conference on Artificial Intelligence*, 209–220. Springer Nature Singapore, Singapore.

Wang Y, D Cao, L Chen, M Li, W Zheng and N Ohkohchi (2021) 'Current Trends in Three-Dimensional Visualization and Real-Time Navigation as Well as Robot-Assisted Technologies in Hepatobiliary Surgery', *World Journal of Gastrointestinal Surgery*, 13(9): 904–922, https://doi.org/10.4240/wjgs.v13.i9.904.

Wegman B, JP Stannard and BS Bal (2012) 'Medical Liability of the Physician in Training', *Clinical Orthopaedics and Related Research*, 470(5): 1379–1385, https://doi.org/10.1007/s11999-012-2244-4.

Weng SF, J Reps, J Kai, JM Garibaldi and N Qureshi (2017) 'Can Machine-Learning Improve Cardiovascular Risk Prediction Using Routine Clinical Data?', *PLoS One*, 12(4): e0174944, https://doi.org/ 10.1371/journal.pone.0174944.

Wexler J (2017) 'Google AI Blog, Facets: An Open Source Visualization Tool for Machine Learning Training Data', https://ai.googleblog.com/2017/07/facets-open-source-visualization-tool.html.

Whalen R (2016) 'Legal Networks: The Promises and Challenges of Legal Network Analysis', *Mich. St. L. Rev.*, 539–565.

Whitney SN, AL McGuire and LB McCullough (2004) 'A Typology of Shared Decision Making, Informed Consent, and Simple Consent', *Ann Intern Med.*, 140(1): 54–9, https://doi.org/10.7326/0003-4819-140-1-200401060-0001.

Whittam S (2022) 'Mind the Compensation Gap: Towards a New European Regime Addressing Civil Liability in the Age of AI', *International Journal of Law and Information Technology*, 30(2): 249–265, https://doi.org/10.1093/ijlit/eaac013.

Whittlestone J, R Nyrup, A Alexandrova, K Dihal, and S Cave (2019) 'Ethical and Societal Implications of Algorithms, Data, and Artificial Intelligence: A Roadmap for Research', Nuffield Foundation, https://www.nuffieldfoundation.org/sites/default/files/files/Ethical-and-Societal-Implications-of-Data-and-AI-report-Nuffield-Foundat.pdf.

Wiley L and N Terry (2016) 'Liability for Mobile Health and Wearable Technologies', *Annals of Health Law*, 25: 62–97.

Williams R and R Yampolskiy (2021) 'Understanding and Avoiding AI Failures: A Practical Guide', *Philosophies*, 6(3), 53, https://doi.org/10.3390/philosophies6030053.

Williams S (2015) 'Tort Law Who's Really Who: Apportioning Liability of Independent Contractors Who Work for Hospitals that Qualify for Sovereign Immunity in Rural State; Campbell County Memorial Hospital V. Pfiefle, 317 p.3d 573 (wyo. 2014)', *Wyoming Law Review*, 15(1): 211–230.

Williamson SM and V Prybutok (2024) 'Balancing Privacy and Progress: A Review of Privacy Challenges, Systemic Oversight, and Patient Perceptions in AI-Driven Healthcare', *Applied Sciences*, 14(2), 675, https://doi.org/10.3390/app14020675.

Witkowski K, R Okhai and SR Neely (2024) 'Public Perceptions of Artificial Intelligence in Healthcare: Ethical Concerns and Opportunities for Patient-Centered Care', *BMC Med Ethics*, 25, 74, https://doi.org/10.1186/s12910-024-01066-4.

Wong TY and NM Bressler (2016) 'Artificial Intelligence with Deep Learning Technology Looks into Diabetic Retinopathy Screening', *JAMA*, 16(22): 2366–2367, https://doi.org/10.1001/jama.2016.17563.

Wood B and H Castro, 'AI as Consulting Physicians: Balance & Innovation', https://open.spotify.com/episode/2TUSWYid8xf9baqGDqawEY?go=1&sp_cid=50368fe3fc60dd0336d862b70a2a03de&utm_source=embed_player_p&utm_medium=desktop&nd=1&dlsi=b3b8b0fd3a044dfa.

Woollacott E (2024) 'What is 'AI Washing' and Why Is It a Problem?', BBC News, https://www.bbc.com/news/articles/c9xx81228930.

World Health Organization (2016) Working for Health and Growth: Investing in the Health Workforce', WHO, http://apps.who. int/iris/bitstream/10665/250047/1/9789241511308-eng.pdf.

World Health Organization (2021) *Ethics and Governance of Artificial Intelligence for Health: WHO Guidance*. World Health Organization, Geneva.

World Health Organisation (2024) 'Sepsis', https://www.who.int/news-room/fact-sheets/detail/sepsis.

World Medical Association (2025) WMA Council Resolution to Uphold the Ethical Framework of Healthcare, Adopted by the 229th WMA Council Session, Montevideo, Uruguay, https://www.wma.net/policies-post/wma-council-resolution-to-uphold-the-ethical-framework-of-healthcare/.

Wright WL (2021) 'The Second Victim in Medical Malpractice Litigation: The Stress of Litigation', in *The Medical-Legal Aspects of Acute Care Medicine*, JE Szalados (Eds.), 645–658. Springer, Cham, https://doi.org/10.1007/978-3-030-68570-6_32.

Wu M, X Du, R Gu and J Wei (2021) 'Artificial Intelligence for Clinical Decision Support in Sepsis', *Frontiers in Medicine*, 8: 665464, https://doi.org/10.3389/fmed.2021.665464.

Wulf AJ and O Seizov (2024) '"Please Understand We Cannot Provide Further Information": Evaluating Content and Transparency of GDPR-Mandated AI Disclosures', *AI & Society*, 39(1): 235–256.

Xie C, Y Xie, Y Wang, P Zhou, L Lu, Y Feng and C Liang (2025) 'Understanding Older Adults' Continued-Use Intention of AI Voice Assistants', *Univ Access Inf Soc*, 24: 1687–1699, https://doi.org/10.1007/s10209-024-01172-5.

Xu C, P Coen-Pirani and X Jiang (2023) 'Empirical Study of Overfitting in Deep Learning for Predicting Breast Cancer Metastasis', *Cancers*, 15(7), https://doi.org/10.3390/cancers15071969.

Xu H and KMJ Shuttleworth (2024) 'Medical Artificial Intelligence and the Black Box Problem: A View Based on the Ethical Principle of "Do No Harm"', *Intelligent Medicine*, 4(1), 52–57, https://doi.org/10.1016/j.imed.2023.08.001.

Xu Y, X Liu, X Cao and J Zhang (2021) 'Artificial Intelligence: A Powerful Paradigm for Scientific Research', *The Innovation*, 2(4), 100179, https://doi.org/10.1016/j.xinn.2021.100179.

Yampolskiy RV (2024) 'On Monitorability of AI', *AI Ethics*, 5: 689–707, https://doi.org/10.1007/s43681-024-00420-x.

Yang G, Q Ye and J Xia (2022) 'Unbox the Black-Box for the Medical Explainable AI Via Multi-Modal and Multi-Centre Data Fusion: A Mini-Review, Two Showcases and Beyond', *Inf Fusion*, 77: 29–52, https://doi.org/10.1016/j.inffus.2021.07.016.

Yau KLA, HJ Lee, YW Chong, MH Ling, AR Syed, C Wu and HG Goh (2021) 'Augmented Intelligence: Surveys of Literature and Expert Opinion to Understand Relations Between Human Intelligence and Artificial Intelligence', *IEEE access*, 9: 136744–136761.

Yeung K (2018) 'A Study of the Implications of Advanced Digital Technologies (Including AI Systems) for the Concept of Responsibility Within a Human Rights Framework', MSI-AUT 5, https://rm.coe.int/a-study-of-the-implications-of-advanced-digital-technologies-including/168096bdab.

Yeung L (2019) 'Empirical Analysis of Judicial Decisions', in *Encyclopedia of Law and Economics*, A Marciano and GB Ramello (Eds.). Springer, New York, NY, https://doi.org/10.1007/978-1-4614-7753-2_692.

Yu KH, AL Beam and IS Kohane (2018) 'Artificial Intelligence in Healthcare', *Nat Biomed Eng*, 2: 719–731, https://doi.org/10.1038/s41551-018-0305-z.

Yuan S, S Coghlan, R Lederman, and J Waycott (2023) 'Ethical Design of Social Robots in Aged Care: A Literature Review Using an Ethics of Care Perspective', *Int J of Soc Robotics*, 15: 1637–1654, https://doi.org/10.1007/s12369-023-01053-6.

Yun JH, E-J Lee, and DH Kim (2021) 'Behavioral and Neural Evidence on Consumer Responses to Human Doctors and Medical Artificial Intelligence', *Psychology & Marketing*, 38(4): 610–625.

Zacharias J, M Von Zahn, J Chen and O Hinz (2022) 'Designing a Feature Selection Method Based on Explainable Artificial Intelligence', *Electron Markets*, 32: 2159–2184. https://doi.org/10.1007/s12525-022-00608-1.

Zamzam AH, K Hasikin, and AKA Wahab (2023) 'Integrated Failure Analysis Using Machine Learning Predictive System for Smart Management of Medical Equipment Maintenance', *Engineering Applications of Artificial Intelligence*, 125: 106715.

Zanfir-Fortuna G (2020) 'Article 21 Right to Object', in *The EU General Data Protection Regulation (GDPR): A Commentary*, C Kuner, LA Bygrave, C Docksey and L Drechsler (Eds.). Oxford Academic, New York, https://doi.org/10.1093/oso/9780198826491.003.0054.

Zapata KAC, R Patil, T Ward, R Loughran and F McCaffery (2023) 'Analysis of the Classification of Medical Device Software in the AI Act Proposal', *HHAI Workshops*, 19–34.

Zech H (2021) 'Liability for AI: Public Policy Considerations', *ERA Forum*, 22: 147–158, https://doi.org/10.1007/s12027-020-00648-0.

Zemmar A, AM Lozano and BJ Nelson (2020) 'The Rise of Robots in Surgical Environments during COVID-19', *Nat Mach Intell*, 2: 566–572, https://doi.org/10.1038/s42256-020-00238-2.

Zhang J and ZM Zhang (2023) 'Ethics and Governance of Trustworthy Medical Artificial Intelligence', *BMC Med Inform Decis Mak*, 23(1), 7, https://doi.org/10.1186/s12911-023-02103-9.

Zhu Y, T Gao, L Fan, S Huang, M Edmonds, H Liu, F Gao, C Zhang, S Qi, YN Wu, JB Tenenbaum and S-C Zhu (2020) 'Dark, Beyond Deep: A Paradigm Shift to Cognitive AI with Humanlike Common Sense', *Engineering*, 6(3): 310–345.

Zimmerman EJ (2017) 'Machine Minds: Frontiers in Legal Personhood' https://papers.ssrn.com/sol3/papers.cfm?abstract_id=2563965.

Ziosi M, J Mökander, C Novelli, F Casolari, M Taddeo, and L Floridi (2023) 'The EU AI Liability Directive (AILD): Bridging Information Gaps', *European Journal of Law and Technology*, 14(3).

Zou J and L Schiebinger (2018) 'AI Can Be Sexist and Racist- It's Time to Make It Fair', *Nature*, 559(7714): 324–326, https://doi.org/10.1038/d41586-018-05707-8.

Index

 | HTTPS://DOI.ORG/10.1163/9789004543980-011

www.ingramcontent.com/pod-product-compliance
Lightning Source LLC
LaVergne TN
LVHW010631300926
849162LV00001B/124
* 9 7 8 9 0 0 4 5 4 3 9 9 7 *